Ordered and Disordered Metamaterials

Design and applications

Online at: https://doi.org/10.1088/978-0-7503-5462-2

IOP Series in Electromagnetics and Metamaterials

Series Editor
Akhlesh Lakhtakia, *Pennsylvania State University*

Series in Electromagnetics and Metamaterials
The Series on Electromagnetics and Metamaterials, published by IOP, is an innovative and authoritative source of information on a fundamental science that has been enabling a multitude of transformative technologies for two centuries and more. The electromagnetic spectrum extends from millihertz waves to microwaves to terahertz radiation to ultraviolet light and even soft x-rays. In each spectral regime, different classes of materials have different kinds of electromagnetic response characteristics. Much has been discovered and much has been technologically exploited, but even more remains to be discovered and even more remains to be put to use for diverse applications.

Electromagnetics is an ever more vibrant arena of techno-scientific research. This is amply exemplified by the huge current interest in metamaterials. By virtue of carefully designed and engineered morphology, metamaterials exhibit response characteristics that are either completely absent or muted in their constituent materials.

Each book in the series offers an extended essay on a foundational topic; an emerging topic; a currently hot topic; and/or a tool for metrology, design, and application. Ranging from 60 to 120 pages, books are written by internationally renowned experts who have been charged with making the content not only authoritative, but also easy to understand, thereby offering more synthesis and depth than a typical review article in a journal.

Illustrated in full color for both ebook and printed copies, these short books are easily searchable in the ebook format. The series is thus more modular and dynamic than traditional handbooks and more coherent than contributed volumes.

This series is edited by Akhlesh Lakhtakia, the Charles Godfrey Binder (Endowed) Professor of Engineering Science and Mechanics at the Pennsylvania State University. Initial topics targeted in the series include symmetries of Maxwell equations, homogenization of bianisotropic materials, metamaterials and metasurfaces, transformation optics, nanophotonics for medicine and biology, single photons, radiation sources, optical bolometry, magnetic resonance imaging, and detection and imaging of buried objects. Additional topic suggestions are welcomed and will be promptly considered and decided upon. As part of the IOP digital library, students and professors at purchased institutions will have unlimited access to the ebooks for classroom and research usage.

A full list of the titles published in this series can be found at: https://iopscience.iop.org/bookListInfo/iop-series-on-electromagnetics-and-metamaterials#series.

Ordered and Disordered Metamaterials

Design and applications

Edited by
Pankaj K Choudhury
College of Optical Science and Engineering, International Research Center for Advanced Photonics (Haining Campus), Zhejiang University, Hangzhou, Zhejiang, People's Republic of China

Tatjana Gric
Department of Electronic Systems, Vilnius Gediminas Technical University, Vilnius, Lithuania

IOP Publishing, Bristol, UK

ISBN 978-0-7503-5462-2 (ebook)
ISBN 978-0-7503-5460-8 (print)
ISBN 978-0-7503-5463-9 (myPrint)
ISBN 978-0-7503-5461-5 (mobi)

DOI 10.1088/978-0-7503-5462-2

Version: 20241101

IOP ebooks

British Library Cataloguing-in-Publication Data: A catalogue record for this book is available from the British Library.

Published by IOP Publishing, wholly owned by The Institute of Physics, London

IOP Publishing, No.2 The Distillery, Glassfields, Avon Street, Bristol, BS2 0GR, UK

US Office: IOP Publishing, Inc., 190 North Independence Mall West, Suite 601, Philadelphia, PA 19106, USA

Those who strive for peace.

Contents

Preface

Metamaterial technology has been in the frontline of research for over a decade because metamaterials offer *on-demand* tailoring of electromagnetic behavior. This means that they have great technological potential in applications ranging from solar cells, perfect absorbers, varieties of miniaturized sensors, diffractive lenses, THz imaging devices, invisibility cloaks, advanced medical diagnostics, etc, thereby opening up the development of sustainable materials. The development of metamaterials primarily involves material scientists and engineers, physicists, and also mathematicians. However, the use of metamaterials requires specialities to be relied on, and it may also even involve expert scientists in the biological, biomedical, environmental, space and military arenas to meet specific design issues.

The study of metamaterials and their subsequent realization involve specific formalisms in solving the different types of technical problems involved. As is known, metamaterials are subwavelength-sized periodic structures, and the periodicity may be of the arrayed and/or layered forms. A common conceptual understanding relating to metamaterial design is to assume that they are in the ordered forms, which is evident from the massive number of research reports in the literature. However, the disordered forms of metamaterials are also attracting interest among the R&D community due to their possible implementation, particularly in the biosensing and/or biomedical arena. With this viewpoint in mind, the editors had the idea for a book—*Ordered and Disordered Metamaterials: Design and Applications*—that incorporates both the ordered and disordered kinds of engineered media. This volume focuses more on an overview of innovative phenomena enabled by both types of metamaterials and their applications, apart from the basic formulation techniques to be exploited for design purposes.

Within this context, the volume contains nine chapters describing aspects of the ordered and disordered engineered media and their applications, contributed by pioneering authors from 10 different countries. In an attempt to obtain a glimpse of the book, one must recall the cloaking applications of metamaterials, which have been very interesting, as these virtually conceal objects placed within. Cloaking structures comprise specific forms of metamaterials having refractive index(es) so that the waves are not allowed to penetrate the medium. Pivoting to the cloaking of objects, chapter 1 focuses on cylindrical invisibility cloaks comprising ordered metamaterials such as the perfect electromagnetic conductor and nihility media. In this chapter, Shahzad *et al* take up different electromagnetic conditions for theoretical analyses and outcomes, emphasizing the co- and cross-polarized scattering coefficients of different orders.

Chapter 2 deals with reconfigurable ordered metamaterials for scattering control in the microwave band of the electromagnetic spectrum. In this chapter, Jiang and Ma discuss different types of methods used for this purpose, which primarily include integrating active devices or materials into passive meta-atoms. Typically, achieving the property of reconfigurability involves electrical, optical, thermal, chemical, mechanical, and microfluidic methods, and the chapter describes some of the most commonly used modulation techniques. It also touches upon the types of reconfigurable metamaterials with all elements controlled uniformly, as well as those with each element controlled independently. Overall, the chapter provides a glimpse of the design issues related to programmable metasurfaces that can be realized and implemented for varieties of technological purposes.

The theme of chapter 3 is programmable hyperbolic metamaterials with layered configurations, wherein Pourmand and Choudhury use an array of phase-change media and dielectrics to generate primary color pixels. They also discuss such layered structures for tunable dual-band nearly perfect absorption in the visible and near-infrared regimes of the electromagnetic spectrum. In particular, they use $GeSe_3$–Al bilayers for structural color generation and Sb_2S_3–Ag bilayers for tunable absorption. In both situations, they implement particle swarm optimization to achieve the best possible geometrical parameters for the respective metamaterial configurations. The reported results are of importance in designing reconfigurable reflective color displays and tunable perfect absorbers.

In chapter 4, Rybin considers two-layer wire metamaterials for creating high-impedance surfaces and generating microwave surface plasmon polaritons. The author discusses analytically the scattering and the effective constitutive properties of the single- and two-layer wire gratings in the microwave band exploiting dipole approximation. The analytical studies carried out in this chapter are confirmed by numerical finite-difference time-domain (FDTD) simulations, wherein the author demonstrates that, in GHz frequencies, the layered wire gratings behave almost like a mirror for the *p*-polarization incidence, whereas they are transparent for the *s*-polarization. The author also demonstrates that a two-layer wire grating backed by a plane conductive surface can be used to design high-impedance surfaces in the microwave band.

Choi *et al* in chapter 5 briefly touch upon machine learning- or artificial intelligence-based models for the analysis, design, and optimization of various properties of optical metamaterials. In this chapter, the authors explore recent studies in the relevant direction considering forward prediction of optical parameters in metamaterials by machine learning, inverse design of optical metamaterials using machine learning, and machine learning-based optimization of optical metamaterials. Following these, they suggest potential models to drive innovation in the development of various optical elements and components in connection with optical metamaterials.

Medical diagnostics is one of the potential areas for the application of metamaterials, which essentially involves biosensing. Within this viewpoint in mind, chapter 6 of this book deals with ordered metamaterials that can be exploited for biosensing applications, using terahertz waves. In this chapter, Sheta *et al* provide in brief the background for metamaterial designs, followed by a review of a few different types of metamaterial-based biosensor designs reported by the investigators to detect biological analytes. They also incorporate some recent research results of their own on certain kinds of polarization-insensitive layered terahertz metamaterial configurations that can be used for sensing gamma radiations. The obtained results indicate the potential of the design in gamma radiation dosimetry.

Chapter 7 of this book amalgamates discussions on the ordered and disordered metamaterials used for biosensing applications exploiting the phenomenon of surface plasmon resonance. So far as the ordered structure is concerned, Ghasemi and Choudhury consider an illustrative example of the nanocomb grating array configuration, which the authors fabricated and experimentally demonstrated for sensing glucose of different concentrations. The reported results show the usefulness of the nanocomb grating array structure in biosensing applications. As for the disordered structure, the authors take up the case of BALB/c rat's dried DNA thin film (of different DNA concentrations) deposited over a gold nanolayer. They evaluate the spectral characteristics under plasmonic conditions, which determine the technique to be very useful in biosensing.

In the context of biosensing applications of disordered metamaterials, chapter 8 describes the theoretical and computational determination of the effective permittivity of the composite disordered media, as this requirement is vital to examine the electromagnetic behavior for assessing *on-demand* needs. In this context, Gric presents the methodology as a perfect tool to analytically evaluate the permittivity tensor of the sensing measurand for analysing biological media. The author claims that the presented technique allows for the creation of phantom tissue models for further use in clinical applications.

Chapter 9 focuses on sensing molecular chirality at the nanoscale—an area that has been quite challenging due to the inherently weak nature of the associated chiro-optical signals. In this chapter, Droulias demonstrates the possibility of detecting weak chiral inclusions with the aid of achiral photonic metasurfaces. The author presents the study of both the isotropic and anisotropic metasurfaces, and demonstrates them enabling enhanced detection, unambiguous enantiomer differentiation and complete measurement of the total chirality.

The topics covered in this book illuminate the underlying theories of the ordered and disordered kinds of engineered metamaterials suitable for excitation wavelengths of different regimes of the electromagnetic spectrum. It also incorporates research advances in the stated relevant fields, which makes the volume useful to both undergraduate researchers venturing into research careers and expert scientists in universities and research organizations, keeping them abreast of developments in the area. In this way, the volume will appeal to professionals, researchers and those involved in the design of photonics-based techniques for various reasons, namely energy harvesting, color displays, health diagnostics and several other relevant technological needs. Finally, the themes incorporated in this volume are also directed to applications of ordered and disordered engineered metamaterials. A few illustrative examples are the range of uses from cloaking to tunable energy harvesters, reconfigurable structural color generation and scattering control to wired structures for possible antenna designs, and biosensing to chirality detection at the molecular level. Apart from these, reviews are also made of metamaterial designs with the aid of machine learning and artificial intelligence.

The editors expect the content of the book to meet the needs of a range of readers from students to early career researchers to established scientists in their fields. They extend sincere thanks to all of the esteemed contributors for happily taking interest in the project, which finally resulted in realizing this volume, which discusses topics relating to both the ordered and disordered forms of engineered structures. However, it took a few more months than expected to complete the project due to unexpected situations; the editors are grateful to the contributors for their patience and continuous support.

Pankaj K Choudhury
Tatjana Gric

Acknowledgements

The editors take the opportunity to acknowledge Emily Tap of the IOP for inviting them to consider this project, Phoebe Hooper and Mia Foulkes for subsequent technical handling of the same. Also, PKC acknowledges the Director of the International Research Center for Advanced Photonics (Zhejiang University, China) for continuous support and encouragement. Finally, PKC is indebted to Professor B N Basu—the series editor—for inviting him to take on the project, and for his profound knowledge and guidance, which have always been illuminating.

Editor biographies

Pankaj K Choudhury

Pankaj K Choudhury received a PhD in Physics in 1992. From 1992 to 1997, he was a research associate at the Department of Electronics Engineering, Indian Institute of Technology, Banaras Hindu University (Varanasi, India). In 1997, he joined the Department of Physics, Goa University (Goa, India) as a lecturer. In late 1999, he became a researcher at the Center of Optics, Photonics and Lasers (COPL), Laval University (Quebec, Canada). From 2000 to 2003, he was with the Faculty of Engineering, Gunma University (Kiryu, Japan), as a researcher. In May 2003, he received the position of professor at the Faculty of Engineering, Multimedia University (Cyberjaya, Malaysia) where he was until late 2009. During that span, he also served Telekom Research and Development (TMR&D, Malaysia) as a consultant for projects on optical devices. In late 2009, he became a professor at the Institute of Microengineering and Nanoelectronics (IMEN), Universiti Kebangsaan Malaysia (The National University of Malaysia, Malaysia), where he was attached until mid-2022. He is now a professor at the College of Optical Science and Engineering, International Research Center for Advanced Photonics, Zhejiang University (China). His current research interests lie in the theory of optical waveguides, which include complex media, fiber and integrated optics, fiber optic devices, optical sensors and metamaterial properties. He has published over 285 research papers in peer-reviewed international journals and conference proceedings; contributed 29 chapters to research level books; and edited and co-edited 12 research level books in these areas. He is a reviewer for over 50 research journals, and the Editor-in-Chief of the *Journal of Electromagnetic Waves and Applications* (Taylor and Francis, UK). He is a fellow of the Institution of Engineering and Technology (IET, UK) and the Society of Photo-Optical and Instrumentation Engineers (SPIE, USA), and a senior member of the Institute of Electrical and Electronics Engineers (IEEE, USA) and Optica (formerly OSA, USA). He is also a chartered engineer (CEng) registered with the Engineering Council (UK).

Tatjana Gric

Tatjana Gric is a professor at the Department of Electronic Systems, Vilnius Gediminas Technical University (Vilnius, Lithuania). She has been engaged (since 2003) in the investigation of waveguide devices (e.g., waveguide modulators, filters, etc), namely on proposing electrodynamical analyses of various complex structures (e.g. tissue-like structures). Another major goal of her studies has been plasmonics as the examination of the interaction between electromagnetic field and free electrons in a metal. She has also been involved in the studies of optically active nanostructures and their fundamental photonic properties. During the past few years she has been working on the investigation of nanostructured composites and their fascinating properties. She has authored and co-authored over 70 articles in refereed journals and conference proceedings. She also holds one Lithuanian patent.

List of contributors

Shakeel Ahmed
Quaid-i-Azam University, Islamabad, Pakistan

Inseop Byeon
School of Electrical and Electronic Engineering, Yonsei University, Seoul, Republic of Korea

Jong-ryul Choi
Medical Device Development Center, Daegu-Gyeongbuk Medical Innovation Foundation, Daegu, Republic of Korea

Pankaj K Choudhury
College of Optical Science and Engineering, International Research Center or Advanced Photonics (Haining Campus), Zhejiang University, Hangzhou, Zhejiang, China

Sotiris Droulias
Department of Digital Systems, ICT School, University of Piraeus, Greece

Masih Ghasemi
Laser and Plasma Research Institute, Shahid Beheshti University, Tehran, Iran

Tatjana Gric
Department of Electronic Systems, Vilnius Gediminas Technical University, Vilnius, Lithuania

A-B M A Ibrahim
Faculty of Applied Sciences, Universiti Teknologi MARA, Selangor, Malaysia

Xinyu Jiang
College of Optical Science and Engineering, Zhejiang University, Hangzhou, China

Donghyun Kim
School of Electrical and Electronic Engineering, Yonsei University, Seoul, Republic of Korea

Yungui Ma
College of Optical Science and Engineering, Zhejiang University, Hangzhou, China

Qaisar Abbas Naqvi
Quaid-i-Azam University, Islamabad, Pakistan

M Pourmand
Department of Applied Physics and Electronics, Umeå University, Umeå, Sweden

Oleg Rybin
School of Radio Physics, Biomedical Electronics and Computer Systems, V N Karazin Kharkiv National University, Kharkiv, Ukraine

Anjum Shahzad
Quaid-i-Azam University, Islamabad, Pakistan

E M Sheta
International Centre for Radio Astronomy Research, Curtin University, Bentley, Australia

Contributor biographies

Shakeel Ahmed graduated from the University of Peshawar (Pakistan) and received a PhD from Quaid-i-Azam University, Islamabad (Pakistan). He is the author of more than 40 international journal publications. His current research interests include electromagnetic scattering by different types of geometries and metamaterials.

Inseop Byeon received a BSc from the Department of Nanoelectronic Physics, Kookmin University (Republic of Korea) in 2022. He is now a PhD candidate at Yonsei University, where he is currently conducting research on plasmonics techniques. His present research interests encompass optical modeling and structural synthesis of metallic nanoparticles, along with the investigation of plasmonic properties of metastructures.

Jong-ryul Choi received a BSc and a PhD from the Department of Electrical and Electronic Engineering, Yonsei University (Seoul, Republic of Korea); the latter one in 2013. Since 2013, he has been a researcher at the Medical Device Development Center, Daegu-Gyeongbuk Medical Innovation Foundation (K-MEDI hub), Daegu (Republic of Korea). His research interests include biomedical optics (especially optical sensing, imaging, manipulation, and stimulation), nanoscale photonics for medical applications, and machine learning-assisted biophotonics.

Sotiris Droulias received a diploma in electrical and computer engineering and a PhD in nonlinear photonics from the National Technical University of Athens, Greece, in 2001 and 2007, respectively. From 2009 to 2012, he was an adjunct lecturer with the University of Patras, Greece, and from 2012 to 2020, he was with the Phononic- and Meta-Materials Group, FORTH-IESL, Crete, Greece. He is currently a research associate with the Department of Digital Systems, ICT School, University of Piraeus, Greece. He is the author of more than 50 articles and two book chapters. He has worked on several EC funded projects. His research interests include electromagnetic modeling, metasurfaces, nanolasers and plasmonics. In 2019, he received the Best Poster Award for his work on metasurface lasers in META 2019, Lisbon, Portugal, and in 2020, he was recognized as an Outstanding Reviewer by the Institute of Physics (IOP). He has received several talk invitations in prestigious conferences and he serves as a reviewer for international scientific journals.

Masih Ghasemi received an MEng in telecommunication from the Faculty of Engineering, Multimedia University (Cyberjaya, Malaysia) in 2012, and a PhD in nanophotonics from the Institute of the Microengineering and Nanoelectronics, Universiti Kebangsaan Malaysia (Bangi, Malaysia) in 2016. Since then he has been actively pursuing his postdoctoral research at Shahid Beheshti University (Tehran, Iran) in areas related to silicon photonics applications, broadband plasmonics data analysis, metasurface engineering, sensory application of 1D and 2D thin films, biochip-based thin films, the design and development of automated reflective, and transmissive optical characterization of nanoengineered thin films. He has over 40 publications in international journals and conference proceedings, 2 book chapters and 1 patent (registered in Iran) to his credit.

Abdel-Baset M A Ibrahim earned a PhD in nonlinear optics from the University Sains Malaysia (Penang, Malaysia) in 2009. Previously, he earned a BSc in physics and an MSc in quantum optics from Al-Zagazig University in Egypt and the University of Malaya (Kuala Lumpur, Malaysia), respectively. From 2000 to 2005, he worked as a research officer at Telekom Research and Development (Malaysia) on optical materials and devices. 2010 marked his arrival at Universiti Teknologi MARA (UiTM, Shah Alam, Malaysia), where he is currently an associate professor at the Faculty of Applied Sciences and the head of the Photonics and Materials Research Group. He was a visiting researcher at Abu Dhabi University (UAE) from September 2018 to December 2020. He has supervised over a dozen MSc and PhD students in quantum optics, nonlinear optics, and condensed matter physics. He has published more than 70 international journal articles and book chapters. He has edited two books and written more than 70 reviews for physics-based journals. He has been a member of Optica (formerly OSA) since 2012. His current research interests include quantum and nonlinear optics, waveguide theory, light–matter interaction, and the optical properties of materials.

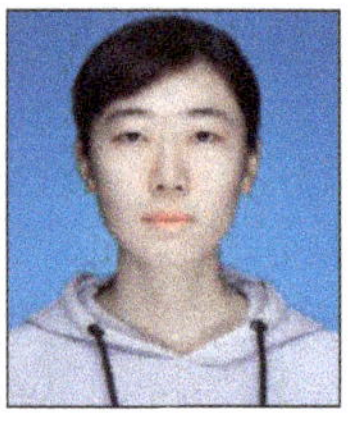

Xinyu Jiang received a BSc degree from the Physics Department, Zhejiang University (China) in 2021. Now she is a master's student at Zhejiang University. Her current research focuses on tunable microwave metasurfaces.

Donghyun Kim received a BSc and an MSc in electronics engineering from Seoul National University, Seoul (Republic of Korea), and a PhD in electrical engineering from the Massachusetts Institute of Technology (USA). His doctoral research focused on innovative multidimensional display technologies and smart optical filters. Following his PhD, he worked as a senior research scientist at Corning, Inc., New York (USA), where he specialized in next-generation fiber optic access communication systems. Subsequently, as a postdoctoral fellow at Cornell University (USA), he explored cellular biophotonic sensors for cell-based assays.

Since 2004, he has been leading the Biophotonics Engineering Laboratory, Yonsei University. His research at Yonsei centers on nanophotonic technology and its applications in biomedical engineering, particularly using plasmonic techniques and metastructures. In addition, he has been serving as an adjunct professor at the Department of Biomedical Engineering of the Chinese University of Hong Kong since 2024. He has published more than 130 journal articles and 60 conference papers and holds more than 40 patents across the Republic of Korea, Japan, and the USA, with additional filings under the PCT. He is a fellow of the SPIE and a senior member of the IEEE.

Yungui Ma received a BSc and a PhD from Lanzhou University (China). He worked as a research fellow and a research scientist at National University of Singapore from 2005 to 2010. Afterwards, he joined the College of Optical Science and Engineering, Zhejiang University (China) as a full professor. His current research interests include metamaterials, nanophotonics and near-field heat transfer. He is the Deputy Director of the Joint Laboratory for International Cooperation in Photonics and Technology of the Ministry of Education (China), Director of the International Research Center for Advanced Photonics, Zhejiang University, and the Secretary General of the 8th Council of the Zhejiang Optical Society. In 2011, he was selected for the New Century Excellent Talents Support Program by the Ministry of Education. In 2015, he received support from the Zhejiang Provincial Natural Science Foundation for Distinguished Young Scholars. In 2020, he was selected as a leading innovative talent in Zhejiang Province of China.

Qaisar Abbas Naqvi received a PhD from the Department of Electronics, Quaid-i-Azam University (Islamabad, Pakistan), where he is now a professor. He has contributed to the areas of fractional calculus, fractional electromagnetics and metamaterials, with 260 notable journal articles, a book and a book chapter. In recognition of his scientific contributions, he was awarded the Diversity Outreach Award for 2022 by the International Society for Optics and Photonic. Also, the Pakistan Academy of Sciences awarded him a Gold Medal in Engineering Sciences.

M Pourmand received a BSc in electrical engineering from the Khajeh Nasirodin Toosi University of Technology (Iran) in 2004. During 2004–2020, he worked as an electronic engineer in the industry sector. During the span, he studied for an MSc in electrical engineering at Tafresh University (Iran), and received the degree in 2013. He received a PhD from the Institute of Microengineering and Nanoelectronics, Universiti Kebangsaan Malaysia (The National University of Malaysia, Malaysia) in 2023. Soon after that he joined the Department of Applied Physics and Electronics, Umeå University (Sweden) as a postdoctoral fellow. His research interests lie in metamaterials, photonic crystals and plasmonics, and he has a few publications in these areas in reputable international journals.

Oleg Rybin received an MSc from V N Karazin Kharkov National University (Ukraine) in 1994, a PhD in radio physics from Kharkov National University of Radioelectronics (Ukraine) in 1999, and a DSc in electrical engineering from V N Karazin Kharkov National University in 2018. He was a postdoc (2005–2006) at Nagoya Institute of Technology (Japan). He was a foreign associate professor at Bahauddin Zakariya University (Multan, Pakistan) in 2006; an associate professor at Kharkov State University of Food Technology and Trade (Ukraine) in 2008; a research professor at Wuhan Textile University (China) in 2013; and a senior researcher at the Department of Theoretical Radio Physics at V N Karazin Kharkiv National University until 2016. He has been a professor at the Department of Quantum Radio Physics at V N Karazin Kharkiv National University since 2019. His research interests are in modeling and simulation in electromagnetics, metamaterials and applications, and invisibility cloaking.

Anjum Shahzad received an MSc and MPhil in electronics from Quaid-i-Azam University (Islamabad, Pakistan) in 2007 and 2010, respectively. He has some international publications to his credit, and is currently pursuing his PhD at Quaid-i-Azam University. His current research interests focus on multiband electromagnetic wave absorbers for microwave and millimeter-wave applications. He received the SPIE Optics and Photonics Educational Award in 2022. He was also the recipient of the IRSIP award from HEC Pakistan to conduct a six-month research collaboration at Zhejiang University (China). Additionally, he has been serving as the president of the SPIE student chapter at Quaid-i-Azam University, and actively participating in various training and outreach activities to promote emerging research trends in optics and photonics.

E M Sheta received a BSc and an MSc, both in electronics and communications engineering, from Ain Shams University (Egypt). Since graduating in 2016, he has been working in academia. He is currently a PhD student at the International Center for Radio Astronomy Research, Curtin University (Bentley, Australia). He has 10 publications so far in peer-reviewed research journals and conference proceedings. His research interests lie in microelectronics, metamaterials, and optical sensing.

IOP Publishing

Ordered and Disordered Metamaterials
Design and applications
Pankaj K Choudhury and Tatjana Gric

Chapter 1

Metamaterials-based cylindrical invisible cloaks

Anjum Shahzad, Shakeel Ahmed, Qaisar Abbas Naqvi and Pankaj K Choudhury

A cylindrical wave expansion method was employed to derive the scattering fields for a two-dimensional cylindrical invisibility cloak incorporating perfect electromagnetic conductor (PEMC) and nihility media in a perturbed void region. A near-ideal model of the cloak was developed to address the boundary value problems at the inner boundary of the cloak shell. The study confirms the perfection of cloaks using the PEMC and nihility media by analysing the transition of scattering coefficients from the near-ideal to the ideal case. Additionally, it was observed that, with the use of nihility medium at a very small distance δ, the convergence rate remains independent of the type of the incident field.

1.1 Introduction

Metamaterials exhibit exotic electromagnetic (EM) properties that arise from their structure rather than their composition. Ordered metamaterials are a class of engineered media with a highly regular and predictable arrangement of their constituent structural units arranged in a periodic pattern, which endows them with unique properties that do not typically exist in natural materials. The periodicity is crucial in determining their interaction with electromagnetic, acoustic, or elastic waves. As such, the key characteristics of ordered metamaterials include their periodic structure, predictable behavior, and tailored properties. The applications of ordered metamaterials encompass a wide range of advanced technological fields, including high-resolution lenses and imaging systems, cloaking devices for rendering objects invisible, efficient waveguides for guiding light with minimal loss, innovative antenna designs with enhanced bandwidth and directivity, and improved signal processing and transmission in telecommunications [1–7]. In this chapter, we focus on cylindrical invisibility cloaks based on ordered metamaterials such as perfect electromagnetic conductor (PEMC) and nihility metamaterials.

doi:10.1088/978-0-7503-5462-2ch1

Disappearance or invisibility has long been attractive to the general public, and is frequently portrayed in science fiction books and movies with specific characters. Progress in science and technology is making realistic what was previously only fiction. Researchers have reported EM invisibility cloaks (or just cloaks), which are devices that can render objects invisible to specific EM waves [7]. Anything positioned inside or beneath these gadgets will be successfully hidden. These devices are generally made of specific metamaterials that block EM waves from passing through because of their variable refractive index. Rather, the EM waves that hit the cloaking material are steered smoothly around it, emerging from the other side without distortion and being in phase with waves that pass through it. Cloaks thereby prevent both shadow casting and wave reflection, making the thing they cover up virtually invisible. This, in theory, enables an observer to view what is hidden behind the covered object as though it was not there at all.

Metamaterials designed for cloaking applications are characterized by a negative refractive index and are often referred to as left-handed (LH) metamaterials [1, 8–10]. J. Pendry [1] demonstrated that negative refraction can create a perfect lens. Smith *et al* [8] further explored the concept of negative refraction in metamaterials, while Shalaev introduced the concept of optical negative-index metamaterials, as discussed in [9]. Experimental work by Shelby *et al* [10] provided evidence for a metamaterial with negative medium properties at microwave frequencies without violating any of the Maxwell equations.

In a groundbreaking theoretical study, Pendry and Schurig [11] demonstrated that an object could be rendered invisible to EM fields by utilizing coordinate transformation within anisotropic and inhomogeneous metamaterials. Following this, Pendry *et al* [2] successfully constructed a metamaterial-based cloak, achieving practical EM cloaking in the microwave range. Cummer *et al* extended this research on EM cloaking structures by performing full-wave simulations [12]. Their work focused on cylindrical cloaking structures, using both ideal and non-ideal EM parameters. The simulations revealed that the low reflection and power flow bending properties of EM cloaking are relatively insensitive to moderate variations in permittivity and permeability. The cloaking performance gradually diminishes as the loss increases. Nonetheless, effective low-reflection shielding is still attainable using a cylindrical shell with eight-layer homogeneous approximation of the ideal continuous medium.

Schurig *et al* [2] were the first to achieve the practical realization of a metamaterial EM cloak operating at microwave frequencies. In their study, a copper cylinder was concealed within the cloak, effectively reducing scattering from the hidden object and minimizing its shadow, making the combined cloak–object system resemble an empty space. Cai *et al* [13] explored optical cloaking using metamaterials. Ruan *et al* [14] investigated a cloak using ideal material parameters, confirming symmetrically perfect invisibility by analysing the scattering coefficients from a near-ideal to an ideal scenario. Yan *et al* [15] examined the inherent visibility of a cylindrical cloak using simplified material parameters. They explained that such a cloak allows the zeroth-order cylindrical wave to pass through it as if it was a homogeneous isotropic medium, thus making it visible to higher-order cylindrical waves. Their numerical simulations indicated that, while this cloak retains some characteristics of an ideal cloak, it still produces some scattering. Greenleaf *et al* [16] worked on enhancing cloak performance by incorporating soft and hard surface (SHS) linings. Their results revealed that the SHS lining significantly enhanced the cloaking performance, reducing the far-field scattered wave and preventing the blow-up of electric and magnetic field densities.

Yan *et al* [15] investigated the scattering properties of a simplified cylindrical version of the invisibility cloak and compared the performance of three variants: the simplified linear cloak, the improved linear cloak, and the simplified quadratic cloak. They found that both the improved linear and quadratic cloaks exhibited superior invisibility compared to the simplified linear cloak. Zhang *et al* [17] examined the cylindrical invisibility cloaks with EM waves, noting that the incoming waves induce both electric and magnetic surface currents at the inner boundary of the cloak, which are not accounted for in coordinate transformation theory. Cai *et al* [18] proposed a nanomagnetic cloak designed to minimize scattering, utilizing higher-order transformations to achieve a smooth outer interface rather than a discontinuous one. Yan *et al* [19] explored coordinate transformations to develop a perfect invisibility cloak for arbitrary shapes. Yan *et al* [20] explored the effects of geometrical perturbations at the inner boundaries of invisibility cloaks. In addition to these examples, investigators reported a substantial amount of research on cloaking devices [20–25].

In this discourse, we commence by exploring the coordinate transformation method, subsequently delving into some metamaterials employed in the realm of cloaking, and then take up a few designs of these to demonstrate certain cloaking applications. In this chapter, we have used time dependence of the form $e^{-j\omega t}$, which is suppressed throughout the analyses.

1.2 Coordinate transformation

Most analyses of cylindrical and spherical invisibility cloaks concentrate on those generated by applying coordinate transformations solely in the radial direction. The effectiveness of these cloaks is validated by deriving the exact EM fields within the cloaking medium using Maxwell's equations. In practical applications, however, it is often preferable to design invisibility cloaks that conform to the shapes of specific objects. Therefore, a comprehensive understanding of invisibility cloaks with random geometries, created through coordinate transformations, is essential [11, 22]. This technique reveals that, when Maxwell's equations are used in a curved space, they retain the same form as in Cartesian coordinates but with modified medium parameters, current density, and electric charge density [11, 22]. Furthermore, [5] demonstrates that, when the EM fields $\mathbf{E}^i$ and $\mathbf{H}^i$ impinge on the cloak, those within the cloaked medium can be determined through coordinate transformation as follows:

$$\hat{E}_r(r, \theta, z) = f'(r)E_r^i(f(r), \theta, z),$$

$$\hat{H}_r(r, \theta, z) = f'(r)H_r^i(f(r), \theta, z),$$

$$\hat{E}_\theta(r, \theta, z) = \frac{f(r)}{r} E^i_\theta(f(r), \theta, z),$$

$$\hat{H}_\theta(r, \theta, z) = \frac{f(r)}{r} H^i_r(f(r), \theta, z),$$

$$\hat{E}_z(r, \theta, z) = E^i_z(f(r), \theta, z),$$

$$\hat{H}_z(r, \theta, z) = H^i_z(f(r), \theta, z),$$

with $[E^i_r, E^i_\theta, E^i_z]$ and $[H^i_r, H^i_\theta, H^i_z]$ as the components of the incident fields represented in cylindrical coordinates.

1.2.1 PEMC medium

The PEMC material was introduced by Lindell and Shivola [26]. Since it prevents the penetration of electromagnetic energy, an interface with this medium acts as an ideal boundary for the electromagnetic field. Consider the boundary between the PEMC medium and air, characterized by a unit normal vector **n**. Given that the tangential components of fields **E** and **H** are continuous across any interface between two media, one of the boundary conditions on the air side of the interface is

$$\mathbf{n} \times (\mathbf{H} + M\mathbf{E}) = \mathbf{0}. \tag{1.1}$$

This occurs because a corresponding term vanishes on the PEMC medium side. The second condition arises from the continuity of the normal components of the **D** and **B** fields, providing an additional boundary condition as

$$\mathbf{n} \times (\mathbf{D} - M\mathbf{B}) = \mathbf{0}. \tag{1.2}$$

Here M represents the admittance parameter of the PEMC material. It is important to note that the PEMC material generalizes both the perfect electric conductor (PEC) and perfect magnetic conductor (PMC) materials. Specifically, the PEC and PMC are the special cases of the PEMC, corresponding to $M = \infty$ and $M = 0$, respectively. For the special case of PEMC, $M = 1/\eta_o$, with η_o being the free space impedance.

Studies related to PEMCs have attracted the attention of many researchers [26–45]. Ruppin [26] developed an analytical theory for EM scattering by a PEMC cylinder. Ahmed and Naqvi [35] investigated the behavior of a buried PEMC, while Shahzad *et al* [44] examined the response of a buried coated PEMC cylinder.

1.2.2 Nihility medium

The idea of nihility medium was introduced by Lakhtakia [46], and has attracted enormous interest among the electromagnetic community [47–52]. A nihility medium has both the relative permittivity and permeability null-valued [46], and is characterized as the electromagnetic nilpotent. Such a medium is described by the constitutive relations as follows:

$$\mathbf{D}(x, y, z, \omega) = 0 \tag{1.3}$$

and

$$\mathbf{B}(x, y, z, \omega) = 0. \tag{1.4}$$

Consequently, the medium prevents the propagation of EM energy in the absence of sources [49–53]. As a result, the directionality of the phase velocity relative to the wavevector in nihility becomes irrelevant. Under the condition of nihility, Maxwell's equations are expressed as follows:

$$\nabla \times \mathbf{E}(x, y, z, \omega) = 0 \tag{1.5}$$

and

$$\nabla \times \mathbf{H}(x, y, z, \omega) = 0. \tag{1.6}$$

The Bruggeman formalism [54] does not accurately predict the occurrence of nihility in particulate composites, whether through the combination of anti-vacuum with vacuum or by blending an LH material with a conventional material of the same impedance, such as Teflon. This limitation arises because the polarizability and magnetizability per unit volume of an isotropic dielectric–magnetic sphere within a nihility medium remain independent of the constitutive parameters of the material of sphere. The Maxwell Garnett formalism [54] indicates that a homogeneous dispersion of electrically small anti-vacuum spheres in vacuum effectively simulates nihility when the volume fraction of anti-vacuum is 0.25. Conversely, if the matter is extracted from anti-vacuum as small spheres, the obtained Swiss cheese-like composite will also replicate nihility, provided that the volume fraction of anti-vacuum is 0.75. Hence, nihility can be emulated by particulate composites at $\omega = \tilde{\omega}$.

1.3 PEMC invisibility cloak

In this section, we discuss an invisibility cloak termed the PEMC cylindrical cloak. Our primary objective is to examine the scattering properties of a PEMC cylindrical cloak and the convergence rate characteristics of the scattered EM fields generated by the cloak, as explored in [55]. In this work, we focus more on the formulation of the PEMC cylindrical cloak and present the corresponding numerical results.

1.3.1 Formulation of the PEMC cloak

A two-dimensional (2D) cylindrical invisibility cloak, which incorporates a PEMC layer with a small perturbation δ, can be designed by compressing EM fields within a cylindrical region $r' < b$ of a concentric cylindrical shell with $a < r < b$, as figure 1.1 depicts. To prevent the medium parameters from approaching infinity at the inner boundary of the cylindrical cloak, a thin layer of width δ has been removed from this boundary. The region outside the shell (i.e. $r > b$) represents free space,

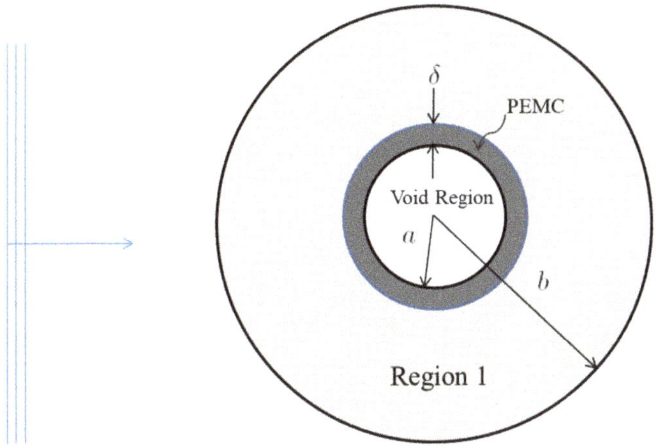

Figure 1.1. Schematic representation of an invisibility cloak incorporating a PEMC tiny perturbation.

characterized by a wavenumber $k_0 = \omega\sqrt{\mu_0\epsilon_0}$ and a medium impedance $\eta_0 = \sqrt{\mu_0/\epsilon_0}$. The region between $(a + \delta) < r < b$ is referred to as region 1.

The medium enclosed by the cloak is characterized by an unknown medium. A coordinate transformation in cylindrical coordinates is considered, as follows:

$$f(r) = r' = \frac{b}{(b - a)}(r - a), \tag{1.7}$$

given that $f(a) = 0$ and $f(b) = b$, while θ and z remain constant [11]. The permittivity and permeability of the cloak region can be derived using the coordinate transformation method [11] as

$$\epsilon_r = \mu_r = \frac{f(r)}{rf'(r)}, \tag{1.8}$$

$$\epsilon_\theta = \mu_\theta = \frac{f(r)f'(r)}{f(r)}, \tag{1.9}$$

and

$$\epsilon_z = \mu_z = \frac{f(r)f'(r)}{r}, \tag{1.10}$$

where the prime ($'$) denotes differentiation with respect to the argument r.

1.3.1.1 Electromagnetic fields in the free space region $(r > b)$

We consider a scenario in which a transverse electric (TE) wave impinges onto the cloak from free space. The incident EM field, expressed in cylindrical coordinates $(r,\ \theta)$, can be described as

$$H_{0z}^{i} = \sum_{n=-\infty}^{\infty} J_n(k_0 r)e^{in(\theta)}. \tag{1.11}$$

Here, $J_n(.)$ denotes a Bessel function of the first kind. Utilizing Maxwell's equations, the θ-component of the incident electric field can be written as

$$E_{0\theta}^{i} = -i\eta_0 \sum_{n=-\infty}^{\infty} J_n'(k_0 r)e^{in(\theta)}. \tag{1.12}$$

The scattered field from the outer boundary of the cylindrical invisibility cloak, in response to the incident field, can be written in terms of an unknown scattering coefficient, as follows:

$$H_{0z}^{s} = \sum_{n=-\infty}^{\infty} a_n\, H_n^{(1)}(k_0 r)e^{in(\theta)}, \tag{1.13}$$

and the θ-component of electric field will be

$$E_{0\theta}^{s} = -i\eta_0 \sum_{n=-\infty}^{\infty} a_n\, H_n^{(1)\prime}(k_0 r)e^{in(\theta)}, \tag{1.14}$$

where $H_n^{(1)}(.)$ denotes a Hankel function of the first kind.

1.3.1.2 Electromagnetic fields in region 1 ($(a + \delta) < r < b$)

The electromagnetic field that is transmitted in region 1 can be expressed using an unknown coefficient as

$$H_{1z}^{i} = \sum_{n=-\infty}^{\infty} b_n J_n[k_0 f(r)]e^{in(\theta)}. \tag{1.15}$$

By utilizing Maxwell's equations, the electric field can be written as

$$E_{1\theta}^{i} = -i\eta_0 f'(r) \sum_{n=-\infty}^{\infty} b_n\, J_n'[k_0 f(r)]e^{in(\theta)}. \tag{1.16}$$

In region 1, the co-polarized component of the field scattered by the cloak is

$$H_{1z}^{s} = \sum_{n=-\infty}^{\infty} c_n\, H_n^{(1)}[k_0 f(r)]e^{in(\theta)}, \tag{1.17}$$

and the corresponding electric field is

$$E_{1\theta}^{s} = -i\eta_0 f'(r) \sum_{n=-\infty}^{\infty} c_n H_n^{(1)\prime}[k_0 f(r)]e^{in(\theta)}. \tag{1.18}$$

For the E-polarized excitation, the scattered field from a PEMC border contains the E-polarized fields in addition to the H-polarized fields [26]. Therefore, the cross-polarized component of the scattered field in region 1 can be written as

$$E_{1z}^{s} = -i\eta_0 \sum_{n=-\infty}^{\infty} d_n\, H_n^{(1)}[k_0 f(r)]e^{in(\theta)}, \tag{1.19}$$

and the θ-component of the scattered field is

$$H_{1\theta}^{i} = f'(r) \sum_{n=-\infty}^{\infty} d_n H_n^{(1)\prime}[k_0 f(r)]e^{in(\theta)}. \tag{1.20}$$

The unknown coefficients a_n, b_n, c_n, and d_n in each of the aforementioned expressions can be found by applying the proper boundary conditions at interfaces. At the interface $r = b$, the boundary conditions are

$$H_{0z} = H_{1z},\ r = b,\ 0 \leqslant \theta \leqslant 2\pi \tag{1.21a}$$

and

$$E_{0\theta} = E_{1\theta},\ r = b,\ 0 \leqslant \theta \leqslant 2\pi. \tag{1.21b}$$

That is, the tangential components of the total fields continue at the interface $r = b$. Similarly, the boundary conditions at $r = a + \delta$ interface are

$$H_{1z} + ME_{1z} = 0,\ r = a + \delta,\ 0 \leqslant \theta \leqslant 2\pi \tag{1.22a}$$

and

$$H_{1\theta} + ME_{1\theta} = 0,\ r = a + \delta,\ 0 \leqslant \theta \leqslant 2\pi, \tag{1.22b}$$

where M is the admittance parameter of the PEMC material. By applying the above boundary conditions at interfaces $r = a + \delta$ and $r = b$, we drive the subsequent set of equations:

$$J_n(k_0 b) + a_n H_n^{(1)}(k_0 b) = b_n J_n[k_0 f(b)] + c_n H_n^{(1)}[k_0 f(b)], \tag{1.23a}$$

$$J_n'(k_0 b) + a_n H_n^{(1)'}(k_0 b) = b_n f'(b) J_n'[k_0 f(b)] + c_n f'(b) H_n^{(1)'}, \tag{1.23b}$$

$$b_n J_n[k_0 f(a+\delta)] + c_n H_n^{(1)}[k_0 f(a+\delta)] - Mi\eta_0 d_n H_n^{(1)}[k_0 f(a+\delta)] = 0, \tag{1.23c}$$

and

$$\begin{aligned} &d_n H_n^{(1)\prime}\big[k_0 f(a+\delta)\big] - Mi\eta_0 b_n J_n'\big[k_0 f(a+\delta)\big] \\ &\quad - Mi\eta_0 c_n H_n^{(1)\prime}\big[k_0 f(a+\delta)\big] = 0 \end{aligned} \tag{1.23d}$$

The unknown scattering coefficients in the aforementioned equations, after simplifications, are

$$a_n = D_n' B_n' - C_n' \tag{1.24a}$$

$$b_n = D_n' \tag{1.24b}$$

$$c_n = D_n' A_n' \tag{1.24c}$$

$$d_n = \frac{D_n' I_n + D_n' A_n' N_n}{\beta N_n}, \tag{1.24d}$$

where

$$A_n' = \frac{\beta^2 N_n L_n - I_n K_n}{K_n N_n - \beta^2 K_n N_n}, \quad B_n' = \frac{C_n' + A_n' H_n}{I_n}, \quad C_n' = \frac{E_n}{I_n},$$

$$D_n' = \frac{C_n' B_n - A_n}{B_n B_n' - C_n - A_n' D_n}, \quad A_n = J_n(k_0 b), \quad B_n = H_n^{(1)}(k_0 b),$$

$$C_n = J_n[k_0 f(b)], \quad D_n = H_n^{(1)}[k_0 f(b)], \quad E_n = J_n'(k_0 b),$$

$$F_n = H_n^{(1)}\left(k_0' b\right), \quad G_n = f'(b) J_n'[k_0 f(b)], \quad H_n = f'(b) H_n^{(1)}[k_0 f(b)],$$

$$I_n = J_n[k_0 f(a + \delta)], \quad K_n = H_n^{(1)}[k_0 f(a + \delta)], \quad L_n = J_n'[k_0 f(a + \delta)],$$

$$N_n = H_n^{(1)}[k_0 f(a + d)], \text{ and } B = iM_0.$$

It is observed that the cloaked medium is free space under the matched condition, as Yan *et al* mentioned in [20]. Taking this into account, it is evident from the fields mentioned above that $a_n = c_n$ and $b_n = 1$. The aforementioned coefficients in equation (1.24) now become

$$c_n = \frac{\beta^2 N_n L_n - I_n K_n}{K_n N_n (1 - \beta^2)} \tag{1.25}$$

and

$$d_n = \frac{\beta N_n L_n - \beta K_n I_n}{K_n N_n (1 - \beta^2)}. \tag{1.26}$$

The scattering coefficients for the PEC and PMC cases may be found, since they represent the limiting cases of PEMC [26]; these are

$$M \to 0(\text{PMC}): c_n = \frac{-I_n}{N_n} = \frac{-J_n[k_0 f(a + d)]}{H_n^{(1)}[k_0 f(a + d)]} \tag{1.27}$$

and

$$M \to \pm\infty(\text{PEC}): c_n = \frac{L_n}{K_n} = \frac{J_n'(k_0 f(a + \delta))}{H_n^{(1)}(k_0 f(a + \delta))}. \tag{1.28}$$

These coefficients are the same as those discussed in [20], and d_n being the cross-polarized coefficient becomes zero for the PMC or PEC case.

1.3.1.3 Results and discussion

We now present the numerical results derived from the analytical formulations for a cylindrical invisibility cloak with a PEMC layer at δ. The plots illustrate the zeroth-, first-, and second-order co- and cross-polarized scattering coefficients for the TE case. We set the inner and outer boundary radii at $a = 0.1$ and $b = 0.2$, respectively, with a perturbation range of $10^{-8}a < \delta < 10^{-2}a$. The frequency of the incident wave is 2 GHz. In all plots, the solid line represents the co-polarized coefficient, while the dotted line corresponds to the cross-polarized scattering coefficient. The zeroth-order coefficients are specifically plotted to highlight the characteristics of the cylindrical invisibility cloak.

Figure 1.2 shows the co- and cross-polarized scattering coefficients corresponding to the PMC case (i.e. $M = 0$). A comparison with the results in [20] shows strong agreement, with the cross-polarized coefficient vanishing and the co-polarized coefficient decreasing as δ decreases. Figure 1.3 exhibits the scattering coefficients for $M = \pm 1$, demonstrating improved convergence as compared to the $M = 0$ case.

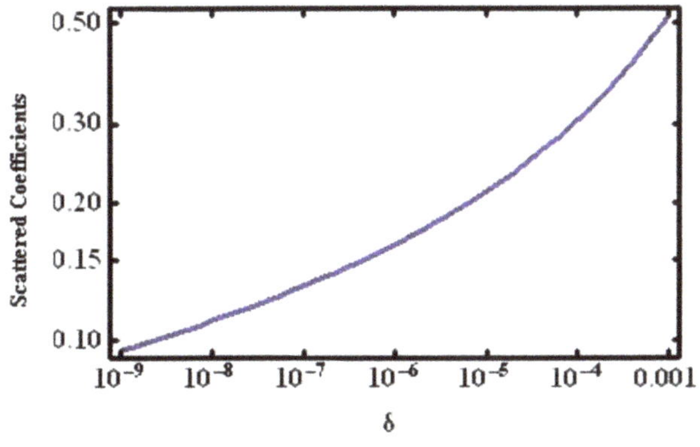

Figure 1.2. Plot of the zeroth-order scattering coefficients with parameters $a = 0.1$, $b = 0.2$, $M = 0$, assuming that the medium within the clock is free space. Adapted from [55], CC0.

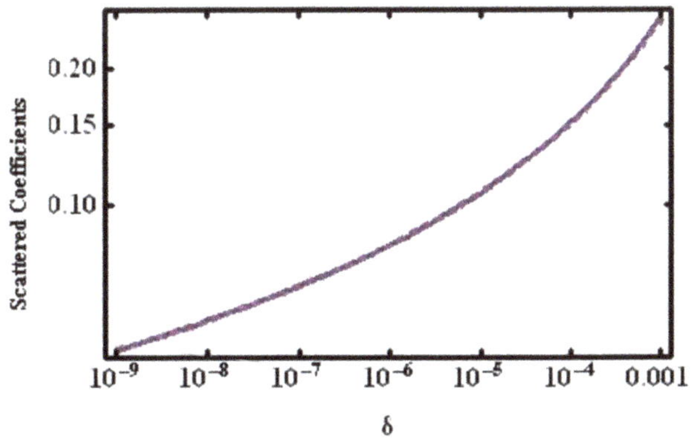

Figure 1.3. Plot of the zeroth-order scattering coefficients with parameters $a = 0.1$, $b = 0.2$, $M = \pm 1$, assuming that the medium within the clock is free space. Adapted from [55], CC0.

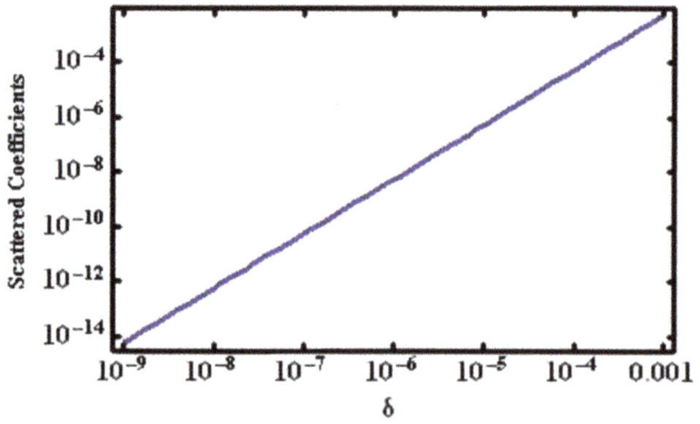

Figure 1.4. Plot of the zeroth-order scattering coefficients with parameters $a = 0.1$, $b = 0.2$, $M \to \pm\infty$, assuming that the medium within the clock is free space. Adapted from [55], CC0.

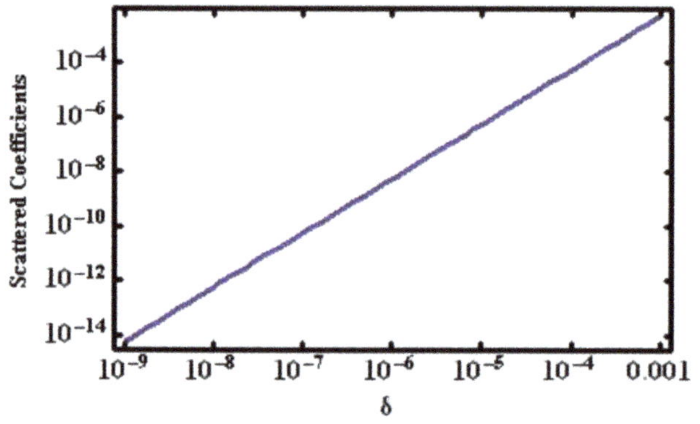

Figure 1.5. Plot of the zeroth-order scattering coefficients with parameters $a = 0.1$, $b = 0.2$, $M = 0$, assuming that the medium within the clock is free space. Adapted from [55], CC0.

This plot exhibits a better convergence rate than that reported in [20]. As δ decreases from $10^{-5}a$ to $10^{-8}a$, the scattering coefficients reduce from 0.0815 to 0.0476.

Figure 1.4 presents the scattered coefficients when $M \to \pm\infty$ (i.e. the PEC case) and we observe the result to be in good agreement with that in [20]. We observe that the cross-polarized coefficient is no longer present, and the convergence rate of the scattering coefficients is significantly improved.

Figures 1.5–1.7 illustrate the first-order co- and cross-polarized scattering coefficients for various values of M. Figure 1.5 shows these coefficients for $M = 0$, corresponding to the PMC case. The plots clearly demonstrate that, as δ decreases, the co-polarized component also diminishes, while the cross-polarized coefficient approaches zero. Figure 1.6 depicts the co- and cross-polarized coefficients for $M = \pm 1$. Here, the results corresponding to the cross-polarized case agree well with those reported in [20], and the convergence rate of the co-polarized field shows significant improvement over the results in [20]. As δ decreases from $10^{-5}a$ to $10^{-8}a$, the scattering coefficients for the co- and cross-polarized components

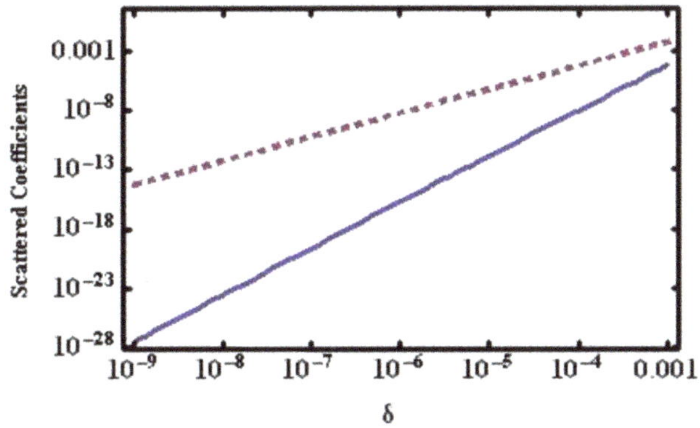

Figure 1.6. Plots of the first-order scattering coefficients with parameters $a = 0.1$, $b = 0.2$, $M = \pm 1$, assuming that the medium within the clock is free space. Adapted from [55], CC0.

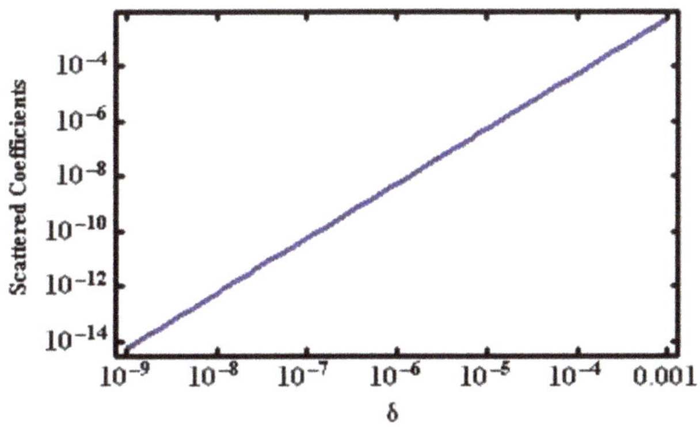

Figure 1.7. Plot of the first-order scattering coefficients with parameters $a = 0.1$, $b = 0.2$, $M \to \pm\infty$, assuming that the medium within the clock is free space. Adapted from [55], CC0.

reduce from 1.81×10^{-16} to 3.12×10^{-28} and 5.50×10^{-9} to 5.50×10^{-15}, respectively. Figure 1.7 exhibits the co- and cross-polarized scattering coefficients for $M \to \pm\infty$, representing the PEC case, which shows excellent agreement with the previously published results [20].

Figures 1.8–1.10 show the co- and cross-polarized scattering coefficients for varying values of M. Figure 1.8 presents these coefficients for $M = 0$, representing the PMC case. The plots clearly indicate that as δ decreases, the co-polarized coefficient also diminishes, while the cross-polarized coefficient becomes negligible. Figure 1.9 displays the coefficients for $M = \pm 1$, where the cross-polarized field aligns well with the results in [20], and the co-polarized field shows a notably improved convergence rate compared to that reported in [20]. Specifically, as δ decreases from $10^{-5}a$ to $10^{-8}a$, the scattering coefficients reduce from 2.81×10^{-27} to 2.95×10^{-45} for the co-polarized field and from 4.82×10^{-15} to 4.82×10^{-30} for the cross-polarized field. Figure 1.10 illustrates the coefficients for $M \to \pm\infty$,

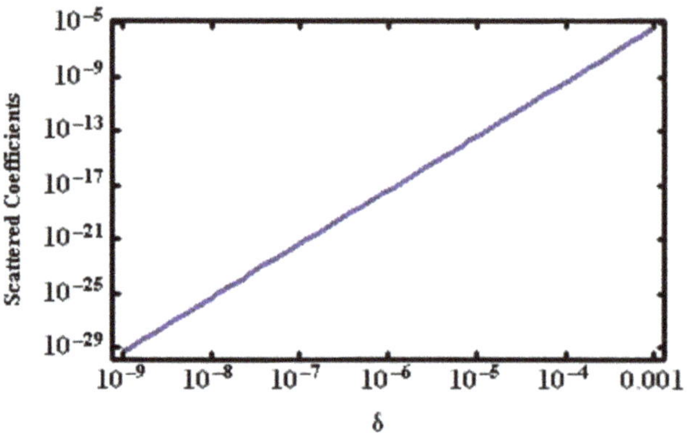

Figure 1.8. Plot of the second-order scattering coefficients with parameters $a = 0.1$, $b = 0.2$, $M = 0$, assuming that the medium within the clock is free space. Adapted from [55], CC0.

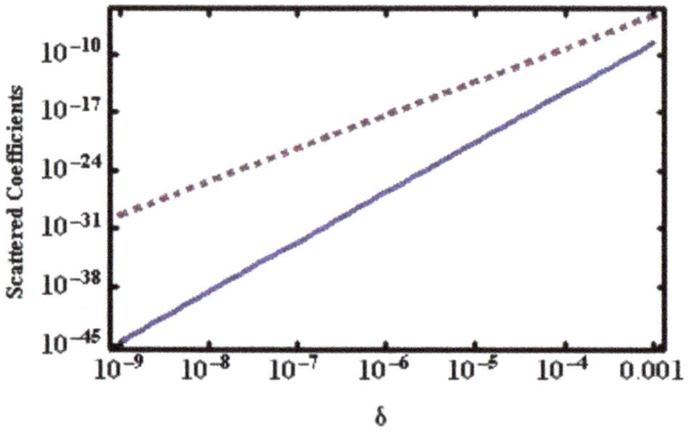

Figure 1.9. Plot of the second-order scattering coefficients with parameters $a = 0.1$, $b = 0.2$, $M = \pm 1$, assuming that the medium within the clock is free space. Adapted from [55], CC0.

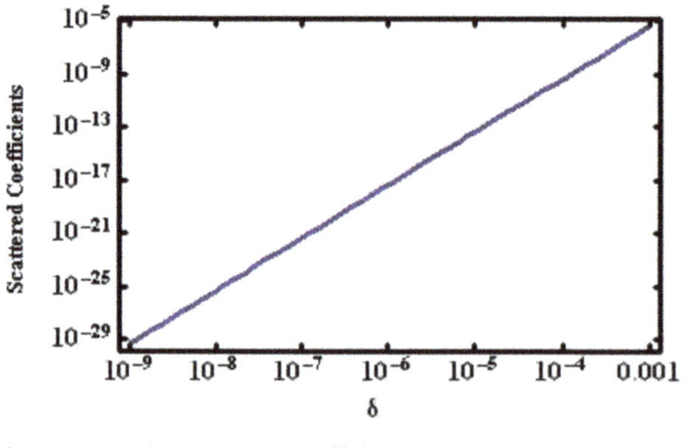

Figure 1.10. Plot of the second-order scattering coefficients with parameters $a = 0.1$, $b = 0.2$, $M \to \pm\infty$, assuming that the medium within the clock is free space. Adapted from [55], CC0.

corresponding to the PEC case, showing excellent agreement with the existing literature in this regard.

The above-presented plots for PEMC material are identical in the case of transverse magnetic (TM) incidence.

In this study, we focus on the convergence rate of the co- and cross-polarized scattered fields in the context of an ideal cloak, where ideally these fields should vanish—a scenario not yet achievable. The observed convergence rate in this work surpasses that reported in previous studies [20] where PEMC materials were employed. The dual nature of PEMC contributes to an improved convergence rate, and notably, this rate is independent of the type of incident field when a PEMC medium is utilized. The superior convergence rate observed here is attributed to the dual nature of PEMCs and their admittance parameter. Optimal results are achieved when $M = \pm 1$, leading to an enhanced convergence rate of the coefficients. Furthermore, adjusting the admittance parameter M allows for further improvement in the convergence rate of the scattered coefficients.

1.3.2 Nihility cloak

In this section, we examine the scattering characteristics of the cylindrical invisible nihility cloak and address the improvement of the convergence rate of scattered fields in free space. Specifically, for the nihility case, the zeroth-order Bessel function contribution shifts to the first-order when $n = 0$, leading to a reduction in scattered fields, as presented in [56]. The subsequent sections focus on the formulation of the problem for both TE- and TM-polarized incident radiation.

1.3.2.1 Formulation of the nihility cloak

Figure 1.11 exhibits the geometry of a 2D cylindrical nihility cloak. An invisibility cloak functions by compressing EM fields within a cylindrical region defined by

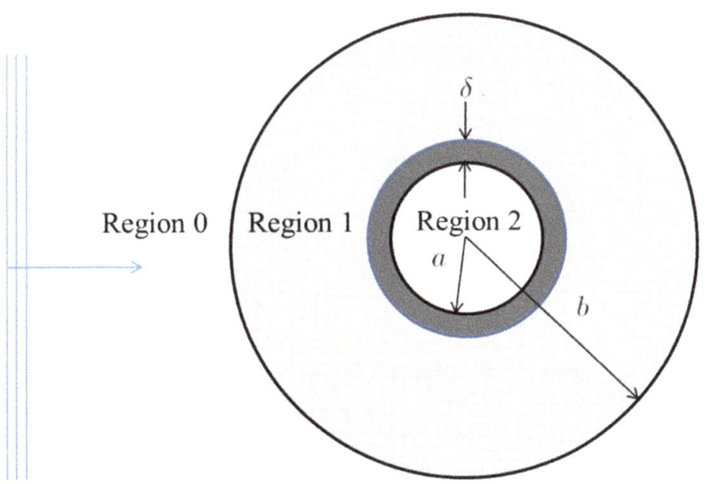

Figure 1.11. Schematic representation of an invisible nihility cloak.

$r' < b$ into a concentric cylindrical shell between $a < r < b$. The area outside the radius b is referred to as region 0, which represents free space. In this region, the wavenumber k_0 and impedance η_0 are given by $k_0 = \omega\sqrt{\mu_0 \epsilon_0}$ and $\eta_0 = \sqrt{\mu_0/\epsilon_0}$, respectively.

Following figure 1.11, region 1 is represented as $(a + \delta) < r < b$, composed of an unknown medium, and region 2 is $r < a + \delta$, which, in the first instance, is assumed to be a dielectric medium with $k_2 = \omega\sqrt{\mu_2 \epsilon_2}$ as the wavenumber and $\eta_2 = \sqrt{\mu_2/\epsilon_2}$ as impedance.

Consider the coordinate transformation in cylindrical coordinates [11] as

$$f(r) = r' = a + \frac{r(b - a)}{b}, \tag{1.29}$$

given that $f(a) = 0$ and $f(b) = b$, while θ and z are kept constant. The permittivity and permeability of the cloak region can be derived using the coordinate transformation method [11] as

$$\epsilon_r = \mu_r = \frac{f(r)}{rf'(r)}, \tag{1.30}$$

$$\epsilon_\theta = \mu_\theta = \frac{f(r)f'(r)}{f(r)}, \tag{1.31}$$

and

$$\epsilon_z = \mu_z = \frac{f(r)f'(r)}{r}, \tag{1.32}$$

where the prime ($'$) denotes differentiation with respect to r.

1.3.2.2 Transverse electric (TE) case

We consider the case where a TE wave incident from free space interacts with a cloak, whose core is assumed to be composed of a dielectric medium. The total electromagnetic field in the region $r > b$ is expressed as

$$H_z = \sum_{n=-\infty}^{\infty} J_n(k_0 r)e^{jn(\theta)} + a_n H_n(k_0 r)e^{jn(\theta)} \tag{1.33}$$

and

$$E_\theta = \frac{k_0}{j\omega\, \epsilon_0}\sum_{n=-\infty}^{\infty} J_n'(k_0 r)e^{jn(\theta)} + a_n H_n'(k_0 r)e^{jn(\theta)}. \tag{1.34}$$

In this context, $J_n(\cdot)$ and $H_n(\cdot)$ denote Bessel and Hankel functions, respectively, while a_n represents the scattering coefficient for the region where $r > b$. The inclusion of these scattering coefficients is crucial due to the dielectric nature of region 1, which necessitates reflection and transmission of the electromagnetic field at the boundary between regions 0 and 1. An ideal cloak is characterized by the absence of scattering, whereas a cloak with minimal scattering closely approximates this ideal condition. Hence, the consideration of scattering coefficients is imperative.

The total electromagnetic field for the region $a < r < b$ is given as

$$H_z = \sum_{n=-\infty}^{\infty} b_n J_n \left[k_1(r - a)\right] e^{jn(\theta)} + c_n H_n \left[k_1(r - a)\right] e^{jn(\theta)} \tag{1.35}$$

and

$$E_\theta = \frac{k_1}{j\omega \ \epsilon_\theta(r)} \sum_{n=-\infty}^{\infty} b_n J_n' [k_1(r - a)] e^{jn(\theta)} + c_n H_n' [k_1(r - a)] e^{jn(\theta)}, \tag{1.36}$$

with $k_1 = k_0 b/(b - a)$ and b_n, c_n as the coefficients for transmission and reflection, respectively.

In the region $r < a$, the fields are

$$H_z = \sum_{n=-\infty}^{\infty} d_n J_n \left(k_2 r\right) e^{jn(\theta)} \tag{1.37}$$

and

$$E_\theta = \frac{k_2}{jw\epsilon_2} \sum_{n=-\infty}^{\infty} d_n J_n' (k_2 r) e^{jn(\theta)}, \tag{1.38}$$

where d_n is the transmission coefficient for region $r < a$.

The boundary conditions necessitate the continuity of the tangential components of the **E** and **H** fields across the interfaces at $r = b$ and $r = a + \delta$. By applying these conditions and leveraging the impedance matching principles discussed in [14–16], the unknown scattering coefficient a_n in free space can be determined as follows:

$$a_n = \frac{k_2 J_n(k_1 \delta) - \frac{k_1}{\epsilon_\theta(a+\delta)} L_n[k_0(a + \delta)] J_n' [k_0(a + \delta)]}{\frac{k_1}{\epsilon_\theta(a+\delta)} L_n[k_0(a + \delta)] H_n' [k_0(a + \delta)] - k_2 H_n(k_1 \delta)}, \tag{1.39}$$

where

$$L_n(\chi) = \frac{J_n(\chi)}{J_n'(\chi)}. \tag{1.40}$$

1.3.2.3 Transverse magnetic (TM) case

Next, we consider the case where a TM wave traveling from free space is incident on the cloak. The total electromagnetic field in the region $r > b$ can be expressed as

$$E_z = \sum_{n=-\infty}^{\infty} J_n \left(k_0 r\right) e^{jn(\theta)} + a_n' H_n \left(k_0 r\right) e^{jn(\theta)} \tag{1.41}$$

and

$$H_\theta = \frac{-k_0}{j\omega \ \mu_0} \sum_{n=-\infty}^{\infty} J_n' (k_0 r) e^{jn(\theta)} + a_n' H_n' (k_0 r) e^{jn(\theta)}. \tag{1.42}$$

The total electromagnetic field within the region $a < r < b$ is expressed as

$$E_z = \sum_{n=-\infty}^{\infty} b_n' J_n\left[k_1(r-a)\right] e^{jn(\theta)} + c_n' H_n\left[k_1(r-a)\right] e^{jn(\theta)} \tag{1.43}$$

and

$$H_\theta = \frac{-k_1}{j\omega\, \mu_\theta(r)} \sum_{n=-\infty}^{\infty} b_n' J_n'\left[k_1(r-a)\right] e^{jn(\theta)} + c_n' H_n'\left[k_1(r-a)\right] e^{jn(\theta)}. \tag{1.44}$$

In the region $r < a$, the fields are

$$E_z = \sum_{n=-\infty}^{\infty} d_n' J_n(k_2 r) e^{jn(\theta)} \tag{1.45}$$

and

$$H_\theta = \frac{-k_2}{j\omega\, \mu_2} \sum_{n=-\infty}^{\infty} d_n'(k_2 r) e^{jn(\theta)}, \tag{1.46}$$

where the scattering coefficients a_n', b_n', c_n', and d_n' correspond to the TM case and are determined by applying the appropriate boundary conditions at $r = b$ and $r = a + \delta$. The free space scattering coefficients a_n' can be expressed as

$$a_n' = \frac{k_2 J_n(k_1\delta) - A^* J_n'\left[k_0(a+\delta)\right]}{A^* H_n'\left[k_0(a+\delta)\right] - k_2 H_n(k_1\delta)}, \tag{1.47}$$

where

$$\mathrm{A} = \frac{k_1}{\mu_\theta(a+\delta)} L_n\left[k_0(a+\delta)\right]. \tag{1.48}$$

1.3.2.4 Limiting process for the nihility cloak

Now, the refractive index of nihility must be null-valued because $\epsilon_2 = \mu_2 = 0$, as discussed by Lakhtahia in [46]. For the function $L_n(\chi)$, we have

$$\lim_{\chi \to 0} \chi\, L_0(\chi) = -2, \; n = 0 \tag{1.49}$$

and

$$\lim_{\chi \to 0} \chi^{-1} L_0(\chi) = n^{-1}, \; n \neq 0. \tag{1.50}$$

Therefore, after taking the limit $k_2 \to 0$, equations (1.39) and (1.47) for a nihility cloak simplify to

$$a_0 = a_0' = \frac{J_1\left[k_0(a+\delta)\right]}{H_1\left[k_0(a+\delta)\right]}, \; n = 0 \tag{1.51}$$

and

$$a_n = a_n' = \frac{J_{|n|}[k_0(a+\delta)]}{H_{|n|}[k_0(a+\delta)]}, \quad n \neq 0. \tag{1.52}$$

The results from the above equations indicate that the nihility medium, characterized by $\epsilon_2 = \mu_2 = 0$, corresponds to a void region due to the absence of any fields.

1.3.2.5 Numerical results and discussion

We now present numerical results derived from the proposed geometry for a cylindrical invisible cloak incorporating a nihility layer at δ. We set the free space wavenumber to $k_0 = 0.064/\pi$. The radius of the inner boundary is defined as $a = 1.2\pi/k_0$, and that of the outer boundary is $b = 2\pi/k_0$. The perturbation parameter δ varies within the range $10^{-8}a < \delta < 10^{-2}a$.

Figure 1.12 presents the scattering coefficient for $n = 0$, comparing it with the existing literature and demonstrating significant improvement. In this figure, the dotted line represents the nihility case, while the solid line indicates the scenario without a PEC [20]. The scattering coefficient approaches 0.1 for $\delta = 10^{-8}$ in the absence of nihility, whereas it approaches 10^{-19} in the nihility case at the same value of δ. This suggests a superior convergence rate for the nihility core. Previous studies [14, 20] suggest that optimal cloaking may be achieved using a PEC core for TM polarization and a PMC core for TE polarization. In these configurations, the zeroth-order scattering coefficient depends on the first-order cylindrical wave function, as nihility provides better impedance matching with free space. Figure 1.13 compares the scattering coefficient for $n = 0$ with the case where a PEC layer is incorporated at δ [20], and the results show good agreement with those reported in [20].

Figures 1.14 and 1.15, respectively, exhibit the first- and second-order scattering coefficients corresponding to $n = 1$ and $n = 2$. We compare these results with the plots presented in [20], and find them to be in good agreement.

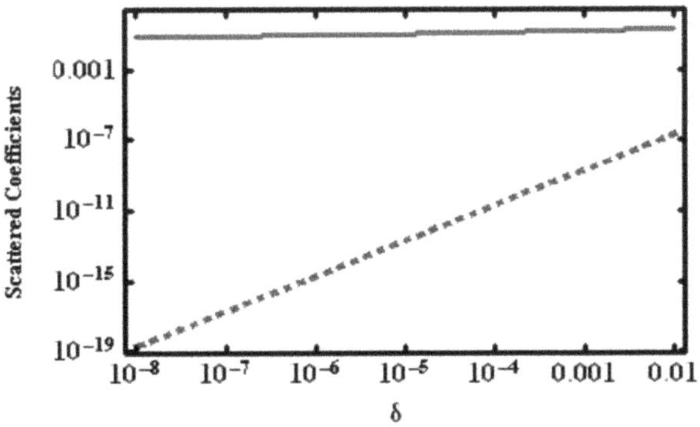

Figure 1.12. Zeroth-order scattering coefficients with parameters $a = 1.2\pi/k_0$ and $b = 2\pi/k_0$. The blue line corresponds to the non-PEC case [20] and the dotted red line represents the nihility case. Adapted from [56], CC BY 4.0.

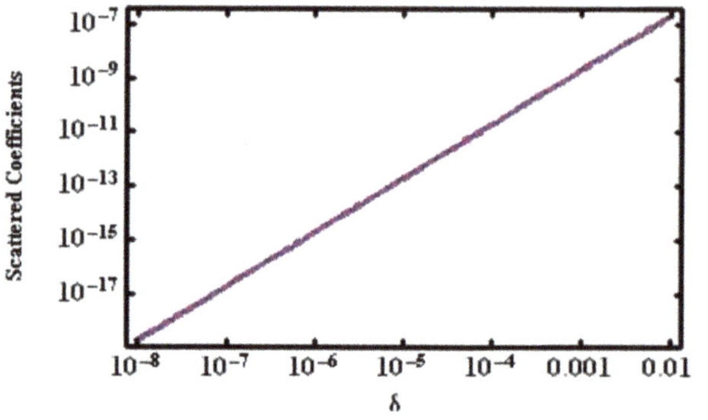

Figure 1.13. Zeroth-order scattering coefficients with parameters $a = 1.2\pi/k_0$ and $b = 2\pi/k_0$. The blue line corresponds to the PEC case [20] and the dotted red line represents the nihility case. Adapted from [56], CC BY 4.0.

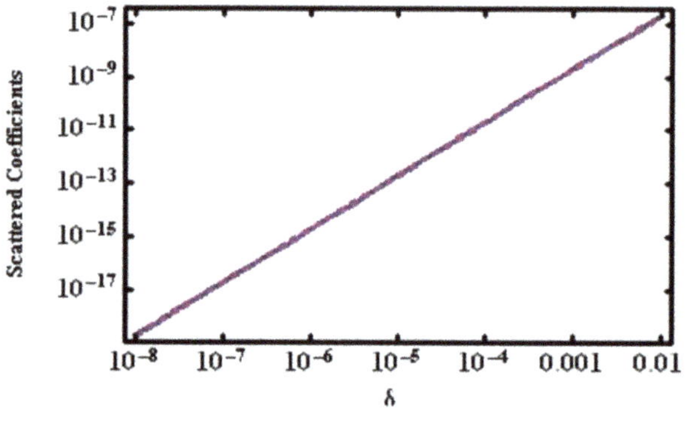

Figure 1.14. First-order scattering coefficients with parameters $a = 1.2\pi/k_0$ and $b = 2\pi/k_0$. The blue line corresponds to the PEC case [20] and the dotted red line represents the nihility case. Adapted from [56], CC BY 4.0.

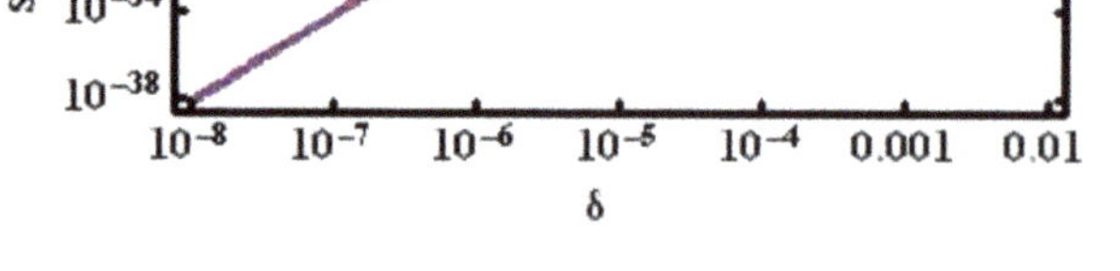

Figure 1.15. Second-order scattering coefficients with parameters $a = 1.2\pi/k_0$ and $b = 2\pi/k_0$. The blue line corresponds to the PEC case [20] and the dotted red line represents the nihility case. Adapted from [56], CC BY 4.0.

For the case of nihility medium at the perturbed region, the same coefficients are obtained corresponding to the TE and TM polarization cases as given in equations (1.51) and (1.52). This indicates that these coefficients are unaffected by the polarization of the incident field. A primary advantage of employing nihility is its approximation to the ideal scenario, which is a crucial feature of cloaking. In the nihility case, the zeroth-order term of Bessel functions transitions to the first order for $n = 0$ (see equation (1.51)), thereby reducing the scattered fields.

1.4 Conclusion

Based on the analysis of the co- and cross-polarized scattering coefficients, it can be concluded that a cylindrical invisibility cloak incorporating PEMC material at a distance δ can be realized, representing a generalization of the PEC and PMC cases. The numerical results lead to the following conclusions.

1. For zeroth-order scattering, PEMC demonstrates a superior convergence rate compared to PMC, while the PEC limiting case of PEMC yields results consistent with those reported in [14, 20].
2. For the first- and second-order scattering, PEMC shows a significantly better convergence rate than both the PEC and PMC, particularly for the co-polarized scattering.
3. The convergence rate remains consistent regardless of the incident field type when PEMC is employed instead of PEC or PMC.

From the analytical calculations performed for the nihility cloak corresponding to both the TM and TE polarization cases, the following conclusions can be drawn.

a. It has been demonstrated that by employing nihility in the δ layer of the geometry, one can achieve a cloaking effect that closely approximates the ideal scenario.
b. Additionally, in the presence of nihility, the scattering coefficients remain invariant with respect to the type of incident field.

References

[1] Pendry J B 2000 Negative refraction makes a perfect lens *Phys. Rev. Lett.* **85** 3966–9
[2] Schurig D, Mock J J, Justice B J, Cummer S A, Pendry J B, Starr A F and Smith D R 2006 Metamaterial electromagnetic cloak at microwave frequencies *Science* **314** 977–80
[3] Xu T, Xu Y, Zhang Y and Li H 2013 Loss compensated broadband metamaterial waveguides *Opt. Express* **21** 15163–71
[4] Yoo Y J and Lim S 2014 Polarization reconfigurable antenna using metamaterial *IEEE Trans. Antennas Propagat.* **62** 5004–10
[5] Zhou B and Cui T J 2009 Directivity enhancement to Vivaldi antennas using compactly anisotropic zero-index metamaterials *Electron. Lett.* **45** 1288–90
[6] Ruan Z and Yan M 2007 Cylindrical invisibility cloak with simplified material parameters is inherently visible *Phys. Rev. Lett.* **99** 113903
[7] Alù A and Engheta N 2005 Achieving transparency with plasmonic and metamaterial coatings *Phys. Rev.* E **72** 016623

[8] Smith D R, Pendry J B and Wiltshire M C K 2004 Metamaterials and negative reflection index *Science* **305** 788–92
[9] Shalaev V M 2007 Optical negative index metamaterials *Nature Photon.* **1** 41–8
[10] Shelby R A, Smith D R and Schults S 2001 Experimental verification of a negative index of refraction *Science* **292** 77–9
[11] Pendry J B and Schuring D R 2006 Controlling electromagnetic fields *Science* **312** 1780–2
[12] Cummer S A, Popa B I, Schurig D, Smith D R and Pendry B 2006 Full-wave simulations of electromagnetic cloaking structures *Phys. Rev.* E **74** 036621
[13] Cai W, Chettiar U K, Kildishev A V and Shalaev V M 2007 Optical cloaking with metamaterials *Nature Photon.* **1** 224–7
[14] Ruan Z, Yan M, Neff C W and Qiu M 2007 Ideal cylindrical cloak: perfect but sensitive to tiny perturbations *Phys. Rev. Lett.* **99** 113903
[15] Yan M, Ruan Z and Qiu M 2007 Scattering characteristics of simplified cylindrical invisibility cloaks *Opt. Express* **15** 17770–82
[16] Greenleaf A, Kurylev Y, Lassas M and Uhlmann G 2007 Improvement of cylindrical cloaking with the SHS lining *Opt. Express* **15** 12717–34
[17] Zhang B, Chen H, Wu B I, Luo Y, Ran L and Kong J A 2007 Response of a cylindrical invisibility cloak to electromagnetic waves *Phys. Rev.* B **76** 121101
[18] Cai W, Chettiar U K, Kildishev A V, Shalaev V M and Milton G W 2007 Nonmagnetic cloak with minimized scattering *Appl. Phys. Lett.* **91** 111105
[19] Yan W, Yan M, Ruan Z and Qiu M 2008 Coordinate transformation makes perfect invisibility cloaks with arbitrary shape *New J. Phys.* **10** 043040
[20] Yan W, Yan M, Ruan Z and Qiu M 2008 Influence of geometrical perturbation at inner boundaries of invisibility cloaks *J. Opt. Soc. Am.* A **25** 968–73
[21] Zolla F, Guenneau S, Nicolet A and Pendry J B 2007 Electromagnetic analysis of cylindrical invisibility cloaks and the mirage effect *Opt. Lett.* **32** 1069–71
[22] Leonhardt U and Philbin T G 2006 General relativity in electrical engineering *New J. Phys.* **8** 247
[23] Yan W, Yan M and Qiu M 2008 Necessary and sufficient conditions for reflectionless transformation media in an isotropic and homogeneous background arXiv:0806.3231v1
[24] Chen H, Wu B I, Zhang B and Kong J A 2007 Electromagnetic wave interactions with a metamaterials cloak *Phys. Rev. Lett.* **99** 063903
[25] Greenleaf A, Kurylev Y, Lassas M and Uhlmann G 2008 Full-wave invisibility of active devices at all frequencies *Commun. Math. Phys.* **275** 749–89
[26] Lindell I V and Sihvola A H 2005 Perfect electromagnetic conductor *J. Electromagn. Waves Appl.* **19** 861–9
[27] Lindell I V 2004 *Differential Forms in Electromagnetics* (New York: Wiley and IEEE Press)
[28] Lindell I V and Sihvola A H 2005 Realization of the PEMC boundary *IEEE Trans. Antennas Propagat.* **53** 3012–8
[29] Lindell I V and Sihvola A H 2005 Transformation method for problems involving perfect electromagnetic conductor (PEMC) structures *IEEE Trans. Antennas Propagat.* **53** 3005–11
[30] Ruppin R 2006 Scattering of electromagnetic radiation by a perfect electromagnetic conductor cylinder *J. Electromagn. Waves Appl.* **20** 1853–60
[31] Lindell I V and Sihvola A H 2006 Losses in PEMC boundary *IEEE Trans. Antennas Propagat.* **54** 2553–8

[32] Sihvola A and Lindell I V 2006 Possible applications of perfect electromagnetic conductor (PEMC) media *Proc. of the 1st European Conf. on Antennas and Propagat. (EuCAP 2006) (Nice, France, 6–10 November 2006)* (ESA SP-626)

[33] Lindell I V and Sihvola A H 2006 The PEMC resonator *J. Electromagn. Waves Appl.* **20** 849–59

[34] Jancewicz B 2006 Plane electromagnetic wave in PEMC *J. Electromagn. Waves Appl.* **20** 647–59

[35] Ahmed S and Naqvi Q A 2008 Electromagnetic scattering from a perfect electromagnetic conductor cylinder buried in a dielectric half-space *Prog. Electromagn. Res.* **78** 25–38

[36] Ahmed S and Naqvi Q A 2008 Electromagnetic scattering from parallel perfect electromagnetic conductor cylinders of circular cross-sections using iterative procedure *J. Electromagn. Waves Appl.* **22** 987–1003

[37] Ahmed S and Naqvi Q A 2008 Electromagnetic scattering from a two-dimensional perfect electromagnetic conductor (PEMC) strip and PEMC strip grating simulated by circular cylinders *Opt. Commun.* **281** 4211–8

[38] Ahmed S and Naqvi Q A 2008 Electromagnetic scattering from a perfect electromagnetic conductor cylinder coated with a metamaterial having negative permittivity and/or permeability *Opt. Commun.* **281** 5664–70

[39] Ahmed S and Naqvi Q A 2008 Electromagnetic scattering of two or more incident plane waves by a perfect electromagnetic conductor cylinder coated with a metamaterial *Prog. Electromagn. Res.* B **10** 75–90

[40] Fiaz M A, Ghaffar A and Naqvi Q A 2008 High frequency expressions for the field in the caustic region of a PEMC cylindrical reflector using Maslov's method *J. Electromagn. Waves Appl.* **22** 358–97

[41] Fiaz M A, Aziz A, Ghaffar A and Naqvi Q A 2008 High frequency expression for the field in the caustic region of a PEMC Gregorian system using Maslov's method *Prog. Electromagn. Res.* **81** 135–48

[42] Illahi A and Naqvi Q A 2009 Scattering of an arbitrarily oriented dipole field by an infinite and a finite length PEMC circular cylinder *Central Eur. J. Phys.* **7** 829–53

[43] Illahi A, Afzaal M and Naqvi Q A 2008 Scattering of dipole field by a perfect electromagnetic conductor cylinder *Prog. Electromagn. Res. Lett.* **4** 43–53

[44] Shahzad A, Ahmed S and Naqvi Q A 2010 Analysis of electromagnetic field due to a buried coated PEMC circular cylinder *Opt. Commun.* **283** 4563–71

[45] Ahmed S, Mannan F, Shahzad A and Naqvi Q A 2011 Electromagnetic scattering from a chiral-coated PEMC cylinder *Prog. Electromagn. Res.* M **19** 239–50

[46] Lakhtakia A 2001 An electromagnetic trinity from 'negative permittivity' and 'negative permeability' *Int. J. Infrared Millimeter Waves* **22** 1731–4

[47] Lakhtakia A and Godden J B 2007 Scattering by nihility cylinder *Int. J. Electron. Commun.* **61** 62–5

[48] Lakhtakia A, McCall M W and Weiglhofer W S 2002 Brief overview of recent developments on negative phase-velocity mediums (alias left-handed materials) *Int. J. Electron. Commun.* **56** 407–10

[49] Ramakrishna S A 2005 Physics of negative refractive index materials *Rep. Prog. Phys.* **68** 449–521

[50] Lakhtakia A 2002 On perfect lenses and nihility *Int. J. Infrared Millimeter Waves* **23** 339–43

[51] Ziolkowski R W 2004 Propagation in and scattering from a matched metamaterial having a zero index of refraction *Phys. Rev.* E **70** 046608
[52] Ahmed S and Naqvi Q A 2010 Electromagnetic scattering from a chiral-coated nihility cylinder *Prog. Electromagn. Res. Lett.* **18** 41–50
[53] Kong J A 1971 Charged particles in bianisotropic media *Radio Sci.* **6** 1015–9
[54] Lakhtakia A 1996 On homogenization of impedance-matched chiral-in-chiral composites *J. Phys. D: Appl. Phys.* **29** 957
[55] Shahzad A, Qasim F, Ahmed S and Naqvi Q 2011 Cylindrical invisibility cloak incorporating PEMC at perturbed void region *Prog. Electromagn. Res.* M **21** 61–76
[56] Shahzad A, Ahmed S, Ghaffar A and Naqvi Q A 2014 Incorporation of the nihility medium to improve the cylindrical invisibility cloak *ACES J.* **29** 9–14

IOP Publishing

Ordered and Disordered Metamaterials
Design and applications
Pankaj K Choudhury and Tatjana Gric

Chapter 2

Reconfigurable ordered microwave metamaterials for smart scattering control

Xinyu Jiang and Yungui Ma

Dynamically controllable reconfigurable metamaterials operating in the microwave bands were reviewed. In particular, to achieve the property of reconfigurability, active devices or materials that may respond to different external stimuli were integrated into passive meta-atoms. Within this context, some of the most commonly used modulation methods were discussed. Further, reconfigurable metamaterials with all elements controlled uniformly and those with each element controlled independently were briefly touched upon. Apart from these, reconfigurable metasurfaces incorporating artificial intelligence or adopting temporal modulation were discussed briefly, as these would further improve the performance of, or extend, applications.

2.1 Introduction

Metamaterials are artificial electromagnetic structures composed of periodically or aperiodically arranged subwavelength elements, which, due to their ability to flexibly manipulate the amplitude, phase and polarization of electromagnetic waves, have aroused tremendous research interest. They can be divided into two general categories according to the structures of elements, namely two-dimensional (2D) and three-dimensional (3D) versions. 3D metamaterials can provide higher design freedom. However, there are also some drawbacks, for instance, complex structure, bulky volume, difficult fabrication and relatively large loss. 2D metamaterials, also known as metasurfaces, consist of one or more layers of periodic or aperiodic arrays of subwavelength elements, and possess a great number of advantages, such as subwavelength profile, small size, light weight, low loss, easy fabrication and potential conformability [1, 2].

Metamaterials have been developed rapidly in the past few decades, and there are various designs and implementations from the microwave band to the optical band.

doi:10.1088/978-0-7503-5462-2ch2

The development of metamaterials operating in the microwave band involves several stages of passive metamaterials with fixed functions, reconfigurable metamaterials based on active materials or active components and programmable metamaterials with each unit cell controlled independently. Moreover, feedback systems have also been ingeniously incorporated into the design of programmable metasurfaces, leading to intelligent metasurfaces that can work adaptively.

This chapter is aimed at giving a brief introduction to reconfigurable microwave metamaterials with identical unit cells arranged periodically. It is worth noting that it is much easier to integrate active components or active materials into meta-atoms when using metasurfaces compared to 3D metamaterials, which facilitates the realization of reconfigurable operations [3]. Therefore, this chapter will focus on the introduction of metasurfaces with 2D structures. The contents mainly include the mechanism of reconfigurability, basic principles of distinct control fashions and diverse functions, followed by the design and analysis methods. Some specific implementations in the literature will also be shown as examples.

2.2 The mechanism of reconfigurability

The function of passive metamaterials is limited and fixed, and cannot be changed after manufacture. Hence, they cannot meet the needs of real time dynamic manipulation of electromagnetic waves, and the application is restricted. Fortunately, by introducing active components or phase change materials into the passive structures, the working modes can be switched or tuned with the help of external stimulus, so as to dynamically manipulate electromagnetic waves. These active metamaterials are also called reconfigurable metamaterials.

There are many kinds of tuning methods that can be used in the design and implementation of reconfigurable metamaterials, such as the electrical, optical, thermal, mechanical, chemical, microfluidic, etc, to mention a few. And these methods may resort to diverse active components and materials. In terms of microwave band, the electrical mode is mostly used. Typically, PIN diodes or varactors are loaded in meta-atoms, and the working states can be reconfigured by applying bias voltages. In the following subsections, some common control mechanisms in the microwave band will be induced briefly, and the basic principles and some examples of reconfigurable microwave metamaterials are presented here.

2.2.1 Electric methods

The most convenient and common way to obtain reconfigurable responses is to use electric modulation. By loading components or materials that are sensitive to electrical stimulation into passive metamaterials, the electromagnetic response of metamaterials can be dynamically controlled by applied voltages. The commonly used active components include varactors, PIN diodes, micro-electro-mechanical systems (MEMSs), transistors and so on, while active materials generally include doped semiconductors, 2D materials, etc.

As for reconfigurable metamaterials operating at microwave frequencies, varactors and PIN diodes are the most commonly used active components. The equivalent circuit models of varactors and PIN diodes in the *on* and *off* states are shown in figure 2.1. The equivalent capacitance of varactors will change with the applied bias voltages, which, in turn, affects the resonance behavior of the meta-atoms. PIN diodes operating in the *on* and *off* states are equivalent to small resistors and capacitors, respectively, and when changing the states of bias the resonance performances of the meta-atoms will also be switchable. Given the extensive use of varactors and PIN diodes, the next sections will refer to many designs based on them, and therefore specific examples in the literature will not be provided. Instead, a simple structure with the corresponding simulated results will be presented for the demonstration of the operating principle of electrically controllable diode-loaded metasurfaces.

Assuming that there is an electrically controlled metasurface, the unit cell is shown schematically in figure 2.2(a), where the periodic metallic structure is placed on an FR-4 substrate with a dielectric constant of 4.3 and a thickness of 0.5 mm, with a square ring-shaped slot across which four PIN diodes are mounted to control the electromagnetic properties of the structure. Taking the parameters of SMP1320-040LF PIN diodes from Skyworks, for example, the simulation results are shown in figure 2.2(b). When the PIN diodes are cut off, each of them is equivalent to an inductor with an inductance of 0.45 nH connected in series with a capacitor with a capacitance of 0.23 pF, and the structure exhibits a band-pass filtering response. When the PIN diodes are switched on, each of them can be regarded as a series

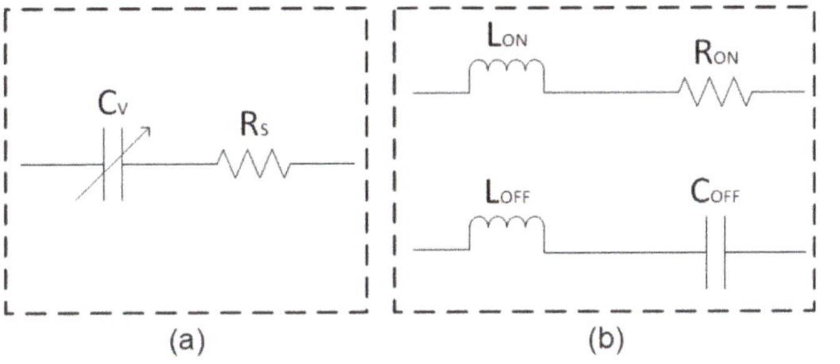

Figure 2.1. The equivalent circuit models of (a) varactors and (b) PIN diodes.

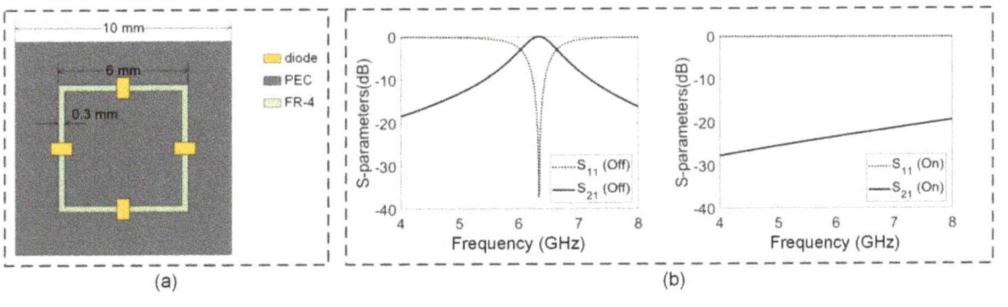

Figure 2.2. (a) The unit cell of a band-pass filter. (b) Simulated scattering parameters when the PIN diodes are switched *off* and *on*.

connection of a 0.75 Ω resistor and a 0.45 nH inductor, and the original pass band around 6.3 GHz is turned off.

If the PIN diodes are replaced with varactors, the increase of the reverse bias voltages will gradually reduce the capacitance of the varactors, and it can be observed that the pass band gradually moves to lower frequency bands.

In addition to varactors and PIN diodes, there are still a great number of other active components and materials that can be applied in reconfigurable microwave metamaterials controlled in electrical manners. For example, the equivalent channel impedance of transistors, controllable by the gate voltages, can be utilized, as in the transistor-based active nonlinear metamaterial absorber presented in [4]. The device realized tunability of the absorption rate under different incident powers when changing the bias voltage of the transistors. Furthermore, a closed-loop structure was proposed, so that the absorber can automatically adjust the absorption rate in real time to adapt to the change of incident electromagnetic wave power. The system used a power meter to measure the power and compared it with the preset threshold, and then automatically controlled the bias voltage of transistors to adjust the absorption rate.

MEMSs are also available for reconfigurable operation, among which electrostatic ones are appropriate for electric modulation. In the electrostatic MEMS-based reconfigurable reflector array presented in [5], two pairs of MEMSs made of polysilicon wires with high resistivity were used to control two linear polarization states in orthogonal directions driven by the electrostatic force. The meta-atoms in the two states have a reflection phase difference of 180°, which was used to achieve the reflection phase coding of 1 bit.

In addition, graphene is one the most commonly used active materials. For instance, a flexible frequency-selective rasorber (FSR) based on graphene was proposed in [6]. A graphene sandwich structure (GSS) was placed between the top lossy layer and the bottom lossless layer of the FSR, which acted as a tunable resistive sheet. By varying the DC voltage applied to the GSS, the tuning of pass band transmittance was realized.

Further, a reconfigurable resonator based on organic electrochemical transistors (OECT) was proposed in [7], which is schematically illustrated figure 2.3. It is noteworthy that researchers designed and prepared the actives devices by themselves, rather than adopting any commercially available transistors. Polyimide substrates, commercial metal nanoparticles and organic mixed ion electron conductors were used, and the metallic resonant structure and depletion mode OECTs were fabricated by inkjet printing. When the gate voltage is 0 V, the OECT is in the *on* state; as the voltage increases, the conductivity decreases due to cation injection, and the transistor is gradually cut off. In this study, the split-ring resonator (SRR) and electric-LC (ELC) resonator structures were used, and the amplitude, phase and frequency of the resonance were adjusted dynamically by changing the gate voltage.

Furthermore, if two or more kinds of active elements or materials are being used simultaneously, more properties can be modulated, such as the combination of graphene and varactors, just like the active metasurface in [8], which used a graphene-based sandwich structure together with a high-impedance surface loaded with varactors. The Fermi level of graphene and the capacitance of varactors were

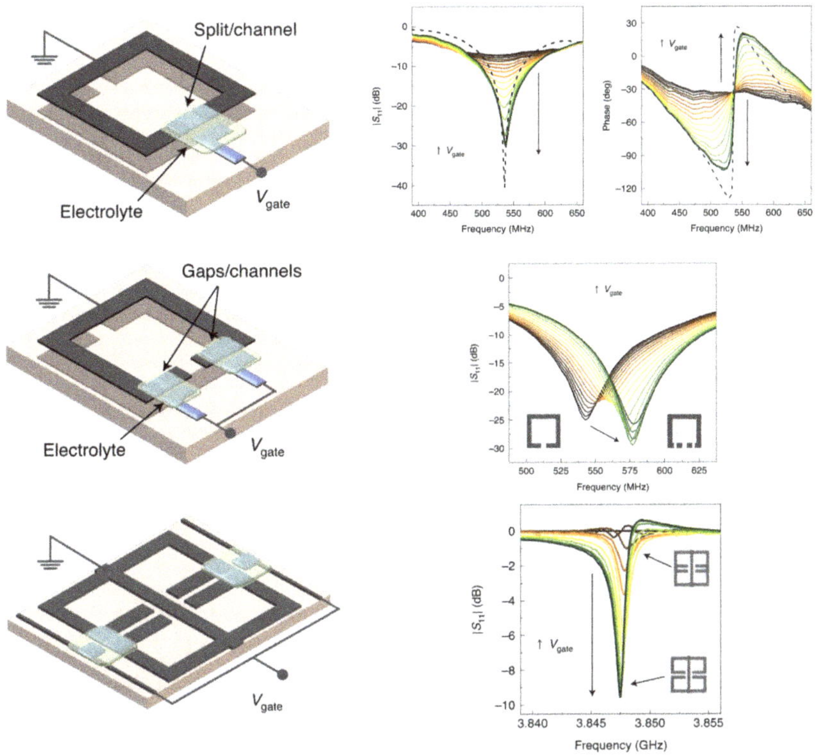

Figure 2.3. Schematics of three kinds of unit cells and the resonance behaviors varying with gate voltages in [7]. Reproduced from [7], with permission from Springer Nature.

controlled independently by external bias voltages, which contributed to the continuous and independent tunability of the magnitude and frequency of resonant absorption, respectively.

Many ingenious designs based on electric modulation are not mentioned here. Overall, it can be concluded that electric modulation features some advantages and disadvantages. On the one hand, the fastest response can be obtained using electrical methods. On the other hand, electrically controlled metasurfaces based on active components, such as diodes and transistors, require additional bias networks to be elaborately designed; the bias networks become more complex when each unit cell needs to be modulated independently, and high power consumption and complicated control circuits will be issues. In addition, some of the active components and materials mentioned above are not available for higher frequencies, for example, the RF diodes, while others such as MEMSs and graphene are applicable for the terahertz band or even higher.

2.2.2 Optical methods

By integrating photosensitive materials into metamaterials, the scattering fields can be modulated by optical stimulus. Commonly used photosensitive materials include semiconductors, photo-induced phase change materials, photoisomerized molecules or polymers, etc [3].

Semiconductors are able to absorb photons with frequencies comparable to their band gaps and generate electron–hole pairs, thereby increasing the concentration of free carriers, and hence the conductivity of the materials, which is called the photoconductivity effect. This feature can be exploited for optical modulation of reconfigurable metamaterials, as in the meta-devices proposed in [9]. The structures of two unit cells and the corresponding scattering parameters are shown in figure 2.4 (a). A low-doped N-type silicon slice was added to an SRR, and the infrared light emitted by a laser diode was illuminated on the silicon slice with the help of a multi-mode fiber. Therefore, pixel-level dynamic modulation of resonance amplitude and frequency was achieved.

The photovoltaic effect of semiconductor photodiodes can also be applied to the optical control of metamaterials. Based on this, a method to realize remote control of metasurfaces by using visible light illumination was illustrated in [10]. The meta-atom consists of an SRR loaded with a varactor, which is supplied by a photodiode. Under different illumination conditions, the DC voltage generated by the photodiode varies, which in turn changes the capacitance of the varactor, causing a variation in the parameters of the resonator, and hence the phase of the reflection coefficient. In the experiment, a single row of meta-atoms was fabricated, and a red LED array was used as the illumination source. The illumination profile was changed by controlling the applied voltage of the LED, and three functions of abnormal reflection, beam focusing and defocusing were realized.

Another reconfigurable metasurface based on the same principle was proposed in [11]; the metasurface controlled by visible light is demonstrated schematically in figure 2.4(b).

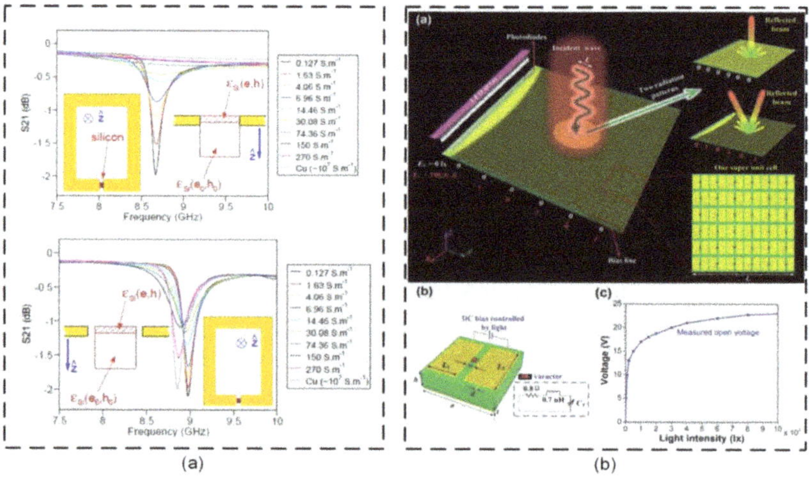

Figure 2.4. (a) The effect of the conductivity of a silicon slice on the resonance performance of SRRs. Reproduced from [9], CC BY 4.0. (b) The operation of a reconfigurable metasurface, schematic of the unit cell, and the DC voltage of photodiode under various illumination conditions. Reproduced with permission from [11], John Wiley & Sons. CC BY 4.0.

The bias voltages of varactors in this study were also provided by the coupling of the LED array and photodiodes, and dynamic control of the reflection phase distribution was realized. Due to the limitations of the hardware, such as low voltage generated by a single photodiode and the relatively high bias voltage required by varactors, extremely complicated bias networks are required for finer control. Therefore, only a few functions were verified experimentally. However, the design idea and prospective application of optical modulation for realizing non-contact remote control of scattered field were demonstrated.

The research also presented a time-varying metasurface transmitter that enabled direct conversion from visible light signal to microwave signal [12]. All elements were driven uniformly by visible light with periodically varied intensity for the generation of harmonics. A photoelectric detection circuit integrated with amplifiers was adopted for higher modulation speed and sensitivity. A dual-channel hybrid wireless communication system was constructed based on the proposed metasurface, and the independent and simultaneous transmission and reception of two different videos were experimentally implemented using the binary frequency shift keying (BFSK) and frequency division multiplexing (FDM) technology.

Additionally, there are other studies utilizing the combination of photodiodes and varactors that fulfilled more sophisticated functions, such as the the visible light-modulated system demonstrated in [13], which was used to obtain cloaked targets, create illusions and dynamically generate vortex beams with different modes. Similarly to [11], several photodiodes were connected in series to provide the bias voltages needed by the varactors, which restricted the miniaturization of the individually addressable sub-arrays.

It can be summarized from the literature that optical methods are expected to enable non-contact remote control over the electromagnetic response of metamaterials. However, the implementation of pixel-wise control is restricted by the hardware, which may be improved with the help of superior active components and optical projection systems, as expected in the literature. Further, due to the requirement for illumination, the optically modulated meta-devices may face the problems of cumbersome systems, bulky sizes, high costs and difficult miniaturization and integration. The response to the change of illumination patterns also takes time.

2.2.3 Thermal methods

The electromagnetic response of metamaterials integrated with materials sensitive to thermal stimulus such as thermosensitive materials and thermally induced phase change materials can be controlled by changing the temperature. The morphological characteristics or electromagnetic properties of thermally controllable active materials are expected to vary with the ambient temperature. For example, the morphology of shape memory alloys (SMAs) may change reversibly with variation in ambient temperature once trained in advance. Based on this, a temperature-controlled metasurface was designed in [14]. When the ambient temperature changes, the morphology of the unit cell made of SMA varies accordingly, resulting

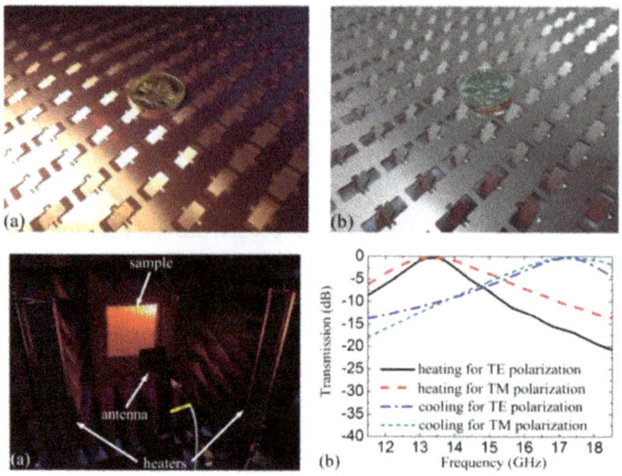

Figure 2.5. Schematic of the unit cell. Variation of the morphology of the metasurface and corresponding electromagnetic response under heating and cooling states. Reproduced from [14], CC BY 4.0.

in the tunability of electromagnetic resonance characteristics, and the position of the resonant peak in the spectrum is shifted, which can be seen in figure 2.5.

The phase change material vanadium dioxide (VO_2) is another commonly used solution for thermal modulation. A VO_2-based metasurface was proposed in [15]. In this study, the VO_2 film was deposited on a sapphire substrate by magnetron sputtering, and later cut into a suitable size with a laser and used for thermistors. The meta-atom was made up of a metallic square ring connected with four thermistors, and polytetrafluorylene with high thermal conductivity was used as the substrate for the bottom metallic ground, with a heating plate connected on the back. Using temperature control, the square resistance of VO_2 film could be significantly changed, and the metasurface was switched between two operating modes, i.e. reflection and broadband absorption.

Water also possesses electromagnetic properties that vary with temperature. For instance, a water-based metasurface rasorber was presented in [16], which was composed of a 3D-printed resin container, a water layer and a band-pass frequency-selective surface (FSS). By controlling the presence or absence of water, the working state could be switched between transmission–absorption and transmission–reflection, while the absorption bandwidth was adjusted by the temperature.

Thermal methods have much wider applicability than the aforementioned electric methods using RF diodes or transistors, and are capable of being extended to higher frequencies. However, it is worth noting that there are some disadvantages of thermal modulation; for instance, it takes time for materials to respond to the change of temperature, and moreover, heating may cause deformation of the meta-devices. Thus, a proper scenario is important for this method.

2.2.4 Mechanical methods

The electromagnetic response of metamaterials can be tuned by mechanically varying parameters such as the geometry of the meta-atoms, the spacing of the adjacent unit cells, or the distance between the adjacent layers. The morphology and period of the meta-atoms both affect the resonance performances of a metamaterial. Based on this, a deformable metasurface with 'kirigami' pattern was presented in [17], as shown in figure 2.6(a). In this research, two kinds of bistable meta-atoms were designed by means of mechanical analysis. By mechanical stretching or compression, the meta-atoms could be switched between the two stable states, so

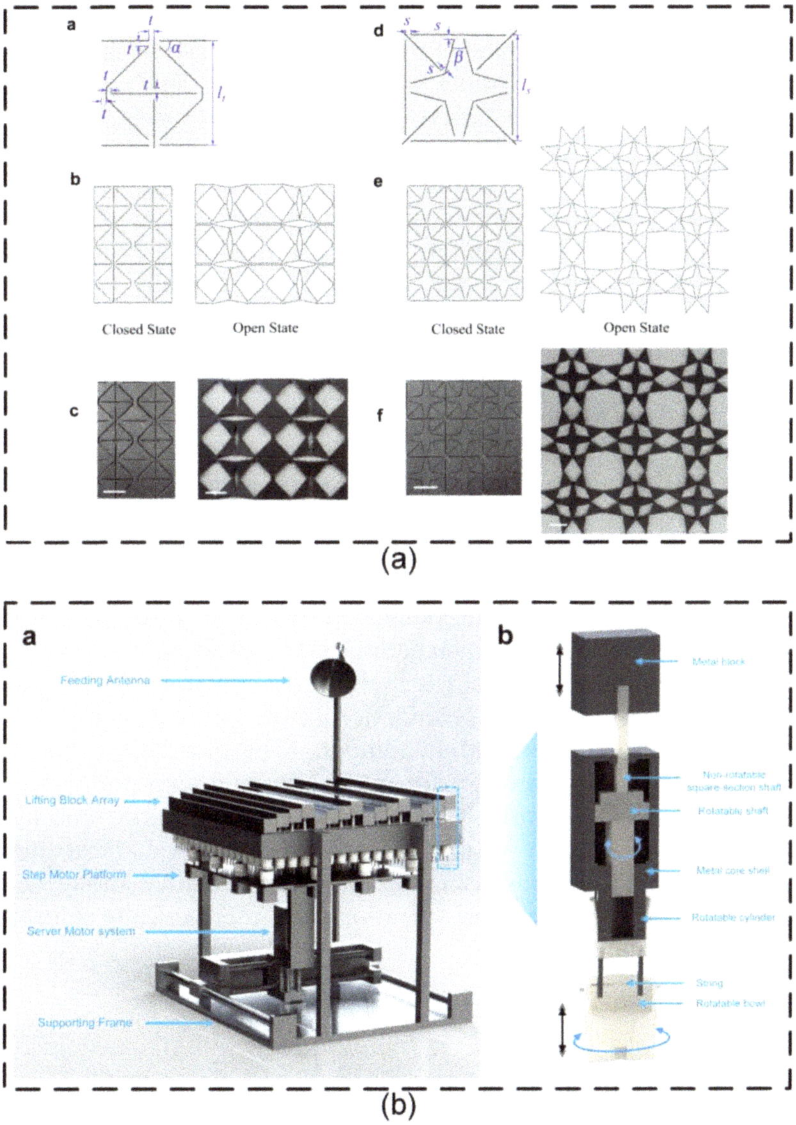

Figure 2.6. (a) Two distinct topologies of the unit cells and the morphology of their bistable states. Reproduced from [17], CC BY 4.0. (b) Structure of the mechanically controllable metamaterial in [21]. Reproduced from [21], CC BY 4.0.

as to dynamically control the electromagnetic response of metasurface. A tunable half-wave dipole and an FSS were presented as verification. The meta-device proposed in [18] is another example of kirigami-based metamaterials, consisting of an array of SRRs with cuts introduced between adjacent unit cells except for the midpoints of sides or the vertices of squares. Deformation between the 2D nonchiral state and the 3D chiral state was allowed by sketching and folding, and handedness could be chosen by the direction of deformation.

Additionally, for metasurfaces consisting of two or more layers, the relation between neighboring layers may also contribute to reconfigurable functions. For example, a polarization-reconfigurable metasurface (PRMS) antenna was proposed in [19], which can be switched among linear polarization, left-hand circular polarization and right-hand circular polarization by mechanically controlling the relative rotation angle of the slot antenna and metasurface.

Apart from the relative orientation between the adjoining layers, the distance between them also exerts a considerable influence on the electromagnetic response of a metasurface, which has been demonstrated in [20] with an FSR consisting of a resistive layer and a band-pass FSS. This design is actually a combination of electrical modulation and mechanical modulation. Schottky barrier diodes were loaded in the bottom layer, and switching between transmission mode and reflection mode at 5.3 GHz was allowed by external bias voltages. By adjusting the distance between the two layers using four stepper motors, the position of absorption band could be tuned.

The system in [21] also utilized motors for mechanical modulation, where a mechanically controlled metamaterial consisting of an array of metal blocks was proposed. The structure of the system is shown in figure 2.6(b), where the height of each block can be independently controlled by motors. By lifting up metal blocks to four different heights, 2-bit reflection phase coding was acquired. Due to the bulky size and heavy weight of the system, experimental verification was restricted, hence only numerical simulation results were given by the researchers. Polarization conversion, anomalous reflection, polarization beam splitting and other functions have been numerically demonstrated. In addition, the group delay coding and phase coding were combined to achieve broadband achromatic reflection.

Typically, mechanical control systems are complex and bulky, and it is usually hard to realize precise and pixelated control with mechanical modulation. Compared with electric methods, using mechanical modulation shows lower speed. Nevertheless, when keeping a certain operation state, mechanically controlled meta-devices consume less power.

2.2.5 Microfluidic technology

Microfluidic technology is also a commonly used modulation method in the design of reconfigurable metamaterials. The functions of a metamaterial can be switched by driving liquid metal, solution or gas to fill the microfluidic channel with pneumatic valves.

An FSR was designed based on microfluidic methods, which took advantage of the fluidity of the liquid metal eutectic gallium indium (EGaIn) [22]. The EGaIn was encapsulated in polymethyl methacrylate (PMMA), onto which the microfluidic channels were laser-etched. There is a transmission window between two absorption bands when EGaIn is injected into the microchannels, while it is tuned to a reflection band when EGaIn is expelled.

Another example is the reflective metasurface for polarization conversion [23]. The microfluidic channel was filled with liquid metal galinstan, which is less toxic than mercury. The lengths of the two arms of the L-type resonator could be adjusted with the pressure of hydrogen chloride vapor, as per the schematic in figure 2.7(a). The device worked as a polarization converter with three polarization states, namely

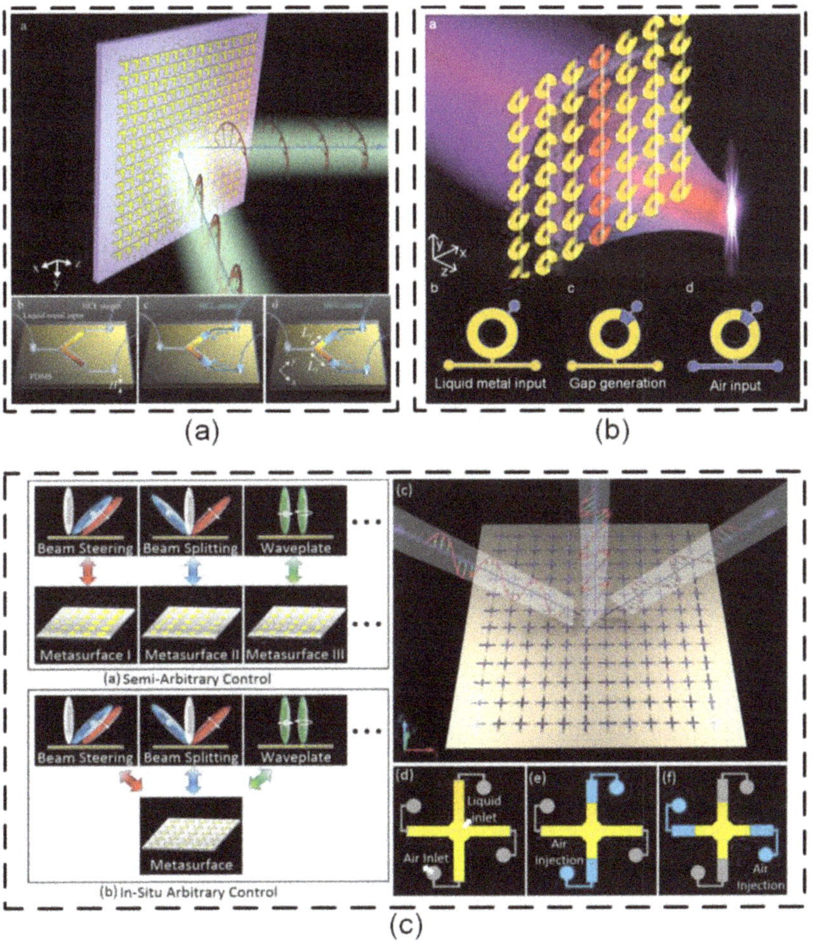

Figure 2.7. (a) Diagram of the reconfigurable polarization converter with L-type resonators. Reproduced with permission from [23], John Wiley & Sons, 2017. (b) Schematic of the tunable meta-lens with circular microcavities. Reproduced with permission from [24], John Wiley & Sons, 2015. (c) Schematic of the multifunctional cross-shaped meta-device. Reproduced with permission from [26] John Wiley & Sons, 2018.

the linear, circular and elliptic polarizations, or acted as an optical attenuator with adjustable intensity.

The meta-atoms were controlled uniformly in the above-mentioned meta-devices, while more complicated microfluidic channels may be designed and fabricated for individually addressable unit cells, such as the flat meta-lens with tunable focal length [24]. In this work, mercury was injected into the annular microcavity first, and then air was injected to form an air gap, as illustrated in figure 2.7(b). The size and direction of the air gap could be changed by controlling the pressure of mercury and air, and thereby modulation of the transmission phase distribution was realized. A similar structure with circular microchannels was also adopted in [25], and dynamic abnormal reflection was achieved.

Further, diverse morphologies of individually addressable meta-atoms have been adopted in other studies, such as a microfluidic metasurface with a cross-shaped microfluidic channel, as shown in figure 2.7(c) [26]. By varying the air pressure, the length of galinstan liquid metal in two orthogonal directions could be controlled independently, which permitted the independent control of the reflection phase gradients of the two orthogonal linear polarization modes. The function of the device could be switched among polarization beam splitting, beam steering and polarization conversion. Another example is the hydrodynamic metasurface in [27], which was used for dynamic beam scanning. Water was used in the microfluidic channels, which is much more economical and user friendly.

Generally, pneumatic valves are used in microfluidic technology to drive liquid metal or gas to fill up microfluidic channels, which gives rise to slow modulation speed. Nevertheless, microfluidic methods are applicable for continuous tunability, as shown in the aforementioned literature, where the lengths of the arms or the orientations and gaps of the circular rings can be controlled smoothly.

2.3 Reconfigurable microwave metamaterials for applications

Metamaterials with exotic electromagnetic responses are capable of freely manipulating the amplitude, phase, polarization state, angular momentum and other properties of the reflected or transmitted electromagnetic waves, enabling absorption, frequency-selective transmission, anomalous reflection or transmission, imaging, holography, beam focusing or defocusing, beam shaping, polarization conversion and other intriguing functions. By dynamically controlling the properties of the scattered waves, reconfigurable metamaterials combined with active components or materials can further realize the switchable operation of two or more functions or the tunability of the original functions, for example, the absorption rate of the incident waves, reflected or transmitted intensity, beam steering angle, hologram patterns, the focal length of a flat lens and the polarization state of the scattered field, to mention a few.

As for general reconfigurable metamaterials, all elements are driven by the same control signal, and the bias networks are usually simple, while the functions achievable are limited. Programmable metasurfaces with each unit cell independently controlled, typically with the help of field-programmable gate arrays (FPGAs) or microcontroller units (MCUs), possess higher degrees of freedom, and are applicable for more functions.

In addition, it can be seen from the above introduction to different modulation methods that control at the unit cell level may also be achieved using microfluidic technology or optical technologies, which contributes to dynamically changing the phase or amplitude locally. Most importantly, for passive metamaterials, the implementation of beam focusing, beam shaping, hologram imaging and other functions usually requires non-periodic arrangement of meta-atoms with different geometric parameters to form a gradient phase distribution. With the help of external control, programmable metasurfaces may achieve the above-mentioned functions simply by periodic arrangement of identical unit cells.

In the following subsections, metamaterials with uniformly controlled meta-atoms will be introduced, and metasurfaces with each unit cell controlled independently shall be presented in the next section. In regard to reconfigurable metasurfaces, the basic principles, design ideas and analysis methods shall be demonstrated, followed by some available functions as illustration. Moreover, since this section is aimed at briefly introducing the controllability of different properties of scattered fields with reconfigurable metamaterials modulated uniformly, almost all the examples presented are based on electrical modulation for convenience.

Reconfigurable metamaterials can dynamically control the scattered field, switching among transmission, reflection and absorption, or dynamically changing the absorption rate or the energy reflected or transmitted, and this brought about reconfigurable devices such as FSSs, absorbers, rasorbers and polarization converters, etc.

2.3.1 Polarization converter

The modulation of polarization of the scattered fields can be realized based on reconfigurable metamaterials. The polarization state of an electromagnetic wave can be defined in terms of the trace of the end of the instantaneous electric field vector, i.e. if the electric field always oscillates along a straight line, the electromagnetic wave is linearly polarized; if the trace is circular or elliptical, it is circularly polarized and elliptically polarized [28]. Circular and elliptical polarizations can be further divided into the left- and right-hand polarizations according to the relation between the moving direction of the end of the electric field vector and the propagation direction of the electromagnetic wave.

Polarization modulation (PoM) possesses significant applications in the field of wireless communication, and may contribute to the increase of channel capacity, the improvement of spectrum utilization and the achievement of higher transmission rate. In addition, because of the vector attribute of polarization and its close correlation with the propagation direction, together with time independence during propagation, information transmission based on PoM is expected to reduce interference, improve reliability, reduce the complexity of transceiver devices and enhance communication security and confidentiality [29].

Theoretically, when combining a reconfigurable metasurface polarization converter with a transmitting antenna with a single fixed polarization state, outgoing waves with any wanted polarization states are supposed to be generated. Thereby, polarization modulation can be realized with a simple hardware structure. In addition, polarization converters based on metasurfaces have the advantages of high efficiency, broadband and subwavelength profile.

Metamaterials composed of anisotropic elements may change the polarization states of the reflected or transmitted waves and operate as polarization converters. Active components such as PIN diodes or varactors can be further integrated, making it possible to dynamically control the polarization states with various applied bias voltages.

Here, the transmission polarization converter shall be taken as an example to illustrate the working principle. We assume that a transmissive metasurface is located in the x–y-plane, with the principal axes of the anisotropic meta-atoms along the x- and y-directions. A linearly polarized plane wave is impacting on the metasurface, with the electric field oscillating along a linear line at an angle of 45° to the x-axis, namely, $\overrightarrow{E_i} = \frac{1}{\sqrt{2}}E_0(\hat{x} + \hat{y})$. The electric field vector of the transmitted wave can be expressed as

$$\overrightarrow{E_t} = \frac{1}{\sqrt{2}}E_0(|T_x|e^{j\delta_x}\hat{x} + |T_y|e^{j\delta_y}\hat{y}), \tag{2.1}$$

where $|T_x|$, $|T_y|$, and δ_x, δ_y denote the magnitudes and phases of the transmission coefficient along the x- and y-directions, respectively. If $|T_x| = |T_y|$ and $\Delta\delta_{xy} = \delta_x - \delta_y = \pm\frac{\pi}{2}$, the transmitted waves will be circularly polarized; if $\Delta\delta_{xy} = 0, \pi$, the transmitted waves will be linearly polarized; all the other cases lead to elliptically polarized waves (the value of $\Delta\delta_{xy}$ is among the internal $(-\pi, \pi)$).

Using anisotropic meta-atoms, distinct transmission phases δ_x and δ_y in the x- and y-directions can be introduced, respectively. At the same time, the amplitude of transmission coefficients in both directions should be approximately similar, and then the polarization state of the transmitted wave can be controlled. Furthermore, PIN diodes or varactors are loaded into the passive structure to change its electromagnetic response with applied external bias voltages, so as to make the phase difference $\Delta\delta_{xy}$ variable and result in the switching of polarization states.

The reflective counterparts can be analysed using a similar method. A reconfigurable transmissive polarization converter with PIN diode-loaded elliptic split rings was proposed in [30]. When the PIN diodes were cut off, the linearly polarized waves were converted into circularly polarized waves. When the diodes were switched on, the polarization state remained the same. An example of reflective polarization converter is the one based on PIN diodes in [31]. The linearly polarized waves are converted into circularly polarized waves when the diodes are off, whereas they are converted to cross-polarized waves when the diodes are switched on.

Instead of PIN diodes, varactors with voltage-controlled capacitance are applicable for the conversion to more polarization states, such as the varactor-based polarization converter proposed in [32]. When the linearly polarized plane waves were incident on the metasurface, the polarization state of the reflected waves could be tuned among linear polarization, elliptical polarization and circular polarization.

Considering another case, that is, the incident wave is circularly polarized, with the phasor form of the electric field vector denoted by $\vec{E_i} = \frac{1}{\sqrt{2}}E_0(\hat{x} \pm i\hat{y})$, the electric field vector of the transmitted wave is formulated as

$$\vec{E_t} = \frac{1}{\sqrt{2}}E_0(|T_x|e^{j\delta_x}\hat{x} \pm i\left|T_y\right|e^{j\delta_y}\hat{y}) = \frac{1}{\sqrt{2}}E_0\left[|T_x|e^{j\delta_x}\hat{x} + \left|T_y\right|e^{j\left(\delta_y \pm \frac{\pi}{2}\right)}\hat{y}\right]. \quad (2.2)$$

If $|T_x| = |T_y|$ and $\Delta\delta_{xy} = \delta_x - \delta_y = 0, \pi$, the scattered waves are still circularly polarized; if $\Delta\delta_{xy} = \pm\frac{\pi}{2}$, the scattered fields will be linearly polarized; and the rest contributes to elliptic polarization.

A reflective reconfigurable polarization converter was demonstrated in [33]. Assuming that the circular-polarized waves were incident, the device worked as a helicity converter and a helicity hybridizer in two separate frequency bands when the PIN diodes were switched on; while when the PIN diodes were switched off, it became a helicity keeper working in a broadband. The diagrams of the unit cell and the operation modes are shown in figure 2.8.

To evaluate the ability for polarization control, one may refer to the reflectivity of co-polarization and cross-polarization. Further, the polarization conversion ratio (PCR) and axial ratio (AR) are also commonly used indicators in the design of polarization converters, and are typically used to appraise the conversion ability of linear cross-polarization and circular polarization, respectively.

PCR is used to measure the ability to convert linear polarization to cross-polarization. Assuming that the metasurface is located in the x–y-plane and the linear polarized plane wave is normally incident, the electric field of which is along the x-axis, the PCR can be calculated using [34]:

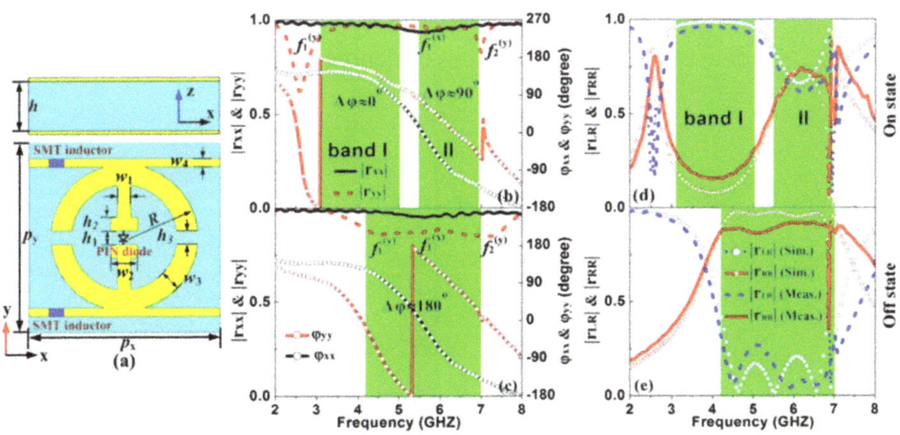

Figure 2.8. Diagram of the unit cell and reflection coefficients in the linear polarization basis and circular polarization basis. Reproduced from [33], CC BY 4.0.

$$PCR_y = \frac{R_{yx}^2}{R_{yx}^2 + R_{xx}^2}, \tag{2.3}$$

where R_{xx} and R_{yx} correspond to the reflection coefficient or transmission coefficient of the co-polarization and cross-polarization, respectively. AR is typically used to evaluate the purity of circular polarization, which denotes the ratio of the major axis to the minor axis of the ellipse, and is formulated as [35]

$$AR = \frac{major\ axis}{minor\ axis} = \frac{\left\{\frac{1}{2}\left[E_x^2 + E_y^2 + \left(E_x^4 + E_y^4 + 2E_x^2E_y^2cos(2\Delta\delta_{xy})\right)^{1/2}\right]\right\}^{1/2}}{\left\{\frac{1}{2}\left[E_x^2 + E_y^2 - \left(E_x^4 + E_y^4 + 2E_x^2E_y^2cos(2\Delta\delta_{xy})\right)^{1/2}\right]\right\}^{1/2}}, \tag{2.4}$$

with $1 \leqslant AR < +\infty$. The dB scale is used most commonly, i.e.

$$AR = 10lg\frac{E_x^2 + E_y^2 + \left(E_x^4 + E_y^4 + 2E_x^2E_y^2cos(2\Delta\delta_{xy})\right)^{1/2}}{E_x^2 + E_y^2 - \left(E_x^4 + E_y^4 + 2E_x^2E_y^2cos(2\Delta\delta_{xy})\right)^{1/2}}. \tag{2.5}$$

In general, the frequency range where AR is lower than 3 dB is considered as the circular polarization bandwidth.

2.3.2 Frequency selective surfaces (FSSs)

Reconfigurable FSSs can also be constructed using uniformly modulated meta-atoms. For example, the aforementioned simple dual-polarized structure with a square ring slot shown in figure 2.2(a) is actually a band-pass FSS, and the transmission window can be switched or tuned by the PIN diodes or varactor diodes welded across the square ring.

In general, microwave metamaterials can be analysed with the help of an equivalent circuit model (ECM). The band-pass FSS is usually regarded as the LC parallel circuit shown in figure 2.9, where Z_0 is the intrinsic impedance of free space. The impedance seen from the input is denoted as Z_{in}, given as

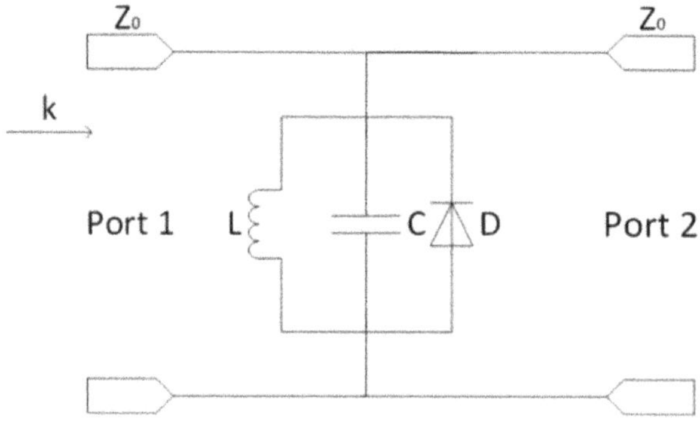

Figure 2.9. ECM of the simple FSS in figure 2.2(a).

$$Z_{in} = j\omega L//\frac{1}{j\omega C}//Z_d//Z_0, \tag{2.6}$$

where Z_d is the impedance of the diode. The reflection and transmission coefficients can be expressed as

$$r = \frac{Z_{in} - Z_0}{Z_{in} + Z_0} \tag{2.7}$$

$$t = \frac{2Z_{in}}{Z_{in} + Z_0}. \tag{2.8}$$

By varying the bias voltages applying to the diodes, the input impedance Z_{in} will be changed and hence the reflectivity and transmissivity.

More sophisticated structures may realize functions such as wider or multiple pass bands, and the corresponding ECMs usually contain more complicated connections or higher order LC resonant circuits. Nevertheless, the basic principle is still to change the resonant properties using PIN diodes or varactors. For example, a reconfigurable second-order band-pass filter with a wide switchable transmission window attained by loading PIN diodes was designed in [36].

2.3.3 Absorber

Another important application of amplitude modulation is the implementation of metamaterial absorbers. Taking metasurface absorbers as an example, one of the simplest structures is composed of a lossy layer, an air or dielectric spacer and a metallic ground. The ECM is shown in figure 2.10, where Z_1 is the intrinsic impedance of dielectric, while θ_1 is the electrical length of spacer. The lossy layer is equivalent to an RLC series circuit, and the spacer, together with the metallic ground, is equivalent to a short-circuit transmission line. The equivalent capacitance of the structure can be dynamically changed by loading PIN diodes or varactors.

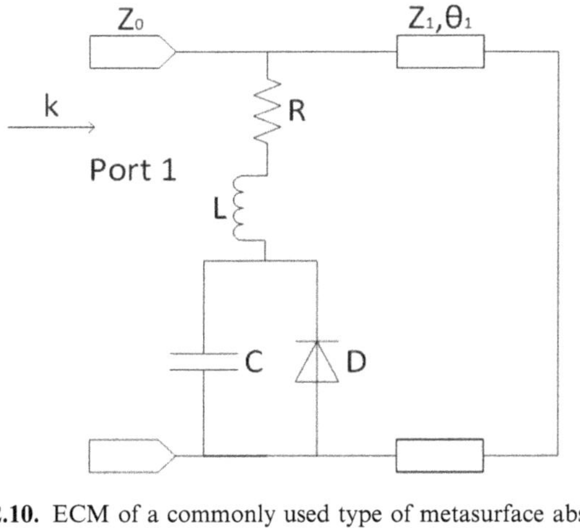

Figure 2.10. ECM of a commonly used type of metasurface absorber.

The reflection coefficient and absorption rate can be analysed using microwave network theory.

In the impedance of the lossy layer is Z_s, the input impedance is calculated as

$$Z_{in} = Z_s//jZ_1 \tan \beta l. \tag{2.9}$$

The reflectivity and absorption rate can be formulated in the following equations:

$$\Gamma = \left| \frac{Z_{in} - Z_0}{Z_{in} + Z_0} \right|^2 \tag{2.10}$$

$$A = 1 - \Gamma. \tag{2.11}$$

Generally, absorption is considered to be realized when the reflectivity is lower than 10%. Theoretically, by changing the resonant properties of structures, the operation modes can be reconfigured, such as switching among absorption, transmission and reflection, tuning the absorption rate, or moving the absorption band.

A switchable absorber based on PIN diodes was proposed in [37]. The amplitudes of the reflection coefficients of the two orthogonal linear polarization states could be modulated independently, which thereby realized the absorption or reflection of single or dual polarization. Another example based on PIN diodes is the reconfigurable absorber presented in [38], where the position rather than the absorption rate of the absorption band was switchable, as shown in figure 2.11(a). Varactors facilitate finer controllability than PIN diodes. For instance, a tunable absorber designed in [39] is shown in figure 2.11(b), where the position of absorption band could be continuously adjusted with varying voltage across varactors.

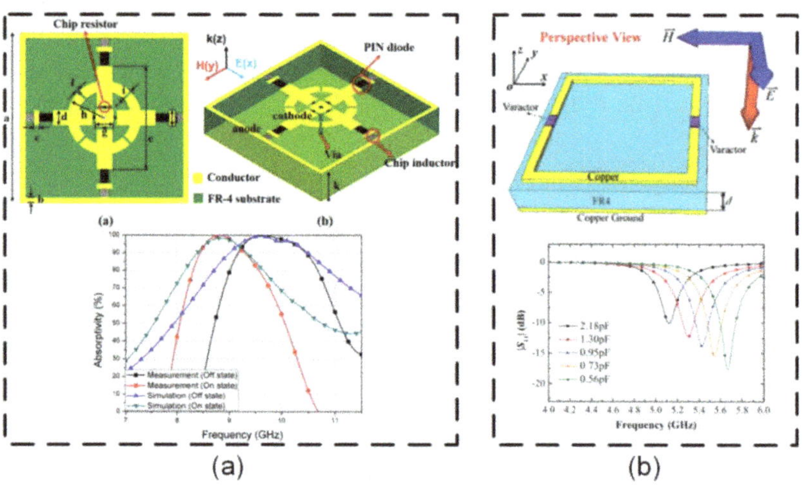

Figure 2.11. (a) Structure of the unit cell, the ECM and the absorption rate. The position of the absorption band is changed when PIN diodes are switched *on* and *off*. Reproduced from [38] CC BY 4.0. (b) Structure of the unit cell, the corresponding ECM, and variation of reflection coefficient according to the capacitance of varactors. Reproduced from [39]. © IOP Publishing Ltd. All rights reserved.

2.3.4 Frequency selective rasorber (FSR)

By replacing the bottom metallic ground of an absorber with an FSS, and introducing additional structures to bypass lossy components in the top resistive layer within the frequency range of the FSS pass band, an FSR can be constructed, where the word rasorber means radome absorber, which indicates the application of FSRs in the design of radar domes. FSRs function as syntheses of FSSs and absorbers, that is, they exhibit absorption or transmission responses to the incoming electromagnetic waves within different frequency ranges, respectively.

In order to reduce the insertion loss of the pass band, a parallel resonant structure with band-pass response is usually introduced in the impedance layer. It should be noted that the pass bands of the upper and lower layers of a FSR must be matched to make the whole structure present band-pass response, while in the frequency range outside pass bands, the responses of both layers are similar to an absorber.

Similar to FSSs and absorbers, the transmission window can be tuned [40] or switched [41, 42] using PIN diodes or varactors. Moreover, the modulation methods mentioned in the last section are all available for the realization of reconfigurable FSRs. However, diode-loaded FSRs shall be taken as examples in this subsection. The ECM of one of the simplest reconfigurable FSRs is shown in figure 2.12(a).

The impedance of the resistive layer and FSS can be denoted as Z_R and Z_F, respectively. Then the scattering parameters S_{21} and S_{11} can be calculated using microwave network theory:

$$S_{21} = \frac{2}{A + \frac{B}{Z_0} + CZ_0 + D} \tag{2.12}$$

$$S_{11} = \frac{A + \frac{B}{Z_0} - CZ_0 - D}{A + \frac{B}{Z_0} + CZ_0 + D}, \tag{2.13}$$

where A, B, C and D are formulated as follows:

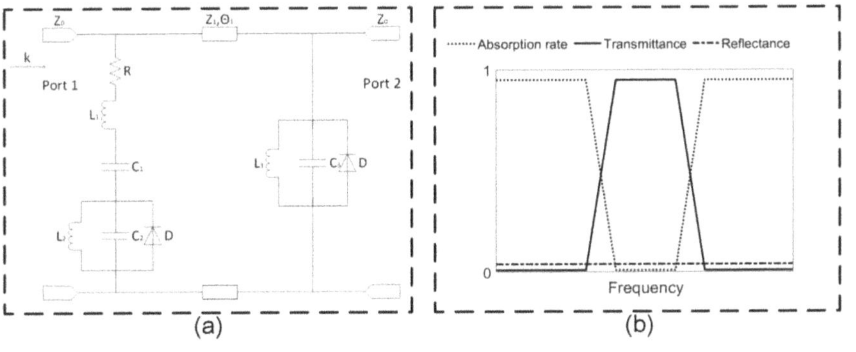

Figure 2.12. (a) ECM of a commonly used type of FSRs. (b) The desirable response of a FSR with a transmission window between two absorption bands.

$$\begin{bmatrix} A & B \\ C & D \end{bmatrix} = \begin{bmatrix} 1 & 0 \\ \frac{1}{Z_R} & 1 \end{bmatrix} \begin{bmatrix} \cos\theta_1 & jZ_1\sin\theta_1 \\ j\frac{1}{Z_1}\sin\theta_1 & \cos\theta_1 \end{bmatrix} \begin{bmatrix} 1 & 0 \\ \frac{1}{Z_F} & 1 \end{bmatrix} \tag{2.14}$$

$$\begin{bmatrix} A\,B \\ C\,D \end{bmatrix} = \begin{bmatrix} 1 & 0 \\ \frac{1}{Z_R} & 1 \end{bmatrix} \begin{bmatrix} \cos\theta_1 & jZ_1\sin\theta_1 \\ j\frac{1}{Z_1}\sin\theta_1 & \cos\theta_1 \end{bmatrix} \begin{bmatrix} 1 & 0 \\ \frac{1}{Z_F} & 1 \end{bmatrix}. \tag{2.15}$$

Then the transmittivity, reflectivity and absorption rate are

$$T = |S_{21}|^2 \tag{2.16}$$

$$R = |S_{11}|^2 \tag{2.17}$$

$$A = 1 - |S_{21}|^2 - |S_{11}|^2. \tag{2.18}$$

FSRs can be analysed with the help of ECMs. In general, the transmittance at the position of pass bands is required to be as close to 1 as possible, while within the absorption bands it is required that both the transmittance and reflectance are close to 0. The desirable spectral response of an absorption–transmission–absorption-type FSR is shown in figure 2.10(b).

The equivalent circuit model of FSRs is not limited to the one mentioned above. Instead, there are also many other ECMs with more complicated topologies representing FSRs with diverse structures. In addition, more complex structures were proposed to improve or extend the function of FSRs, for example, multi-pass band switchability [43], expansion of transmission bandwidth with high-order FSSs [44], improvement of absorption bandwidth by increasing the number of lossy layers [45, 46], adjustable absorption rate [47], and so on.

A multi-band FSR is proposed in [43], whose transmission response can switch between a single pass band at 6 GHz and dual pass bands at 3.5 and 8 GHz. However, the structure only acted on a single polarization state. In [44], a switchable third-order band-pass FSS was adopted, which consisted of a three-layer metallic structure loaded with PIN diodes. An inter-digital resonator connected with a PIN diode was used as a parallel LC resonant structure in each unit cell of the impedance layer. This device exhibited a wide pass band of 3.9–4.34 GHz (fractional bandwidth of 10.7%) in the rasorber mode. This structure also operated in a single polarization state.

If an additional impedance layer is added on the top of a reconfigurable lossy layer and the reconfigurable FSS, the absorption bandwidth may be further expanded. The switchable FSRs in [45, 46] both adopted this method to improve the absorption performance. It is worth mentioning that both of these two FSRs displayed a wide pass band, and the two orthogonal polarization states could be controlled independently without affecting each other. The absorption rate could also be tuned [47]. Since a PIN diode in the forward bias state is equivalent to the series connection of a voltage-controlled resistor and an inductor, the meta-devices in [47] used PIN diodes as variable resistors, and a dynamic absorptive frequency-selective surface integrated with PIN diodes was adopted as the top layer. Thereby, adjustment of the absorption rate was achieved. That is, when the forward bias voltage changed from 0 to 25 V, the absorption rate of the two absorption bands varied from close to 0 to 0.9, while the pass band remained stable during the adjustment.

The above description briefly introduced several common applications of reconfigurable metamaterials with all elements uniformly controlled. Generally, such metamaterials have the advantage of relatively simple bias networks. However, the functions are usually limited due to having the same amplitude and phase of scattered fields for all elements. In the next section, reconfigurable metasurfaces with each unit cell controlled independently shall be briefly illustrated. These structures can apply different control signals to each element, thereby forming certain local phase distributions, and be applicable for more functions.

2.4 Programmable metasurface

Reconfigurable metamaterials with identical control signals applied to all meta-atoms can achieve uniform changes to electromagnetic properties and realize dynamic modulation of working states, which possess relatively simple bias networks but result in limited operation modes. Luckily, if each meta-atom is controlled independently, higher design freedom can be obtained, thereby bringing about more available functions, such as anomalous reflection, beam focusing, beam shaping, holography, imaging, etc. The realization of these functions usually requires meta-atoms to have non-uniform electromagnetic response, forming certain phase distributions.

Typically, one or more PIN diodes are loaded into every meta-atom, and one or more bits of reconfigurable digital coding metasurfaces can be constructed using various combinations of their *on* and *off* states. A digital coding metasurface with N bits has a total of 2^N phase responses of meta-atoms, namely (to avoid ambiguity, in the context of this chapter, phases are all denoted by δ, and azimuth angles are all denoted by φ) [48]:

$$\delta_i = \frac{2\pi}{2^N} \cdot i + \delta_0 \; (i = 0,\ 1,\ \cdots,\ 2^N - 1.\ \ \delta_0 \text{ is the phase of the initial state}). \tag{2.19}$$

The capacitance of varactors varies within a certain range with applied bias voltage. If varactors are used, the meta-atoms may have more phase responses. By applying an appropriate bias to each meta-atom, a phase gradient distribution can be achieved.

In addition to phase coding, amplitude coding, or a combination of phase coding and amplitude coding, are also available. However, encoding the amplitudes may result in lower efficiency. Therefore, only phase coding shall be presented in this section.

2.4.1 Anomalous refection

Programmable metasurfaces can be used to realize abnormal reflections, including beam steering, beam splitting, diffuse reflection, etc. When a plane wave is normally incident, the far-field pattern can be calculated approximately by the following equation:

$$F(\theta,\ \varphi) = \sum_{m=1}^{M}\sum_{n=1}^{N} A_{mn} e^{j\delta_{mn} - j\left[kD_x\left(m-\frac{1}{2}\right)\sin\theta\cos\varphi + kD_y\left(n-\frac{1}{2}\right)\sin\theta\sin\varphi\right]}, \tag{2.20}$$

where A_{mn} and δ_{mn} are the magnitude and phase of the reflection coefficient of the coding element located in the mth row and the nth column, respectively. As for phase modulation, A_{mn} is supposed to be a constant close to 1. D_x and D_y are the sizes of the coding elements. The time factor of the harmonic electric field is $e^{j\omega t}$. It is worth noting that the coupling between adjacent elements is not taken into account, and approximations have been made in the scattered field generated by individual unit cells.

When the scale of the array is small and the coding sequence is regular, the coding matrix of the programmable metasurface can be acquired directly using the above equation. Moreover, the generalized laws of reflection and refraction can also be used [49]. When the array is large and more complex radiation patterns with better performance are desired, such as holograms, RCS reduction and beam steering with reduced sidelobe level, simple and regular coding sequences cannot meet the requirements, and optimization algorithms are usually adopted for reverse designs, such as particle swarm optimization (PSO) and genetic algorithm (GA). One may also apply artificial intelligence techniques to further improve the performance or calculation speed.

One of the examples is the 1-bit digital coding metasurface with each unit cell being loaded with a PIN diode controlled by FPGA presented in [50], as shown in figure 2.13(a). Opposite phases of reflection coefficients are achieved at 8.9 GHz when the diodes are switched *on* and *off*. Four-beam reflection was realized using this device, and chessboard coding matrices with various lattice sizes were adopted to redirect the beams in different elevations and azimuths.

Only two phase responses can be acquired using 1-bit digital coding, which shows low efficiency and limited functions. If 2-bit coding is adopted, four distinct phase responses are available, so that greater freedom can be obtained. For example, two PIN diodes were used in each unit cell in [51], and single-beam deflection, dual- and four-beam splitting and beam diffusion were realized, as shown in figure 2.13(b).

Compared with PIN diodes, whose working states can only be switched between forward bias and reverse bias, varactors are able to provide more phase responses and higher degrees of freedom, due to the fact that the capacitance of a varactor varies continuously with changing reverse bias voltage in a certain range. A gradient phase metasurface was demonstrated using varactors in [52]. The generalized Snell's law was used as a guideline to realize deflection of the reflection beams, and the simulated near-field distribution is shown in figure 2.13(c), together with the

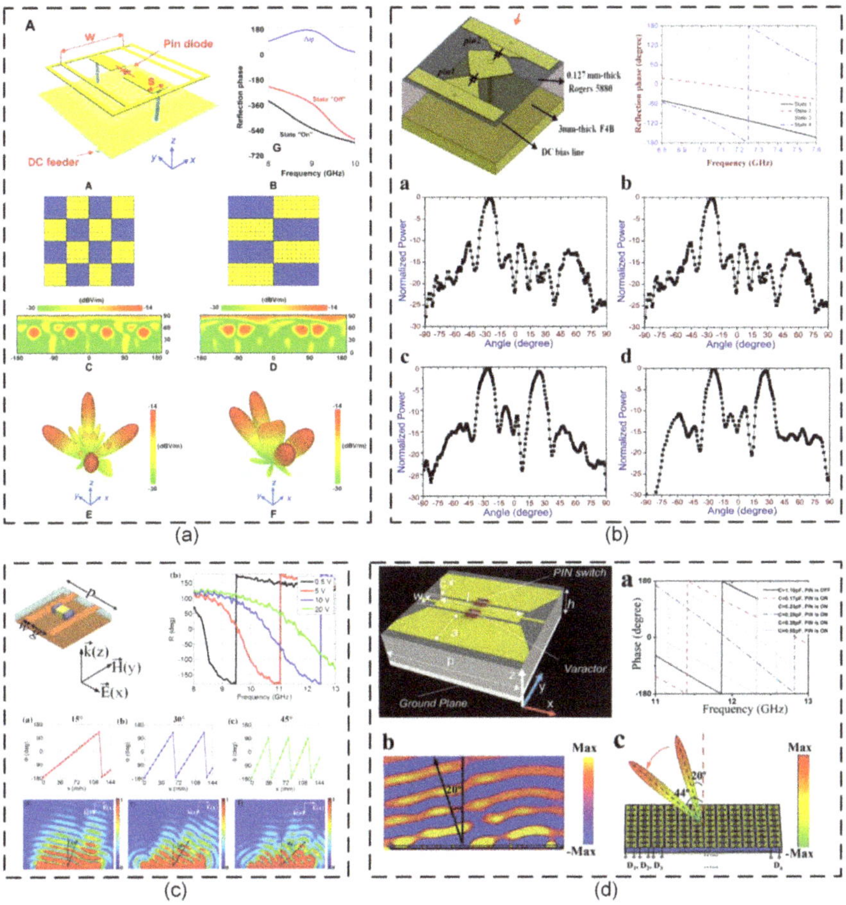

Figure 2.13. (a) One-bit digital coding metasurface based on PIN diodes; the binary phase response; four-beam splitting realized by chessboard coding pattern. Reproduced from [50], CC BY 4.0. (b) Two-bit digital coding with four different phase responses. Measured single-beam deflection and dual-beam splitting performances. Reproduced from [51], CC BY 4.0. (c) Gradient phases achieved by varactor-loaded unit cells, three distinct phase distributions and the corresponding simulated results of beam steering towards different directions [52]. Reproduced from [52], with permission from Springer Nature. (d) Programmable metasurface using both PIN diodes and varactors. A phase step of 60° realized with six capacitance configurations of the two diodes, and beam redirected to the angle of 20° using these six reflection elements. Reproduced with permission from [53], John Wiley & Sons, 2017.

structure of the unit cell and measured phase responses. By changing the bias voltage of varactors, the same structure can be used for frequency agility.

Moreover, PIN diodes and varactors can be used simultaneously [53], as shown in figure 2.13(d). In this work, the *on–off* states of PIN diodes and the capacitance of varactors were controlled jointly for different configurations. Hence, for example, binary phase coding was acquired in various frequency bands, so as to realize dual-beam splitting with frequency agility. The experimental measurement of RCS reduction has confirmed that, by using three different kinds of diode bias configuration, RCS reduction can be achieved at three different frequencies. In addition, due to the high degrees of freedom provided by the combination of the two kinds of diodes, larger phase coverage and more different phase steps can be achieved by using different configurations. Single-beam deflection toward different directions at 12 GHz was verified experimentally.

2.4.2 Holographic imaging

Another available application is metasurface holography, which is significant in a great number of areas, such as 3D display, biomedical imaging, data storage, etc. Holography was proposed by Gabor in 1948 and was first used in microscopy [54]. Traditional holography recorded information for objects by using interference and reconstructed the objects with diffraction. In 1966, Brown and Lohmann proposed computer-generated holograms, which can realize holographic display of objects that do not even exist in reality with the help of spatial filters [55].

Because of their subwavelength structures and exotic ability to manipulate electromagnetic fields, metasurfaces are superior candidates for spatial filters in computer-generated holograms. An increasing number of studies are focused on metasurface holography, involving various wavebands in the electromagnetic spectrum, such as visible light holograms [56], infrared holograms [57], terahertz holograms [58] and microwave holograms [59, 60]. Additionally, researchers have been devoted to figuring out diverse methods to improve metasurface holography, such as using various multiplexing technologies to increase information capacity, realize switchable functions, or display 3D scenes, etc. Some active components or active materials may also be used for reconfigurable holographic display.

Programmable metasurfaces in the microwave band can realize rapid modulation at pixel level by loading PIN diodes or varactors, and can be used for dynamical holographic display. This can also be achieved in other wavebands by using multiplexing technologies or active materials. However, diode-based microwave programmable metasurfaces may show faster response and have the potential to display more different holograms.

Holography can be classified into three types, namely, phase-only holograms, amplitude-only holograms and complex amplitude holograms [61]. Generally, programmable metasurfaces based on phase coding are able to realize phase-only holograms. The phase distributions on metasurfaces are typically calculated using some optimization algorithms, such as the well-known Gerchberg–Saxton (GS) algorithm as well as the algorithms derived from it. A 1-bit digital coding metasurface was used to realize reconfigurable holographic display in [59]. As shown in figure 2.14, the metasurface was used as a programmable holographic phase plate, and the state of the PIN diode in each unit cell was controlled by FPGA in parallel for binary phase coding. This study has provided a new application field for microwave programmable metasurface. However, it used binary coding with only two phases of 0 and π, which may face the problem of low efficiency. A programmable metasurface based on mechanical modulation was designed in [60]. Motors and gears were used to control the rotation angle of each meta-atom and quasi-

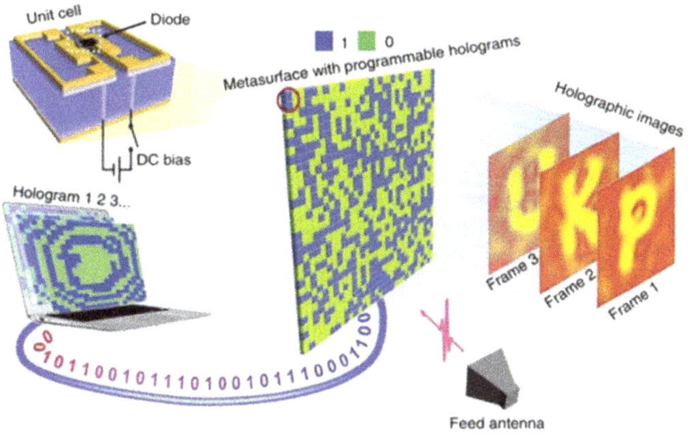

Figure 2.14. Sketch of the operation of the 1-bit programmable metasurface used for holographic imaging in [59]. Reproduced from [59], CC BY 4.0.

continuous Pancharatnam–Berry (PB) phase control was realized; hence the device can be used for high-efficiency holographic display.

In fact, programmable metasurfaces are not restricted to the realization of a certain function. Instead, the digital coding metasurfaces loaded with PIN diodes or varactors for anomalous reflection mentioned in the last subsection can also be used to achieve microwave holographic display. In addition, several reconfigurable metasurfaces based on microfluidic technology are mentioned in the introductory section on reconfiguration mechanisms that are able to achieve pixel-level control; they may also be used for reconfigurable holographic display. Moreover, they are expected to obtain higher phase quantitative levels compared with diode-based structures, so as to achieve more efficient holographic display, but the speed of switching may be much slower than for those using electrical modulation.

2.4.3 Computational imaging

Programmable metasurfaces may also be applied to microwave computational imaging, which performs image reconstruction by solving electromagnetic inverse scattering problems. In microwave computational imaging systems, programmable metasurfaces can dynamically change the radiated field patterns and function as spatial light modulators (SLMs).

A transmissive programmable metasurface was proposed in [62], where two PIN diodes are integrated into each unit for the 2-bit phase coding, which is schematically illustrated in figure 2.15(a). Controlled by FPGA, the device is able to generate a variety of different transmitted modes under the radiation of a feed horn antenna, which can be used for single-sensor and single-frequency imaging. Another implementation is the metasurface imager presented in [63], which used PIN diodes to realize 2-bit reconfigurable coding, and reduced the number of measurements required for image reconstruction with the help of machine learning, as

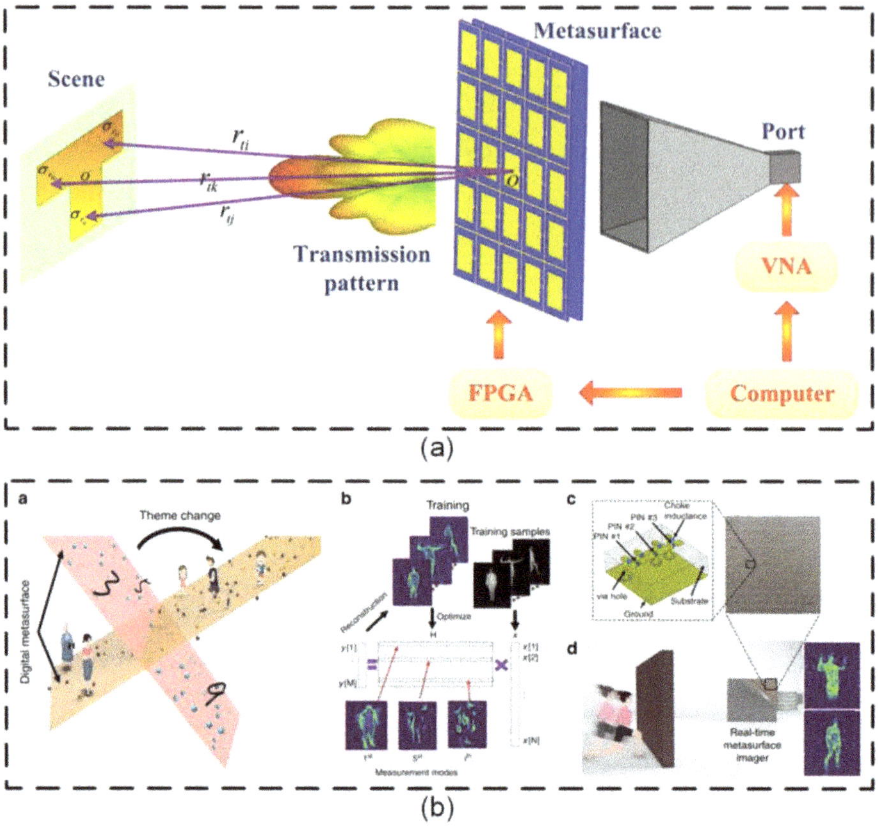

Figure 2.15. (a) Schematic of the 2-bit imaging system in [62]. Reproduced from [62], CC BY 4.0. (b) Sketch of the operation of the metasurface imager assisted by machine learning in [63]. Reproduced from [63], CC BY 4.0.

demonstrated in figure 2.15(b). This design realized the imaging and object recognition of handwritten digits and human poses.

2.4.4 Beam focusing

Assuming that a plane wave is normally incident on a metasurface located in the x–y-plane, if the focal length is f, the phase compensation at (x, y) on the flat metalens is calculated by the following formula according to geometric optics:

$$\delta(x, y) = k(\sqrt{x^2 + y^2 + f^2} - f), \tag{2.21}$$

where k is the wavenumber of free space. The above equation applies to foci on the axis. As for off-axis foci, provided that the coordinate is (x_f, y_f, z_f), the phase compensation is

$$\delta(x, y) = k[\sqrt{(x - x_f)^2 + (y - y_f)^2 + {z_f}^2} - f]. \tag{2.22}$$

A parallel-plate waveguide (PPW)-loaded metasurface lens based on a multi-layer structure loaded with varactors for collimation and beam steering was proposed in [64]. The phase of meta-atoms covered ~337° at 5.67 GHz. Large-angle dynamic beam scanning was verified experimentally with the bias voltages of varactors controlled via 8-channel digital-to-analog converters.

For finer modulation, varactors can be adopted, such as the metasurface mirror shown in figure 2.16(a) [65]. Eight discrete phases were available by applying different bias voltages to the varactors. The dynamic tunability of the position of

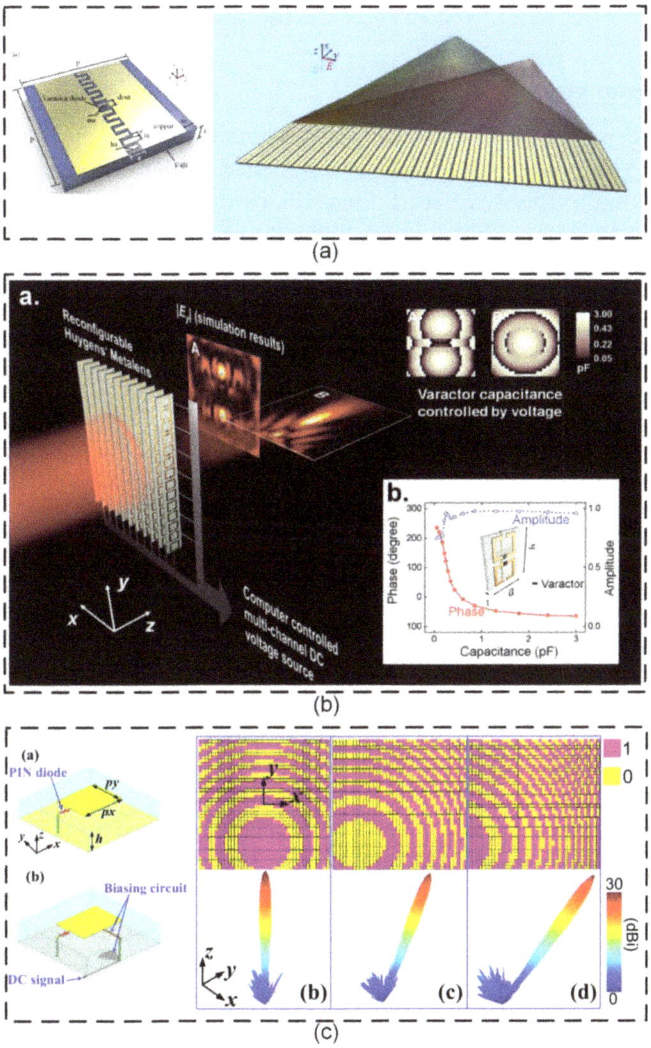

Figure 2.16. (a) Structure of the unit cell and conceptual illustration of the focus-tunable metasurface mirror. Reproduced from [65], CC BY 4.0. (b) Sketch of the reconfigurable meta-lens for dynamic focusing in [66]. Reproduced with permission from [66], John Wiley & Sons, 2017. (c) Unit cell of the multifunctional programmable metasurface in [67] and the beam focusing performance. Reproduced from [67], CC BY 4.0.

foci was proven with the phase distribution controlled using an MCU. Additionally, a reconfigurable meta-lens using varactor diodes was demonstrated in [66] to realize dynamic focusing in a 2D plane and dynamic control of two foci. A sketch of the meta-lens is provided in figure 2.16(b).

Again, it is noted that a programmable metasurface has the potential to realize multifunctional operation, such as the programmable metasurface with 1-bit reflected phase coding based on PIN diodes in [67]. This multifunctional meta-device was used for anomalous reflection, diffusion, beam steering and beam forming, while it has also been proved that the structure was applicable for beam focusing, which is demonstrated in figure 2.16(c).

2.4.5 Vortex beam generation

Electromagnetic waves have spin angular momentum (SAM) and orbital angular momentum (OAM), where the SAM is associated with the polarization state, and the OAM is related to the wavefront phase distribution. Digital coding metasurfaces can generate OAM beams that feature helically phased wavefront distribution, and the electric field is related to the azimuth with a factor $e^{jl\varphi}$, where l is the topological charge and takes an integer value (the positive and negative values depend on the direction of phase variation).

Again, assuming that a plane wave is normally incident on a metasurface located in the x–y-plane, in order to generate OAM beams with topological charges, the reflected or transmitted phase distribution on the metasurface is supposed to satisfy the following relationship [68]:

$$\delta(x, y) = l \text{ Arctan}\frac{y}{x}. \tag{2.23}$$

If spherical waves are fed by the sources, phase compensation also needs to be taken into account. In addition, the phases of the elements on digital coding metasurfaces are only allowed to take discrete values, and therefore, the above equation needs to be further quantized according to the available phases.

A 1-bit coded OAM vortex beam generator loaded with PIN diodes was designed in [69]. The quantization value is 0 when the phase is located in $[0, \pi)$, whereas the quantization value is π when the phase is located in $[\pi, 2\pi)$. A schematic diagram of the phase quantization procedure and the radiation patterns is shown in figure 2.17 (a). Other 1-bit PIN diode-based digital coding metasurfaces were also proposed for the generation of OAM beams [70, 71]. Varactors can be used for higher quantitative levels; for instance, a varactor-loaded vortex beam generator was presented in [72], and is shown in figure 2.17(b). When the bias voltage of varactors changed within a range of 0–5 V, a phase coverage of 320° could be achieved at 5 GHz.

2.4.6 Adaptive metasurfaces

The switching or tuning of the functions of reconfigurable metasurfaces need manual instructions. However, if combining the reconfigurable metasurface with a

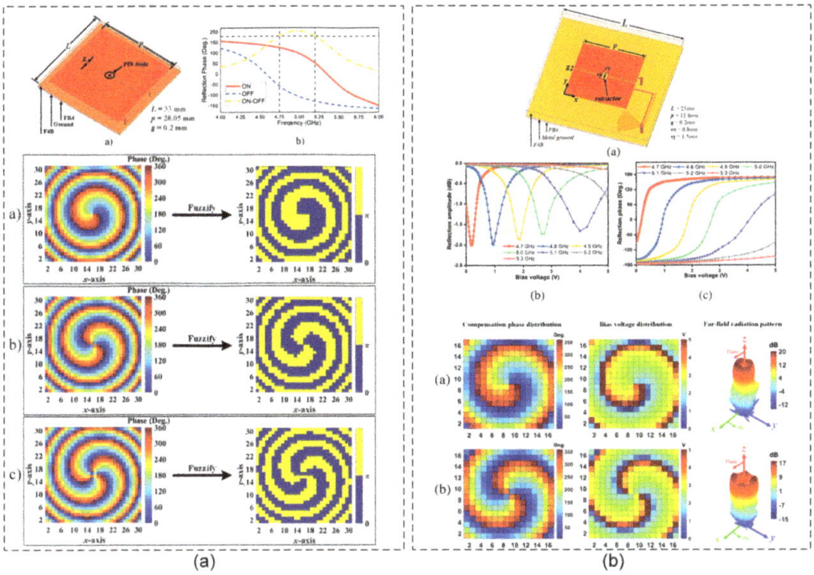

Figure 2.17. (a) Phase distributions required for the generation of vortex beams with different topological charges and the corresponding discrete phase coding matrices; calculated radiation patterns. Reproduced from [69], CC BY 4.0. (b) Unit cell of the varactor-based OAM metasurface, and the reflection response varying with the bias voltage of varactors. Phase distributions and corresponding bias voltage distributions, and the resultant far-field radiation patterns for OAM beam generation. Reproduced from [72], © IOP Publishing Ltd. All rights reserved.

feedback system, one may spare human intervention. The change of external environment can be sensed and transmitted to an MCU in real time using a sensor, and the MCU will automatically process the sensing data and send control signals to the metasurface, and through this an adaptive metasurface without human intervention can be constructed. An example of an adaptive programmable metasurface is the intelligent cloak based on deep learning in [73], which can automatically and quickly respond to ever-changing surrounding environments and incident fields and realize real-time stealth without human intervention. A sketch of the system is provided in figure 2.18(a). This design used two detectors to monitor changes in the incident field and the surrounding environment. A pre-trained artificial neural network was used to rapidly calculate the required bias voltage distribution of varactors. The electromagnetic fields scattered by the metasurface always mimic the scattered fields of the bare surroundings without the hidden object, so as to realize the stealth of the target.

Another adaptive programmable metasurface is that illustrated in figure 2.18(b) [74]. The system consisted of a 2-bit digital coding metasurface, together with a gyroscope, an MCU and an FPGA for closed-loop control. The gyroscope could sense the rotation angle of the metasurface, and the MCU was used to process and analyse the sensing signals, and send instructions to the FPGA to control its output voltages for real time adjustment of the phase distribution on the metasurface. Single-beam steering and dual-beam control were realized.

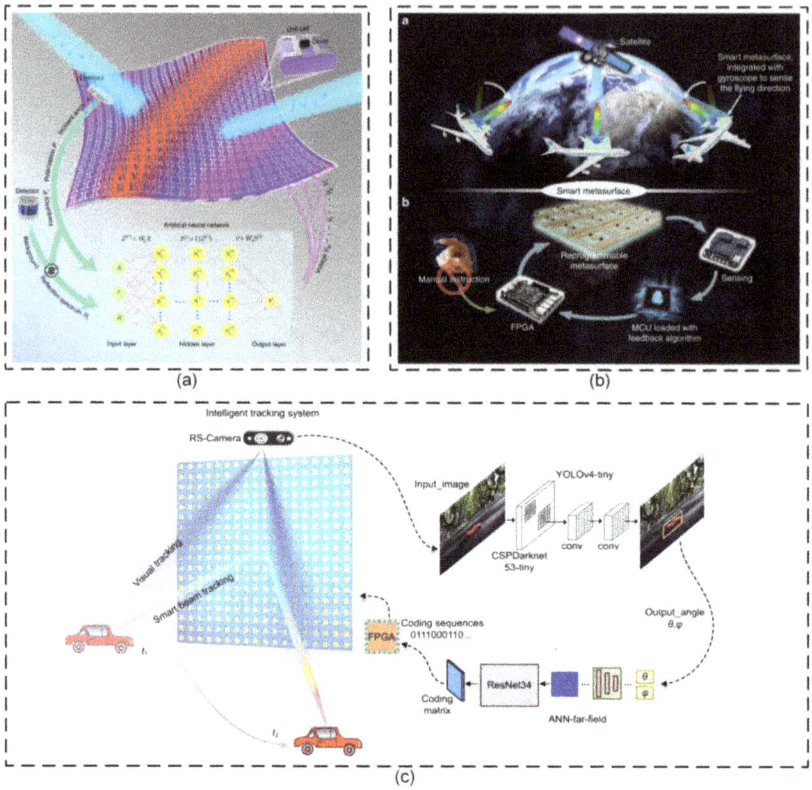

Figure 2.18. (a) Schematic of the self-adaptive metasurface cloak based on deep learning in [73]. Reproduced from [73], with permission from Springer Nature. (b) Sketch of the smart metasurface with a closed-loop system. Reproduced from [74], CC BY 4.0. (c) Diagram of the intelligent target tracking system in [75]. Reproduced from [75], CC BY 4.0.

In addition, an adaptive metasurface for moving target tracking and wireless communication was proposed based on computer vision in [75]. The system is illustrated in figure 2.18(c); it used the Intel Realsense camera as a vision sensor to shoot the moving targets, and the YOLOv4-tiny algorithm for target detection and feature extraction to acquire the position and attitude of the moving target in real time. A pre-trained artificial neural network was used to obtain the required coding sequences to control the steering angle of the reflected beam, so that it was always toward the moving target for real-time tracking. The researchers also proved the feasibility of wireless communication.

2.4.7 Space–time coding metasurfaces

In addition to dynamic control of spatial phase distribution, temporal control of the phase of each meta-atom may further increase the functions of programmable metasurfaces. For example, if periodic time modulation is applied to each element, with a period T_0, the scattered field will contain components with frequency

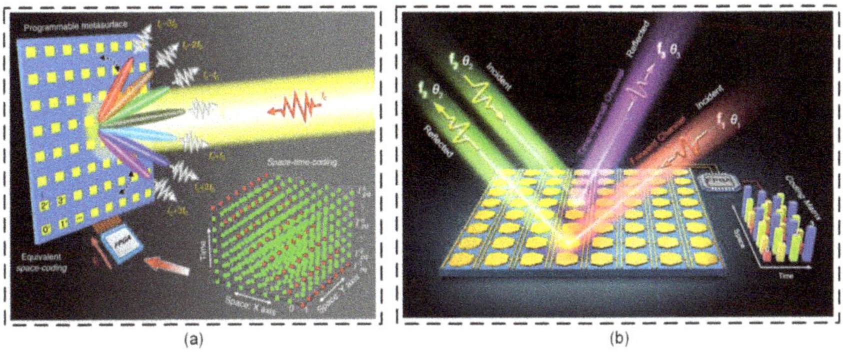

Figure 2.19. (a) Conceptual scheme of the space–time-coding metasurface in [76]. Reproduced from [76], © CC BY 4.0. (b) A sketch of nonreciprocal reflection realized by using a space–time-coding metasurface. Reproduced with permission from [2], John Wiley & Sons, 2019.

$(f_c + mf_0)$ according to a Fourier analysis [76], with f_c being the frequency of the incident wave, $f_0 = \frac{1}{T_0}$, and m being an integer.

A spatiotemporal coding metasurface loaded with PIN diodes was presented in [76], as shown in figure 2.19(a). This study presented several possible functions, such as harmonic beam steering with harmonics at different frequencies steered to different directions. Moreover, the equivalent 3-bit coding was realized by combining spatial coding with temporal modulation, which, compared with the 2-bit spatial coding, exhibited better sidelobe suppression in the function of beam steering at the center frequency. RCS reduction was also realized, and spatial coding contributed to the redistribution of energy in the spectrum. The devices showed superior behavior to metasurfaces that were only coded spatially in reducing backscatter. Further, the generation of vortex beams was also achieved.

Another example is the digital coding metasurface in [2], which broke Lorentz reciprocity using 2-bit spatiotemporal coding, so that the propagation direction and frequency of the time-reversed reflected wave are different from those of the original incident wave, that is, the angular separation and frequency shift are realized. A schematic of the space–time-coding metasurface is illustrated in figure 2.19(b).

2.5 Discussion and outlook

This chapter briefly introduced reconfigurable metamaterials operating in the microwave band. The key to achieving reconfigurability is to integrate active components or materials into passive meta-atoms. With the help of the sensitivity of active components or materials to external stimuli, the electromagnetic response of meta-atoms can be controlled dynamically. Different active components or materials may respond to different external stimuli. According to the categories of external stimuli, the modulation methods of reconfigurable metamaterials can be divided into the electrical, optical, thermal, chemical, mechanical and microfluidic methods, and so on. This chapter introduced some of the most commonly used modulation methods.

In addition to classifying reconfigurable metamaterials according to external excitation, the way in which the control signals are applied may also be used as a basis for classification, according to which the reconfigurable metamaterials can be divided into two types, namely reconfigurable metamaterials with all elements controlled uniformly, and those with each element controlled independently. The former possess simpler bias networks, while the latter are able to achieve more functions. In this chapter, these two types of reconfigurable metamaterials have been briefly introduced, with the basic principles, design ideas and examples of several specific functions demonstrated. The reconfigurable metasurfaces with independently controlled meta-atoms usually use PIN diodes or varactor diodes as active components, and control the bias voltage with the help of FPGA or MCU to realize phase coding. Hence, they are also called programmable metasurfaces.

In combination with closed-loop control systems, the variation of the surrounding environment is detectable by sensors and fed back to control units, so that they can automatically analyse and process the detected signals and control the coding matrices for the purpose of responding in real time to ambient changes. These structures can adaptively change the electromagnetic characteristics without human intervention. Hence, they are also called adaptive metasurfaces.

In addition, the design of reconfigurable metasurfaces may also incorporate artificial intelligence or adopt temporal modulation for further improvement of performance or extension of applications. The matured printed circuit board (PCB) technique greatly facilitates the fabrication of reconfigurable microwave metasurfaces. While using appropriate active components or materials and corresponding manufacturing methods, reconfigurable or even reprogrammable metasurfaces operating in higher frequency bands can also be implemented.

References

[1] Zahra S *et al* 2021 Electromagnetic metasurfaces and reconfigurable metasurfaces: a review *Front. Phys.* **8** 615

[2] Zhang L, Chen X Q, Shao R W, Dai J Y, Cheng Q, Castaldi G, Galdi V and Cui T J 2019 Breaking reciprocity with space-time-coding digital metasurfaces *Adv. Mater.* **31** 1904069

[3] He Q, Sun S and Zhou L 2019 Tunable/reconfigurable metasurfaces: physics and applications *Research* **2019** 1849272

[4] Li A *et al* 2017 High-power transistor-based tunable and switchable metasurface absorber *IEEE Trans. Microwave Theory Tech.* **65** 2810–8

[5] Debogovic T and Perruisseau-Carrier J 2014 Low loss MEMS-reconfigurable 1-bit reflectarray cell with dual-linear polarization *IEEE Trans. Antennas Propag.* **62** 5055–60

[6] Chen H *et al* 2020 Flexible rasorber based on graphene with energy manipulation function *IEEE Trans. Antennas Propag.* **68** 351–9

[7] Bonacchini G E and Omenetto F G 2021 Reconfigurable microwave metadevices based on organic electrochemical transistors *Nat. Electron.* **4** 424–8

[8] Zhang J *et al* 2020 Electrically tunable metasurface with independent frequency and amplitude modulations *ACS Photon* **7** 265–71

[9] Degiron A, Mock J J and Smith D R 2007 Modulating and tuning the response of metamaterials at the unit cell level *Opt. Express* **15** 1115–27

[10] Shadrivov I V *et al* 2012 Metamaterials controlled with light *Phys. Rev. Lett.* **109** 083902
[11] Zhang X G *et al* 2018 Light-controllable digital coding metasurfaces *Adv. Sci.* **5** 1801028
[12] Zhang X G *et al* 2022 A metasurface-based light-to-microwave transmitter for hybrid wireless communications *Light: Sci. Appl.* **11** 126
[13] Zhang X G *et al* 2020 An optically driven digital metasurface for programming electromagnetic functions *Nat. Electron.* **3** 165–71
[14] Chen X, Gao J and Kang B 2018 Achieving a tunable metasurface based on a structurally reconfigurable array using SMA *Opt. Express* **26** 4300–8
[15] Wang Z *et al* 2022 Design of thermal-switchable absorbing metasurface based on vanadium dioxide *IEEE Antennas Wirel. Propag. Lett.* **21** 2302–6
[16] Yan X *et al* 2020 Water-based reconfigurable frequency selective rasorber with thermally tunable absorption band *IEEE Trans. Antennas Propag.* **68** 6162–71
[17] Yang Y *et al* 2023 A new class of transformable kirigami metamaterials for reconfigurable electromagnetic systems *Sci. Rep.* **13** 1219
[18] Jing L *et al* 2018 Kirigami metamaterials for reconfigurable toroidal circular dichroism *NPG Asia Mater.* **10** 888–98
[19] Zhu H L *et al* 2014 Design of polarization reconfigurable antenna using metasurface *IEEE Trans. Antennas Propag.* **62** 2891–8
[20] Wu Z *et al* 2023 Active frequency selective rasorber with switchable transmission band and tunable absorption band *IEEE Microw. Wireless Technol. Lett.* **33** 1247–50
[21] Liu S *et al* 2019 Flexible controls of broadband electromagnetic wavefronts with a mechanically programmable metamaterial *Sci. Rep.* **9** 1809
[22] Kong X *et al* 2023 Wide-passband reconfigurable frequency selective rasorber design based on fluidity of EGaIn *IEEE Antennas Wirel. Propag. Lett.* **22** 1922–6
[23] Wu P C *et al* 2017 Broadband wide-angle multifunctional polarization converter via liquid-metal-based metasurface *Adv. Opt. Mater.* **5** 1600938
[24] Zhu W *et al* 2015 A flat lens with tunable phase gradient by using random access reconfigurable metamaterial *Adv. Mater.* **27** 4739–43
[25] Yan L B *et al* 2017 Adaptable metasurface for dynamic anomalous reflection *Appl. Phys. Lett.* **110** 201904
[26] Yan L *et al* 2018 Arbitrary and independent polarization control *in situ* via a single metasurface *Adv. Opt. Mater.* **6** 1800728
[27] Naqvi A H and Lim S 2022 Hydrodynamic metasurface for programming electromagnetic beam scanning on the azimuth and elevation planes *Microsyst. Nanoeng.* **8** 43
[28] Kong J A 1986 *Electromagnetic Wave Theory* (Toronto: Wiley) p 696
[29] Huang C X *et al* 2021 Polarization modulation for wireless communications based on metasurfaces *Adv. Funct. Mater.* **31** 2103379
[30] Li W *et al* 2016 A reconfigurable polarization converter using active metasurface and its application in horn antenna *IEEE Trans. Antennas Propag.* **64** 5281–90
[31] Yang Z *et al* 2021 Reconfigurable multifunction polarization converter integrated with PIN diode *IEEE Microwave Wirel. Compon. Lett.* **31** 557–60
[32] Gao X *et al* 2018 A reconfigurable broadband polarization converter based on an active metasurface *IEEE Trans. Antennas Propag.* **66** 6086–95
[33] Xu H-X *et al* 2016 Dynamical control on helicity of electromagnetic waves by tunable metasurfaces *Sci. Rep.* **6** 27503

[34] Hassan A G *et al* 2023 Reconfigurable absorptive and polarization conversion metasurface consistent for wide angles of incidence *Sci. Rep.* **13** 18209

[35] Balanis C A 2016 *Antenna Theory Analysis and Design* 4th edn (Hoboken, NJ: Wiley)

[36] Zhao Y *et al* 2022 Design of a broadband switchable active frequency selective surfaces based on modified diode model *IEEE Antennas Wirel. Propag. Lett.* **21** 1378–82

[37] Li H *et al* 2020 A wideband multifunctional absorber/reflector with polarization-insensitive performance *IEEE Trans. Antennas Propag.* **68** 5033–8

[38] Lee D, Jeong H and Lim S 2017 lectronically switchable broadband metamaterial absorber *Sci. Rep.* **7** 4891

[39] Zhu J *et al* 2015 Tunable microwave metamaterial absorbers using varactor-loaded split loops *Europhys. Lett.* **112** 54002

[40] Wu L *et al* 2019 Broadband frequency-selective rasorber with varactor-tunable interabsorption band transmission window *IEEE Trans. Antennas Propag.* **67** 6039–50

[41] Bakshi S C, Mitra D and Teixeira F L 2021 Wide-angle broadband rasorber for switchable and conformal application *IEEE Trans. Microwave Theory Tech.* **69** 1205–16

[42] Qian G *et al* 2019 Switchable broadband dual-polarized frequency-selective rasorber/absorber *IEEE Antennas Wirel. Propag. Lett.* **18** 2508–12

[43] Sun M *et al* 2023 Reconfigurable multiband rasorber with identical metallic-strip-based configuration layers for coordinated absorbing/transmitting design *IEEE Antennas Wirel. Propag. Lett.* **22** 2556–60

[44] Li R *et al* 2021 Switchable rasorber with high-order frequency selective surface for transmission bandwidth extension *Microwave Opt. Technol. Lett.* **63** 1705–11

[45] Jiang H *et al* 2024 Independently switchable rasorber with wide transmission and low-reflection bands under dual polarization *IEEE Trans. Microwave Theory Tech.* **72** 863–77

[46] Jiang B *et al* 2023 A switchable A-T-A rasorber with polarization selectivity and high roll-off characteristics *IEEE Antennas Wirel. Propag. Lett.* **22** 2075–9

[47] Yu D *et al* 2023 High-selectivity frequency-selective rasorber with tunable absorptivity *IEEE Trans. Antennas Propag.* **71** 3620–30

[48] Zhang Z *et al* 2022 Macromodeling of reconfigurable intelligent surface based on microwave network theory *IEEE Trans. Antennas Propag.* **70** 8707–17

[49] Yu N *et al* 2011 Light propagation with phase discontinuities: generalized laws of reflection and refraction *Science* **334** 333–7

[50] Wan X *et al* 2016 Field-programmable beam reconfiguring based on digitally-controlled coding metasurface *Sci. Rep.* **6** 20663

[51] Huang C *et al* 2017 Dynamical beam manipulation based on 2-bit digitally-controlled coding metasurface *Sci. Rep.* **7** 42302

[52] Ratni B *et al* 2018 Active metasurface for reconfigurable reflectors *Appl. Phys.* A **124** 104

[53] Huang C *et al* 2017 Reconfigurable metasurface for multifunctional control of electromagnetic waves *Adv. Opt. Mater.* **5** 1700485

[54] Gabor D 1948 A new microscopic principle *Nature* **161** 777–8

[55] Brown B R and Lohmann A W 1966 Complex spatial filtering with binary masks *Appl. Opt.* **5** 967–9

[56] Ni X, Kildishev A V and Shalaev V M 2013 Metasurface holograms for visible light *Nat. Commun.* **4** 2807

[57] Wang L *et al* 2016 Grayscale transparent metasurface holograms *Optica* **3** 1504–5

[58] Wang Q *et al* 2016 Broadband metasurface holograms: toward complete phase and amplitude engineering *Sci. Rep.* **6** 32867
[59] Li L *et al* 2017 Electromagnetic reprogrammable coding-metasurface holograms *Nat. Commun.* **8** 197
[60] Quan X *et al* 2022 Mechanically reprogrammable Pancharatnam–Berry metasurface for microwaves *Adv. Photon.* **4** 016002
[61] Huang L, Zhang S and Zentgraf T 2018 Metasurface holography: from fundamentals to applications**7** 1169–90
[62] Li Y B *et al* 2016 Transmission-type 2-bit programmable metasurface for single-sensor and single-frequency microwave imaging *Sci. Rep.* **6** 23731
[63] Li L *et al* 2019 Machine-learning reprogrammable metasurface imager *Nat. Commun.* **10** 1082
[64] Li H *et al* 2022 Experimental validation of an active parallel-plate-waveguide-loaded metasurfaces lens for wide-angle beam steering *IEEE Antennas Wirel. Propag. Lett.* **21** 2156–60
[65] Wang Z *et al* 2019 Metasurface-based focus-tunable mirror *Opt. Express* **27** 30332–9
[66] Chen K *et al* 2017 A reconfigurable active Huygens' metalens *Adv. Mater.* **29** 1606422
[67] Yang H *et al* 2016 A programmable metasurface with dynamic polarization, scattering and focusing control *Sci. Rep.* **6** 35692
[68] Zhang L *et al* 2017 Spin-controlled multiple pencil beams and vortex beams with different polarizations generated by Pancharatnam–Berry coding metasurfaces *ACS Appl. Mater. Interfaces* **9** 36447–55
[69] Han J *et al* 2018 1-bit digital orbital angular momentum vortex beam generator based on a coding reflective metasurface *Opt. Mater. Express* **8** 3470–8
[70] Bai X *et al* 2020 High-efficiency transmissive programmable metasurface for multimode OAM generation *Adv. Opt. Mater.* **8** 2000570
[71] Nadi M, Sedighy S H and Cheldavi A 2023 Multimode OAM beam generation through 1-bit programmable metasurface antenna *Sci. Rep.* **13** 15603
[72] Han J *et al* 2018 Versatile orbital angular momentum vortex beam generator based on reconfigurable reflective metasurface *Jpn. J. Appl. Phys.* **57** 120303
[73] Qian C *et al* 2020 Deep-learning-enabled self-adaptive microwave cloak without human intervention *Nat. Photonics* **14** 383–90
[74] Ma Q *et al* 2019 Smart metasurface with self-adaptively reprogrammable functions *Light: Sci. Appl.* **8** 98
[75] Li W *et al* 2023 Intelligent metasurface system for automatic tracking of moving targets and wireless communications based on computer vision *Nat. Commun.* **14** 989
[76] Zhang L *et al* 2018 Space-time-coding digital metasurfaces *Nat. Commun.* **9** 4334

Chapter 3

Phase change media-assisted programmable metamaterials for structural color generation and absorption of light waves

M Pourmand and Pankaj K Choudhury

Light–matter interaction at the nanoscale dimension is discussed, emphasizing phase change media (PCM)-based reconfigurable metamaterials for structural color generation in the visible regime and nearly perfect tunable absorption in the visible and near-infrared (NIR) regimes. After a superficial treatment of the basics of hyperbolic metamaterials (HMMs), studies are reported of metamaterial configurations implementing PCM-embedded bilayer stacks that exhibit HMM behavior. The discussed structures can be used to dynamically modulate the reflected primary colors by the layered forms of embedded pixels, so that these have potential for designing reflective color displays. Further, the possibility of realizing tunable absorbers in the visible and NIR regimes is also discussed, focusing on adjusting the absorption band from 1.310 to 1.570 μm wavelength.

3.1 Introduction

Surface plasmonics has been one of the important phenomena in the context of light–matter interaction at the sub-wavelength scale [1]. According to this, the light illuminating the interface of a metallic film and bulk dielectric medium excites free electrons (of metal) to oscillate while the total electric field inside the metal remains zero. The plasmonic oscillations can travel along the interface or on the thin metallic film surface with a broad spectrum of frequencies from 0 to $\omega_p/\sqrt{2}$, with ω_p being the plasma frequency, given as [1]

$$\omega_p = \sqrt{\frac{ne^2}{\epsilon_0 m}},$$

doi:10.1088/978-0-7503-5462-2ch3

where e is the electronic charge, n is the number density of electron gas, m is the effective optical mass of an electron, and ϵ_0 is the free space permittivity. The physics of plasmons is extended by the concept of surface plasmon polaritons (SPPs), which are coherently oscillating electrons with the incidence wave at a metal–dielectric interface [1]. These SPPs are significantly confined at the surface of metal, giving rise to the strong light–matter interaction. Moreover, sub-wavelength nanoparticles or metallic nanowires embedded in a dielectric medium support localized surface plasmons (LSPs) at the surface of metallic components [1, 2]. The formation of LSPs plays a determining role in the realization of active plasmonic or nanophotonic applications, ranging from biosensing to super-resolution imaging, optical data processing, active filters, and absorbers [3–11].

In the context of SPPs, discussions on metamaterials [11] are obvious, as these have been reported for multifarious applications exploiting the concept of wave propagation at the metal–dielectric and/or dielectric–dielectric interface, which excites the generation of SPP waves. The uncommon electromagnetic features of metamaterials enable them to be vital in developing varieties of artificial media exhibiting negative refraction for applications that include perfect lensing [12], imaging [13], cloaking [14, 15], sensing [16–23], and waveguiding [24, 25].

Though there are several kinds of metamaterials, hyperbolic metamaterials (HMMs) form a class with the ability to demonstrate metallic-to-dielectric behaviors upon altering the incidence wave properties [26–28]. These are extremely anisotropic in nature, which is defined by the relative permittivity and permeability tensors. In certain operational conditions, these exhibit a hyperbolic dispersion relation, thereby allowing the propagation of waves with infinitely large wavenumbers (i.e. high-k) and broadband density of states [27]. Such exotic features of HMMs make them suitable for a broad range of applications [29, 30], including directional waveguiding, imaging, sensing [24, 31–33], negative refraction [12, 34–36], sub-wavelength modes [36, 37], sub-diffraction imaging [38], spontaneous and thermal emission engineering [39–41], high-speed optical communications [42], etc.

This chapter discusses the design of active nanophotonic devices comprising phase change media-embedded HMMs. With this thematic view in mind, we first briefly discuss the fundamentals of HMMs and phase change media, and touch upon comprehensive design procedures for two different kinds of active functional devices to be operated in the visible light regime.

3.2 Fundamentals of hyperbolic metamaterials

It has been reported that in anisotropic media the electric displacement $\overline{D}$ and magnetic displacement $\overline{B}$ vectors can be expressed by the relative permittivity and permeability tensors, $\overline{\overline{\epsilon}}_r$ and $\overline{\overline{\mu}}_r$ [27]. For nonmagnetic materials, $\overline{\overline{\mu}}_r$ is defined by a unit tensor, whereas the tensor $\overline{\overline{\epsilon}}_r$ assumes the form [27]

$$\overline{\overline{\epsilon}}_r = \epsilon_0 \begin{bmatrix} \epsilon_x & 0 & 0 \\ 0 & \epsilon_y & 0 \\ 0 & 0 & \epsilon_z \end{bmatrix} \tag{3.1}$$

in a Cartesian coordinate system, with ϵ_0 being the free space permittivity, and the other permittivity components being along the x-, y- and z-directions. For a uniaxial medium, considering the optical axis to be along the z-direction (of the Cartesian coordinate system), one defines the normal and the parallel permittivity components as $\epsilon_z = \epsilon_\perp$ and $\epsilon_x = \epsilon_y = \epsilon_\parallel$, respectively.

It has been shown that, considering the propagation of harmonic plane waves through the medium, the use of source-free Maxwell's equations finally provides the eigenvalue equation of the form [27]

$$\bar{k}(\bar{k} \cdot \bar{E}) = \left(|k|^2 - k_0^2 \overline{\overline{\epsilon_r}}\right)\bar{E}, \tag{3.2}$$

with $\bar{k}$ being the wavevector (given as $|k|^2 = k_x^2 + k_y^2 + k_z^2$ in the form of combination of wavevector components) of the propagating wave, k_0 being the free space wavevector and $\bar{E}$ being the electric field. Further, the solution to the eigenvalue equation (3.2) gives the dispersion relation as [27]

$$\left(|k|^2 - k_0^2 \epsilon_\parallel\right)\left(\frac{k_x^2 + k_y^2}{\epsilon_\perp} + \frac{k_z^2}{\epsilon_\parallel} - k_0^2\right) = 0. \tag{3.3}$$

That is, equation (3.3) reveals that either

$$k_x^2 + k_y^2 + k_z^2 = k_0^2 \epsilon_\parallel \tag{3.4}$$

$$\text{or } \frac{k_x^2 + k_y^2}{\epsilon_\perp} + \frac{k_z^2}{\epsilon_\parallel} = k_0^2. \tag{3.5}$$

Equation (3.4) describes the dispersion relation as a spherical equation that corresponds to the ordinary waves. This term describes waves with the electric field in the xy-plane (i.e. the transverse electric (TE) waves) that can propagate with dispersion defined by the spherical isofrequency surface (figure 3.1(a), left). However, equation (3.5) describes an ellipsoidal isofrequency surface (figure 3.1 (a), center and right) in the k-space relating to the extraordinary waves; these can be excited by the transverse magnetic (TM) modes of polarization with the magnetic field being in the xy-plane. Equation (3.5) becomes a hyperbolic isofrequency surface if either $\epsilon_\parallel$ or $\epsilon_\perp$ is negative. Figure 3.1(a) determines the isofrequency surfaces for the isotropic (figure 3.1(a), left) and anisotropic (figure 3.1(a), center and right) media for different values of $\epsilon_\parallel$ and $\epsilon_\perp$.

HMM media can be classified into two categories—type I and type II—based on the sign of the permittivity tensor components. Type I HMMs have only one negative component of the permittivity tensor, i.e. $\epsilon_\parallel > 0$ and $\epsilon_\perp < 0$, while type II have two negative components, i.e. $\epsilon_\parallel < 0$ and $\epsilon_\perp > 0$. Type I HMMs support both the TE and TM modes, whereas type II HMMs allow only the TM mode to exist [27]. Moreover, type II HMMs show more metallic behavior and these are more reflective in nature, whereas type I HMMs exhibit high transmission [27, 34].

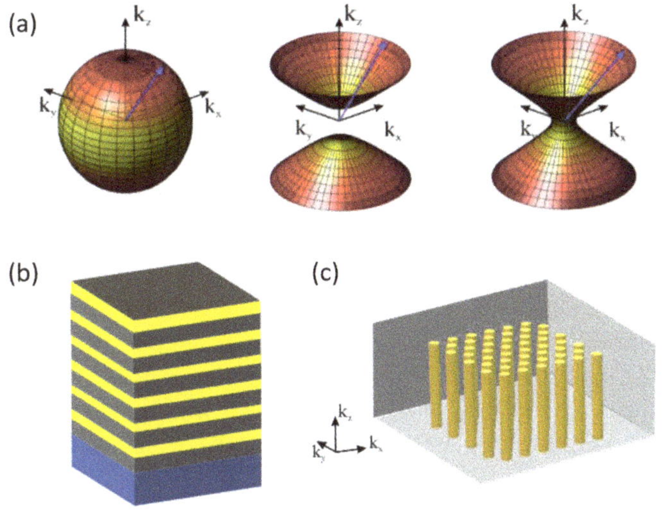

Figure 3.1. (a) Isofrequency surfaces in the k-space for isotropic ($\epsilon_{\parallel}$, $\epsilon_{\perp} > 0$), type I HMM ($\epsilon_{\parallel} > 0,\ \epsilon_{\perp} < 0$), and type II HMM ($\epsilon_{\parallel} < 0,\ \epsilon_{\perp} > 0$) media. Corresponding schematic of HMM (b) multilayered metal–dielectric thin films, and (c) metallic nanowires arranged periodically inside a dielectric host medium.

HMMs may assume two different structural configurations, namely multilayered (figure 3.1(b)) and arrayed (figure 3.1(c)). A thin film multilayer structure comprising alternative layers of metal and dielectric media would exhibit anisotropy. However, the layer thickness must be comparably less than the operating wavelength to comply with homogenization theory (which will be discussed at a later stage). On the other hand, the arrayed HMMs comprise periodically arranged metallic nanowires in a dielectric host, and the periodicity in structure is far below the operating wavelength to fulfill the medium homogenization property. The arrayed HMMs would show high transmission, low loss, and broad bandwidth. The low reflection of these allows type I characteristics to be achieved fairly easily compared to layered HMMs, which mostly exhibit type II behavior [27].

The effective medium theory (EMT) method is widely employed in evaluating the optical response of HMMs. It is a qualitative approach to homogenize an infinite periodic composite with sub-wavelength unit cells, and follows a generalized Maxwell Garnett approach to obtain the $\epsilon_{\parallel}$- and $\epsilon_{\perp}$-components of the effective permittivity tensor [43]. In the case of layered HMMs (figure 3.1(b)), these permittivity components are defined as [27]

$$\epsilon_{\parallel} = \rho\epsilon_m + (1 - \rho)\epsilon_d;\ \epsilon_{\perp} = \frac{\epsilon_m \epsilon_d}{\rho\epsilon_d + (1 - \rho)\epsilon_m}, \tag{3.6}$$

where ρ denotes the filling fraction of metal, given as $\rho = d_m/d_z$, with d_z ($=d_m + d_D$) as the total thickness of the bilayer, and d_m and d_D, respectively, being the thickness of metal and dielectric layers having their respective permittivity values as ϵ_m and ϵ_d. As for the arrayed HMM (figure 3.1(b)), the permittivity tensor components are [27]

$$\epsilon_{\parallel} = \frac{[(1+\rho)\epsilon_m + (1-\rho)\epsilon_d]\epsilon_d}{[(1-\rho)\epsilon_m + (1+\rho)\epsilon_d]}; \ \epsilon_{\perp} = \rho\epsilon_m + (1-\rho)\epsilon_d, \tag{3.7}$$

with the filling fraction given by $\rho = A/A_0$, where A and A_0 denote the cross-sectional areas of metalllic wires and the host dielectric medium, respectively.

Interestingly, the hyperbolic dispersion in HMMs allows wave propagation with large wavevectors [29]—the modes that can propagate within or on the HMM surface. The hyperbolic characteristic of composites can be inferred from the existence of poles and zeros in the plots of effective permittivity ϵ_{eff} (against wavelength λ or frequency f). The zeros and poles refer to the certain wavelengths at which a component of the dielectric tensor either passes through zero (epsilon-near-zero, ENZ) or has a resonant pole (epsilon-near-pole, ENP) [44, 45]. ENZ takes place in the direction of free electron motion, which is parallel to the thin film in the layered configuration or along the length of nanowires in the arrayed configuration. In contrast, ENP happens in the direction in which the motion of free electrons is blocked, meaning perpendicular to the thin film in layered structures and perpendicular to the nanowires in arrayed structures [44]. The behavior of HMMs can be defined by the ENZ and ENP resonances where the reflection and transmission spectra undergo changes; we will present the relevant analyses in the following sections.

3.3 Phase change media-based metamaterials

The use of phase change media (PCMs) in designing metamaterials is crucial, as these offer tunability to photonic devices. The electrical and optical properties of these can be modified by applying an external stimulus such as an electrical or optical pulse and even by adjusting the ambient temperature, and therefore, metamaterials comprising PCMs exhibit adjustable electromagnetic behavior in a wide spectral range. As such, employing PCM-based metamaterials provides a scalable and cost-effective platform to develop wave-controllable structures, thereby allowing the possibility to realize programmable (or reconfigurable) devices.

Early studies reported adjustable electrical resistors based on MoS_2 (molybdenum disulfide—a transition metal dichalcogenide) and ternary glasses comprising As–Te–Br or –I [46], but later works in the 1990s brought the PCM-based devices into prominence by introducing rewritable optical disks [47]. Investigations on PCM-based structures continued to develop tunable devices, such as multi-level photonic memories [48, 49], photonic synapses for neuromorphic photonic devices [50], nonvolatile reflective displays [51], and reflective modulators and switches [52].

PCMs can be categorized into (but not restricted to) three main kinds: (i) metal oxides such as vanadium dioxide (VO_2), (ii) chalcogenides (including MoS_2, stibnite (Sb_2S_3), germanium–antimony–tellurium (GST or $G_2Se_2Te_5$), germanium selenide (GeSe), indium antimonide (InSb)), and (iii) perovskites such as strontium titanate ($SrTiO_3$) and methyl ammonium lead iodide ($MAPbI_3$) [53–56]. However, the phase transition mechanisms differ in each of these classes of PCMs. The phase switching property in metal oxide PCMs is usually reversible. The chalcogenide PCMs (such as

GST) exhibit versatile optical properties and can sustain the phase change even in the absence of stimuli. These turn to the stimulated state with a higher speed of a few pico-seconds, thereby allowing them to be used in high-speed devices [49]. The main drawback of chalcogenide PCMs is their relatively large loss in the optical regime.

Vanadium dioxide (VO_2) exhibits both the insulator-to-metal and crystalline–crystalline phase transitions, from a resistive monoclinic M-VO_2 to conductive rutile structure R-VO_2 [57–59]. At room temperature and ambient pressure, M-VO_2 is an electrical insulator and optically transparent (the atomic structure consists of four vanadium ions per unit cell). On heating above the Curie temperature $T_c = 68$ °C, VO_2 undergoes a phase transition to the metallic state, exhibiting conductive properties. The phase transition from the M-VO_2 state to R-VO_2 occurs after breaking the V–V dimers (in M-VO_2) and straightening V atoms along the *c*-axis, shrinking the unit cell twofold, as there are two V atoms in the R-VO_2 unit cell [59]. Figure 3.2(a) depicts the R-VO_2 (left) and M-VO_2 (right) unit cells. This phase transition happens by changing the ambient temperature [57], applying an electrical voltage [58] or with an ultrafast optical pump [59]. A virtue of metal oxide PCMs is the fast phase transition of states at low temperatures. On the other hand, the nonvolatile characteristic of M-VO_2 and the high optical absorption due to a relatively high extinction ratio ($k \sim 0.5$) make them less suitable for reconfigurable applications, such optical memory. However, these properties make them potentially suitable for versatile uses, such as optical modulators.

A typical chalcogenide PCM, such as GST, undergoes amorphous–crystalline transition on a sub-nanosecond time scale, which can be done reversibly billions of

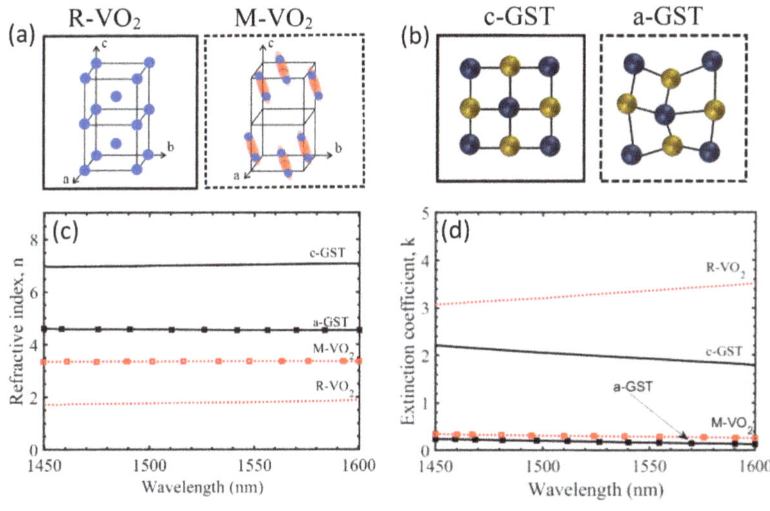

Figure 3.2. Schematic of atomic structures and optical properties of VO_2 and GST. (a) VO_2 structure for rutile (left) and monoclinic (right) states where the vanadium atoms appear in blue and the V–V dimers are presented in the monoclinic structure with an orange shadow [56]. (b) 2D atomic structure of c-GST (left) and a-GST (right), where the golden and black balls represent the Ge, Sb, and Te atoms, respectively [48]. (c, d) The respective real and imaginary parts of the refractive indices of VO_2 and GST.

times [42]. Yet, the electrical and optical properties of GST states change drastically from an optically transmissive, electrical insulator, amorphous state (the *a*-state; figure 3.2(b)) to an optically opaque, electrically conductive, crystalline state (the *c*-state, figure 3.2(b)) [60, 61]. This is attributed to a pronounced change in the resonant bonding of GST in the transition from the *a*-state to the *c*-state, which can be achieved by heating the material above the crystallization temperature (573 K) followed by a slow quenching period. In contrast, the *a*-state is accessed by intensively heating the alloys above the melting temperature (823 K) followed by rapid quenching [49, 61]. The rapid cooling down freezes the liquid-like PCM into the disordered amorphous phase. The required huge temperature alteration can be achieved by utilizing external heaters, highly intensive locally focused optical pulses (the photo-thermal effect) [49] or applying electrical pulses to embedded micro-heaters (electro-thermal effect) [62]. GST exhibits a pronounced change in the real part of the refractive index ($\Delta n = 3.56$) in switching from the *a*-state to the *c*-state in the infrared (IR) regime; however, a high absorption property in the visible light regime limits potential applications.

Figure 3.2 illustrates the schematics of the two states of VO_2 (figure 3.2(a)) and GST (figure 3.2(b)) [47, 63, 64] and their associated optical properties in the wavelength range of 1.45–1.6 μm (figure 3.2(c) and (d)) [65–67]. The depicted drastic change in optical properties is attributed to a fully driven phase transition in the respective PCM. In the partially driven transition, only a fraction of pronounced change will be accessible; the feature can be utilized for programmable nano-photonic devices.

Among many chalcogenide PCMs, Sb_2S_3 exhibits low optical loss and tunable absorption edge [65]. The significant feature of Sb_2S_3 is its large real part of refractive index in both the amorphous and crystalline phases—a feature that makes it more compatible with conventional deposition techniques [56]. In addition, its phase switching can be stimulated by imposing a laser beam or embedded micro-heaters [56, 68]. $GeSe_3$ is another chalcogenide PCM with high refractive index contrast, low optical absorption, and wide bandgap in the visible regime. Unlike the other chalcogenide PCMs (such as GST), the bandgap of $GeSe_3$ can be increased by increasing the concentration of Ge molecules [61, 69]. Employing thermal annealing rearranges the configurations of $GeSe_3$ atoms, thereby affecting the density of $GeSe_3$ molecules, and hence decreases both its refractive index and extinction coefficient. These features can be utilized to develop innovative photonic devices, such as UV absorptive coatings, photodetectors, solar cells, programmable reflectors, and lenses [11, 69, 70].

This chapter emphasizes designing programmable HMM-based nanostructures employing PCMs. In particular, we focus on the use of PCM-based multilayer HMM configurations in developing programmable structural color generation systems and dual-band nearly perfect absorbers. The following sections discuss these two different types of application scenarios for complex-structured layered HMMs, which we touch upon one by one.

3.4 Programmable structural color generation systems

Dynamic color generation structures provide higher resolution and scalability compared to traditional pigmentation-based displays [71, 72]. However, their fixed optical response after practical realization greatly affects their use. To address this issue, several kinds of tunability mechanisms have been introduced, including the implementation of plasmonic nanoantenna structures enabled by liquid crystals [73] and plasmonic resonators with stretchable materials [74]. Chalcogenide PCM-assisted plasmonic structures offer advantages, including flexibility, color durability, high resolution, cost-effectiveness, and complementary metal–oxide–semiconductor (CMOS) compatibility [49]. In this section, we discuss an optimized layered metamaterial structure integrated with PCMs to generate a wide spectrum of colors.

3.4.1 Structure, constitutive properties, and spectral response

Figure 3.3(a) depicts the proposed structure consisting of n bilayers of $GeSe_3$ (of thickness d_D^n) and Al (of thickness d_m^n). To heat the structure, we use a tri-layer graphene sheet of 1 nm thickness beneath the structure. The SiO_2 layers on the top and bottom (of the structure) serve two purposes—the top capping layer of SiO_2 of thickness $d_c = 100$ nm protects the $GeSe_3$ medium from evaporation; a buffer layer (of thickness $d_s = 10$ nm) above the substrate isolates the graphene sheets from the metallic layer above, and provides sufficient adhesion between the graphene sheet and SiO_2 substrate during fabrication process, which can be accomplished by using

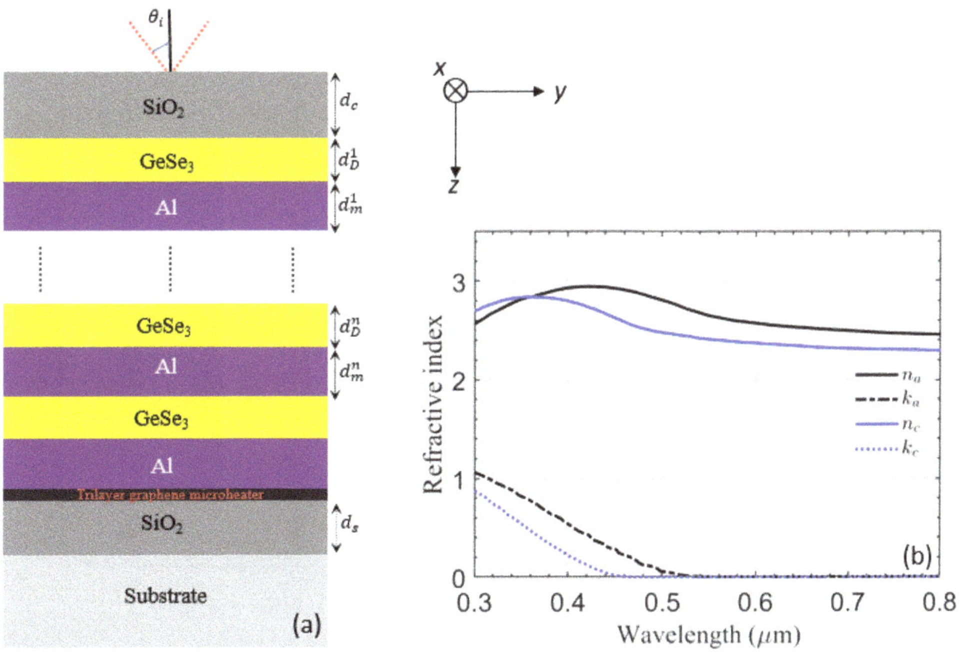

Figure 3.3. (a) 2D illustration of the proposed structure, and (b) wavelength dependence of the optical properties (n and k) of $GeSe_3$ before and after thermal annealing at 360 °C for 1 h.

lithography-free techniques such as physical vapor deposition (PVD) techniques, namely thermal evaporation and/or RF sputtering [69].

The optical properties of chalcogenide PCMs would alter as the state of material is changed from amorphous to crystalline. In the same manner, the optical properties of $GeSe_3$ will be altered as the bulk material is heated for a short time and then cooled down in the laboratory. Figure 3.3(b) illustrates the effect of thermal annealing (at 360 °C for 1 h) on the optical properties of 25 nm thick $GeSe_3$ sputtered on the Si wafer [69]. It is clear from this figure that the extinction coefficient k of the annealed $GeSe_3$ is wavelength-dependent. In particular, both the states of $GeSe_3$ exhibit nominal loss at wavelengths beyond 450 nm.

3.4.1.1 Analysis of spectral response

To examine the optical response of the multilayered structure, we implement the transfer matrix method (TMM) by using the effective permittivity approximation. The unit cell of the stack is homogenized by applying the generalized Maxwell Garnett technique. As such, equation (3.9a) defines the effective permittivity of the bilayer stack in the parallel $\epsilon_{\|}$ and perpendicular $\epsilon_{\perp}$ directions. Regarding the use of ρ in equation (3.9a), it must be noted that the ρ-values of 0 and 1 mean the purely dielectric (i.e. $GeSe_3$) and purely metallic (i.e. Al) layers, respectively. Further, following equation (3.9a), d_z is the total thickness of the stack, i.e. $d_z = d_m + d_D$.

Figure 3.4(a) illustrates the wavelength dependence of the real ($\Re$) and imaginary ($\Im$) parts of the $\epsilon_{\|}$ and $\epsilon_{\perp}$ permittivity components upon exploiting the homogenization approach for the bilayer stack with a ρ-value of 0.4 and a thickness of 50 nm. Figure 3.4(b) shows the corresponding reflectance R, transmittance T, and absorptance A spectra of the same bilayer upon applying the TMM. As we see in figure 3.4 (a), the bilayer stack exhibits type II HMM behavior in a wavelength span of 0.38–0.78 μm because the real parts of the effective permittivity components are $\Re(\epsilon_{\perp}) > 0$, $\Re(\epsilon_{\|}) < 0$ in this regime. It is also evident from the optical response in figure 3.4(b), where the reflectance obtains higher values, $\Im(\epsilon_{\|})$ plunges to the minimum value (figure 3.4(a)) near 500 nm. The transmittance is, however, low in the stated

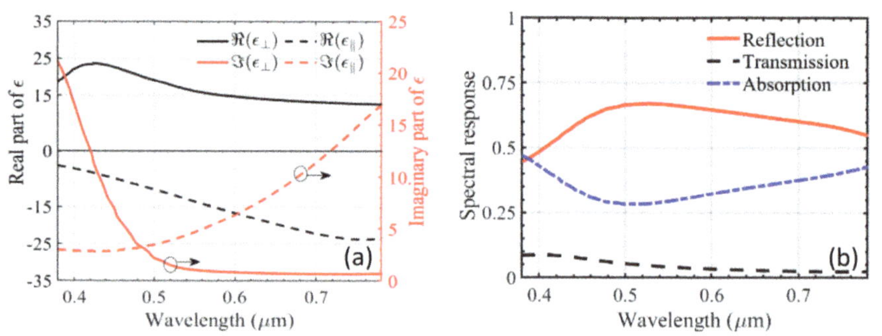

Figure 3.4. Optical characteristics of the $GeSe_3$–Al bilayer stack in the visible regime; (a) the real and imaginary parts of $\epsilon_{\|}$- and $\epsilon_{\perp}$-components as obtained implementing the Maxwell Garnett approach, and (b) the reflectance R, transmittance T, and absorptance A deduced by applying the TMM.

wavelength span, but the absorptance is slightly enhanced for wavelengths above 500 nm (figure 3.4(b)).

We must note that the characteristics of the $GeSe_3$–Al bilayer are different from those of the single layer made of either of the constituent materials. The optical response of the bilayer can be enhanced to attain specific reflection at the desired wavelength by optimizing the operational and geometrical parameters, such as the $GeSe_3$ layer crystallinity state (i.e. crystalline or amorphous states), incidence angle, layer thicknesses, and filling fractions of the structure.

3.4.1.2 Effect of design parameters on the optical response

We now investigate the effect of the geometrical parameters on the spectral response of the proposed structure. We mentioned before that the $GeSe_3$–Al bilayer with a ρ-value of 0.4 exhibits type II HMM behavior in the wavelength window of 0.38–0.78 μm, where the $\epsilon_\perp$- and $\epsilon_\parallel$-permittivity tensor components have opposite signs, i.e. $\Re(\epsilon_\perp) \times \Re(\epsilon_\parallel) < 0$. With the bilayer thickness being fixed at 50 nm, we examine the effect of different ρ-values for the stacks comprising $GeSe_3$ media in the crystalline and amorphous states. Figure 3.5(a) depicts the type II HMM regions for different ρ-values for the bilayer stacks comprising the α- (green) and c-states (red) of $GeSe_3$. We clearly see in figure 3.5(a) that ρ-values between 0.125 and 0.9

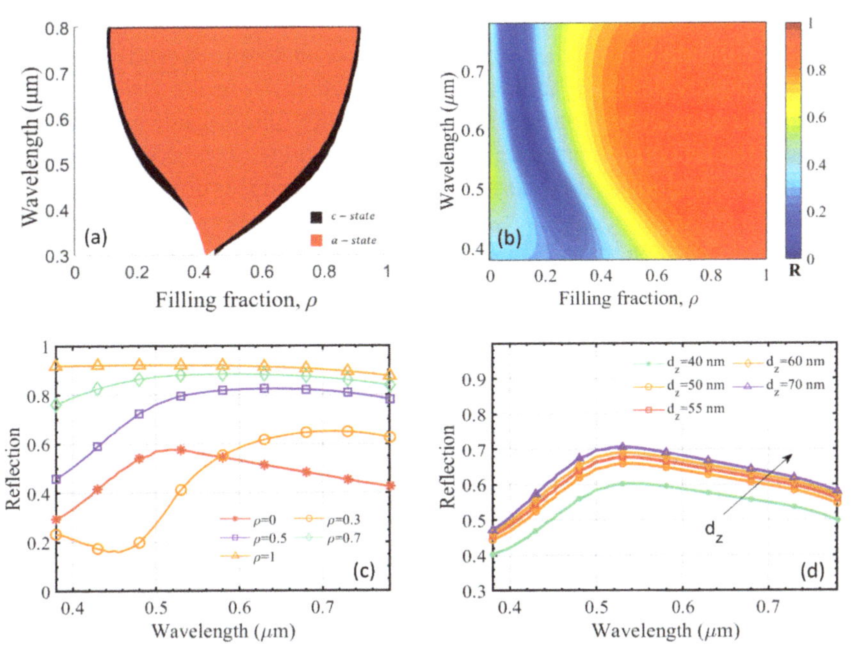

Figure 3.5. Illustrations of the effect of geometrical parameters on the spectral properties of a $GeSe_3$–Al bilayer with a thickness of $d_z = 50$ nm. (a) Map of the type II HMM region deduced for different filling fractions between 0 to 1 and to phases of $GeSe_3$ α- and c-states; (b, c) reflection spectra deduced for a range of ρ-values with $GeSe_3$ in the α-state; (d) reflection spectra obtained for different values of d_z with the α-state of $GeSe_3$ and a ρ-value of 0.4.

provide HMM regions in the visible regime, which is further expanded in the case of crystalline $GeSe_3$ medium.

The filling fraction values and $GeSe_3$ states can modify the spectral response of the structure. Figure 3.5(b) illustrates the effect of changing the ρ-values and phase of the $GeSe_3$ medium. We see that the resonance wavelength of the reflection spectrum follows the left-side border of the hyperbolic region (as depicted in figure 3.5(a)) as the wavelength increases and the area of the lowest reflectance (depicted in dark blue in figure 3.5(b)) is determined. In order to better illustrate the effect of ρ on the reflection, figure 3.5(c) depicts reflection spectra for the p-polarized incidence illumination. This figure shows the reflectance increases for the large ρ-values. But the hyperbolic behavior is evident in 0.5 μm as the reflection drops upon transition from $\rho = 0$ to $\rho = 0.3$ due to the hyperbolic behavior of the stack.

So far we have considered the thickness of the bilayer stack to be fixed at $d_z = 50$ nm. However, the bilayer thickness d_z can also alter the reflection spectra, as figure 3.5(d) depicts. For a fixed ρ-value, a larger d_z increases the reflectance peak, which is attributed to the presence of a larger portion of metal in the stack. Figure 3.5(d) illustrates the results obtained for different values of d_z for the bilayer stack with an α-state of $GeSe_3$ and a ρ-value of 0.4. In this study, the value of d_z is limited to a range of 40–70 nm to comply with the conditions of homogenization. These observations determine that each of the crystallinity states of $GeSe_3$, the filling fraction ρ, and the thickness of the bilayer d_z can independently affect the reflection due to the bilayer stack.

A possible candidate to realize an HMM-based structure would be a stack of 10 bilayers meticulously designed to have a specific reflection in the visible regime. First, we consider the geometrical parameters (namely the filling fraction and layer thickness) to study the reflection characteristics. Then we take into account the effect of the other parameters, including the operational ones. Within this context, the metal filling fraction is defined as $\rho^n = d_m^n/d_z^n$, with n being the number of stacks ($1 \leqslant n \leqslant 10$), and the thickness of each stack (i.e. $d_z^n = d_m^n + d_d^n$) is chosen to be within a range of 40 nm $\leqslant d_z^n \leqslant$ 70 nm, which is far less than the operating wavelength and higher than that required for the fabrication constraints of thin films.

We must note that the structures comprising stacks with different thicknesses and filling fractions allow the formation of Fabry–Perot cavities, thereby enhancing the plasmonic resonances in non-periodic structures, which may not be found in periodic configurations. However, the traditional techniques for achieving the optimal design of such structures are exhausting. For instance, brute force searching of 10 stacks with at least 10 steps of d_z^n would require 10^{10} simulations. To reduce the computational costs, the number of simulation runs can be reduced by accommodating optimization algorithms providing a good solution for problems with limited computational capacity.

3.4.1.3 Designing the structure for color generation

A well-known technique for optimal design of optical devices is the particle swarm optimization (PSO), which is inherently a stochastic evolutionary approach that updates a population of candidates (or particles in the swarm) by evaluating a cost function (CF) [75]. The PSO algorithm assigns an n-dimensional array of design parameters and corresponding gradient vectors to a candidate (i.e. particle). The elements of the arrays are randomly selected to cover all of the design space. At each iteration, the algorithm evaluates the cost function for each particle and then updates the particle parameters to adapt the population to find G_{best} (with the least distance from the target). It is useful to consider a target reflection spectrum with a Gaussian profile. The target is chosen as having a maximum of η at the central wavelength of λ_{center} and the standard deviation of σ, which pertains to the full-width half-maximum (FWHM) characteristic. As such, the target reflection profiles for the blue, green, and red pixels are considered to have center wavelengths of 460, 530, and 670 nm, respectively, with efficiency $\eta = 0.9$ and an FWHM value of 80 nm. The CF is defined as

$$CF = \frac{1}{N_s}\left(\sqrt{\sum_{\lambda_{\min}}^{\lambda_{\max}}\left|R(\lambda) - \eta\exp\left[-\frac{(\lambda - \lambda_{\text{center}})^2}{2\sigma^2}\right]\right|^2}\right). \tag{3.8}$$

Here, N_s is the number of sampling wavelengths in the wavelength window $\lambda_{\min} \leqslant \lambda \leqslant \lambda_{\max}$ and $R(\lambda)$ represents the reflectance spectrum for each particle as deduced by the TMM.

It is worth mentioning that optimized designs for the red and green pixels have been found for the α-state, but the c-state of $GeSe_3$ has been used for the blue pixel. Table 3.1 incorporates the geometrical parameters of the optimized structures for the three primary color pixels. We note that the $GeSe_3$ films experience thickness reduction upon transition to an amorphous state [70]. Thus, in practical investigations, the $GeSe_3$ layer thickness must be scaled accordingly to avoid possible shifts in optical response caused by geometrical alterations.

Figure 3.6(a) illustrates the reflection spectra of the optimized structures. Herein, we demonstrate the primary colors (red, green, and blue) by the optimized structures for the amorphous and crystalline states of $GeSe_3$. It is obvious that the peak reflectance keeps the values above 0.6 in all structures. Moreover, the reflectance undergoes a blueshift upon the transition of the $GeSe_3$ phase from amorphous to crystalline. To investigate the origin of this blueshift, we study the magnetic field

Table 3.1. Design parameters for the $GeSe_3$–Al-based structural color pixels.

Pixel color	Structural parameters	Optimized parameters for 10 stacks from bottom to top									
Red	ρ	1	0.4	0.1	0	0.2	0	0	0	0	0.15
	d_z**(nm)**	50	50	50	65	70	60	70	50	50	50
Green	ρ	1	0	0.4	0.6	0.7	0.45	0	0	0	0.2
	d_z**(nm)**	50	45	55	50	50	50	65	50	40	60
Blue	ρ	1	0	0	1	0.5	0	0.8	0.7	0	0.2
	d_z**(nm)**	50	55	45	50	50	50	50	60	50	55

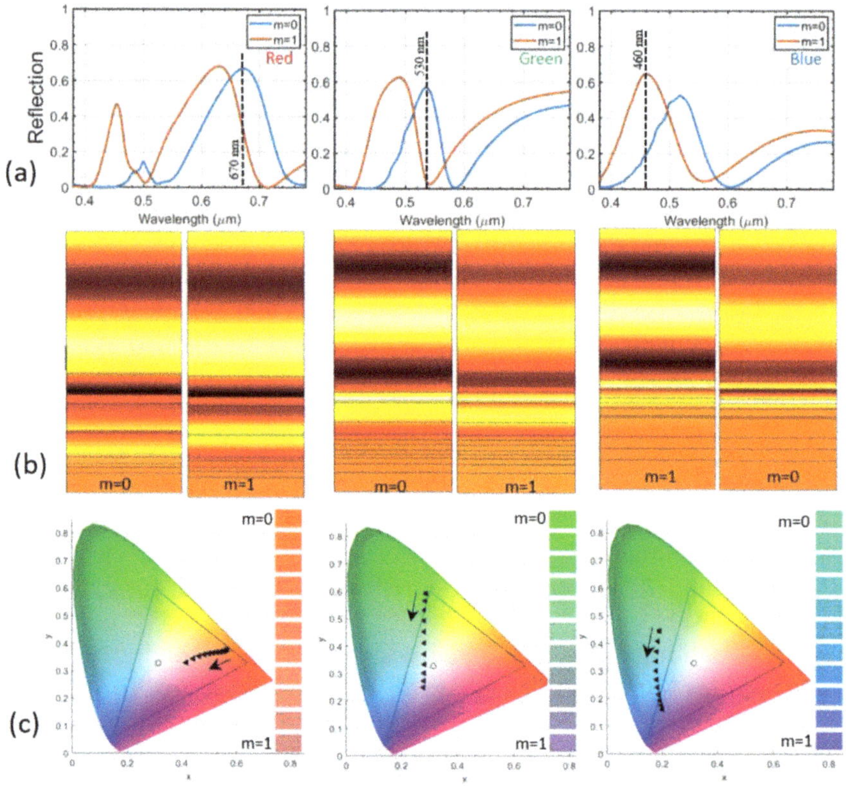

Figure 3.6. Spectral properties of the optimized structure for different m-values. (a) Reflection spectra for the primary color pixels corresponding to the α- and c-states of $GeSe_3$; dashed lines denote the target wavelengths of primary colors considered in the optimization process. (b) Distribution of magnetic field at the target wavelengths of primary colors. (c) Trajectory of reflected colors as a function of m, depicted according to the CIE1931 color space; the white color is illustrated as a circle at the center of the trajectory map and the triangle dissects the NTSC color region.

distributions by illuminating the pixels with linearly polarized plane waves with the wavelengths shown by vertically dashed lines in figure 3.6(a). The results reveal that the magnetic field distribution along the structures shifts through the layers upon phase transition of the $GeSe_3$ state from amorphous to crystalline, which is deemed to be the origin of blueshifts, as figure 3.6(b) depicts.

No less importantly, the intermediate states of $GeSe_3$ may also add more tunability to the optimized structure. Interestingly, the intermediate states can take values of the crystallinity ratio m between 0 (amorphous or α-state) to 1 (crystalline or c-state). Different mechanisms can be used to justify the level of crystallinity [61, 75, 76] by applying external stimuli to heat the PCM layer. For instance, micro-heating systems consisting of thin metal sheets are compact, low-cost mechanisms that can be easily implemented during the fabrication process. Figure 3.3(a) depicts the microheater made of graphene sheets beneath the structure, which provides adequate heat to control the optical properties of PCMs at the nanoscale size by applying external electrical pulses. A wide range of m-values (i.e. the crystallinity levels) is achievable by applying a series of electrical pulses to the graphene sheets. The duty cycles of pulses can be precisely adjusted to obtain exact sequences of the melting and cooling periods [77]. As such, the effect of annealing on the PCM layer can be modeled solely by implementing the crystallinity ratio m in the TMM algorithm; thereby, the level of m is sufficient to simulate the effect of annealing procedure on the spectral response of the structure.

Figure 3.6(c) depicts color trajectories generated for a range of crystallinity ratios by mapping each corresponding reflection spectrum to a point in the CIE1931 color space. As this figure illustrates, the generated colors gradually shift by increasing the m-values at different paces. The color trajectory of the red pixel (the left inset in figure 3.6(c)) shifts toward the central white point at a slow pace, whereas the green pixel (the central inset in figure 3.6(c)) passes a larger area of the NTSC color region. The blue pixel moves with moderate speed from the more green-like colors outside of the triangular dissect toward the dark blue region (the right inset in figure 3.6(c)). The simulation results reveal the predictable and repeatable shifts in color trajectory, which essentially demonstrate the dependence of color generation on the m-values.

Apart from the crystallinity ratio, the angle of incidence radiation also has a pivotal impact on the reflected color from the multilayer structure under investigation. To study this effect, we examine the spectral response of the primary color pixels as a function of incidence angle under the s- and p-polarized illuminations; figure 3.7 exhibits the obtained results, wherein figure 3.7(a) corresponds to the

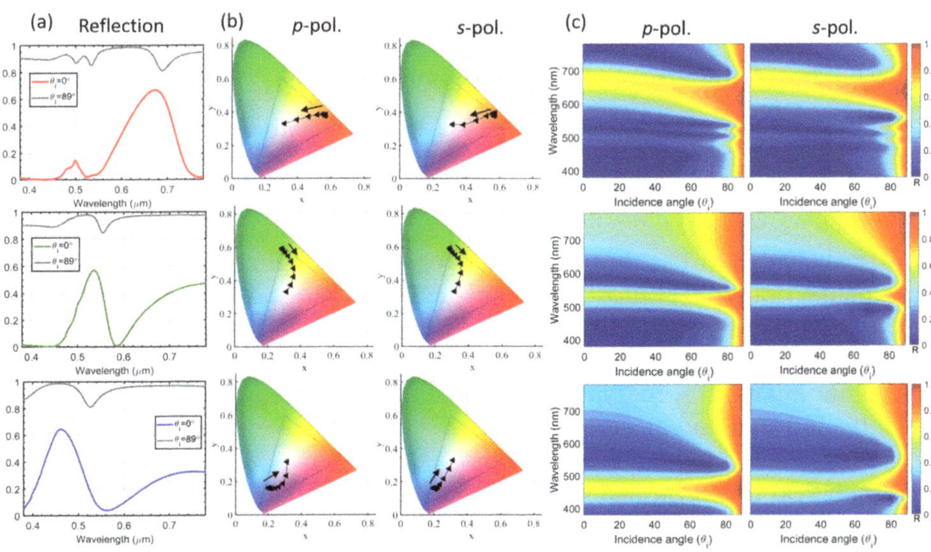

Figure 3.7. (a) Reflection spectra as a function of incidence angle (top to bottom) for red, green, and blue pixels, with $GeSe_3$ being in the α-state for the red and green pixels, but the c-state for the blue pixel. (b) Color trajectory maps of pixels under various incidence angles (shown by a black arrow) of the p-polarized (left column) and s-polarized (right column) light. (c) 2D map of reflectance as a function of incidence angle for the p-polarized (left column) and s-polarized (right column) illuminations.

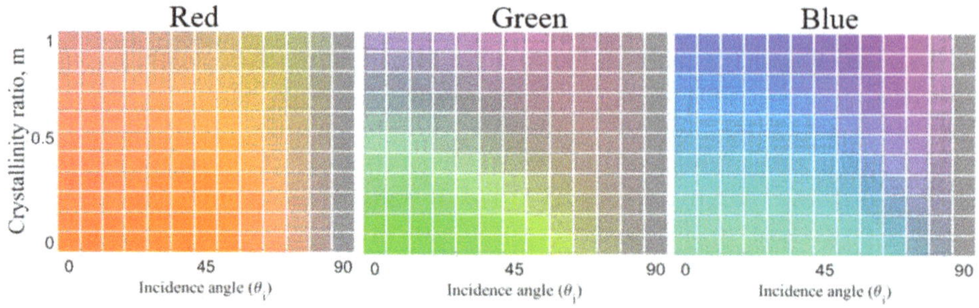

Figure 3.8. Color variations of the red, green, and blue pixels as a function of m-value and incidence angle under p-polarized illumination.

reflection spectra of the primary color pixels for the normal ($\theta_i = 0°$) and grazing ($\theta_i = 89°$) angles. Moreover, figure 3.7(b, c), respectively, depicts maps of the trajectory colors and the corresponding reflection spectra for the s- and p-polarized excitations. We use the α-state of $GeSe_3$ medium for the red and green pixels, and the c-state for the blue pixel. Figure 3.7(c) demonstrates that the s- and p-polarization transitions leave minor discrepancies upon altering the incidence angle.

Notably, the color variations in the trajectory map and the corresponding reflection peak demonstrate the angle-insensitive property of the pixels for a wide range of incidence angles (i.e. 0°–45°), as figure 3.7(b, c) depicts. Figure 3.8 confirms this characteristic, wherein the reflected color of each pixel is deduced under various m-values and incidence angles for the p-polarized excitation. The consistency of the reflected colors for angles $<45°$ is confirmed.

3.5 Dual-band nearly perfect absorber

This section discusses the investigation of a perfect tunable absorber operating in both the visible and near-infrared (NIR) regimes of the electromagnetic spectrum. The proposed structure is composed of HMM configurations of silver and Sb_2S_3 thin films, which show diverse spectral response upon altering the operational and geometrical characteristics. In the design process, we employ the stochastic PSO to determine the optimum configurations of the HMM composites, which is followed by the TMM to investigate the effects of varying the incidence angle and polarization on the spectral properties. The resultant device is capable of being used as a light programmable absorber or an optical filter, or as a major part of sensors to be used in the infrared and/or visible regimes.

3.5.1 Analytical treatment and spectral response

The high loss contributed to the constituent materials limits the application of HMM-based devices in the visible regime. Such a constraint can be resolved by employing either noble metals such as gold or silver with very low loss or by incorporating dielectrics with a high refractive index to compensate for metal-associated loss (in the visible light regime). As for the dielectric medium, stibnite (Sb_2S_3) is a light-harvesting material that is widely used in solar cells and

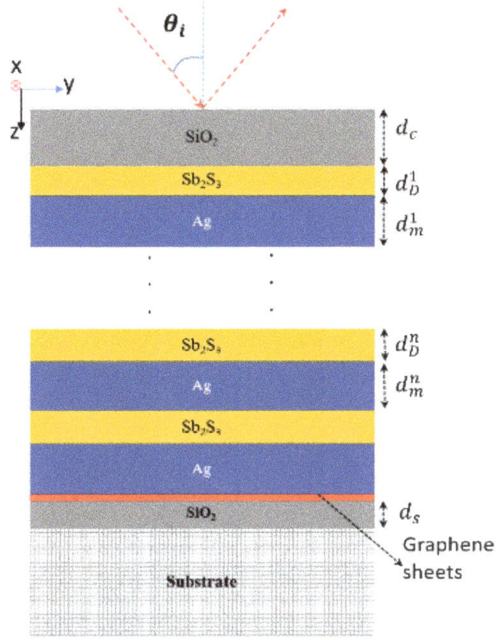

Figure 3.9. Schematic of the multilayer structure comprising the dielectric–metal Sb_2S_3–Ag bilayers and graphene microheater.

photodetectors [56, 60], and is a potential candidate for use in visible spectrum devices. It not only demonstrates a low optical loss and high refractive index in the visible regime, but also is a phase-changing dielectric with a tunable absorption edge between 1.7 and 2.0 eV [56]. Notably, silver and Sb_2S_3 are widely used in the solar energy sector, employing thin film deposition techniques properly established by conventional deposition methods. As such, we chose silver and Sb_2S_3 as the constituent materials to design HMM-based absorbers in the visible regime.

Figure 3.9 represents the proposed structure consisting of Sb_2S_3–Ag bilayers. The capping layer of SiO_2 (of thickness $d_c = 100$ nm) is located on the top of the structure to protect the Sb_2S_3 medium from evaporation. The structure is supposed to comprise n stacks of Sb_2S_3 (of thickness d_D^n) and Ag (of thickness as d_m^n) bilayers. As mentioned in section 3.4, we employ a microheater system embedded beneath the stack of bilayers to generate sufficient heat for phase transition of Sb_2S_3 media. In the configuration we use graphene sheets (with a thickness of 1 nm) and two Ti/Au metal pads with thickness of 100 nm [76].

We evaluate the optical response of the structure (in figure 3.9) by exploiting the TMM in the 0.2–1.6 μm wavelength range. To this end, we first deduce the effective permittivity components of the homogenized dielectric–metal Sb_2S_3–Ag bilayer with a filling fraction of 0.5 by applying equation (3.9a). Figure 3.9 illustrates the results of this study—figure 3.10(a) shows the dispersion plots, including the real ($\mathfrak{R}$) and

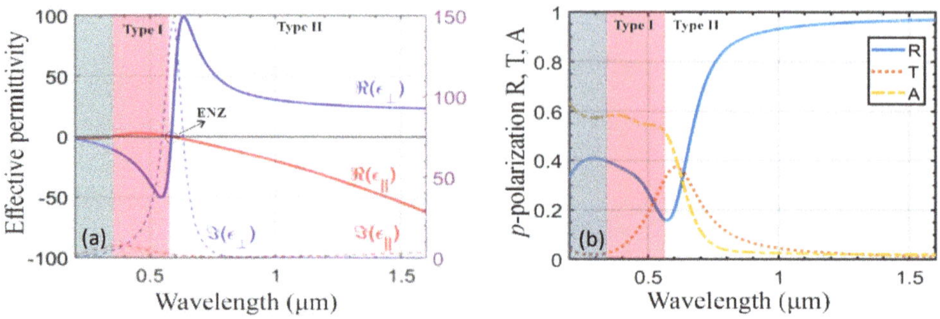

Figure 3.10. Optical characteristics of the dielectric–metal Sb_2S_3–Ag bilayer; plots of the (a) real and imaginary parts of $\epsilon_{\|}$- and $\epsilon_{\perp}$-components, and (b) reflectance R, transmittance T, and absorptance A against wavelength.

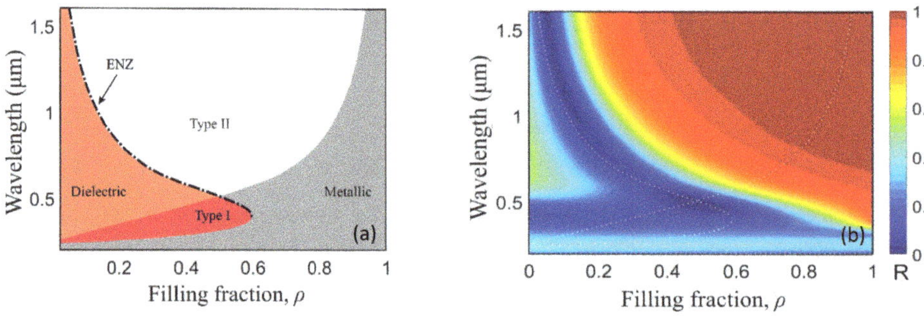

Figure 3.11. Wavelength-dependence of (a) $\Re(\epsilon_{\perp}) \times \Re(\epsilon_{\|})$, and (b) reflectance of the Sb_2S_3–Ag bilayer with a thickness of 50 nm corresponding to different ρ-values.

imaginary ($\Im$) parts of the effective permittivity components, and figure 3.10(b) depicts the spectral response of the Sb_2S_3–Ag bilayer deduced by using the TMM.

As is evident in figure 3.10(a), the bilayer demonstrates type I HMM characteristics in the wavelength span of 0.34–0.58 μm, wherein the effective components are such that $\Re(\epsilon_{\perp}) < 0$, $\Re(\epsilon_{\|}) > 0$, thereby promising more transmission in this region; this is confirmed in figure 3.10(b). The type II HMM region begins at the wavelength wherein the component $\Re(\epsilon_{\|})$ undergoes ENZ, which is marked with a circle in figure 3.10(a) and a vertical dashed line in figure 3.10(b). The type II HMM is a highly reflective medium, as confirmed in figure 3.10(b), and also in figure 3.10(a), where the effective permittivity components exhibit features such that $\Re(\epsilon_{\perp}) > 0$, $\Re(\epsilon_{\|}) < 0$ in the 0.58–1.6 μm wavelength span.

Notably, the ENZ wavelength attributed to the hyperbolic dispersion characteristic is a function of effective permittivity components, and therefore, it can be adjusted by the metal filling fraction ρ of the bilayer. Figure 3.11 shows a 2D map projection of $\Re(\epsilon_{\perp}) \times \Re(\epsilon_{\|})$ and the reflectance as a function of wavelength and ρ. It is evident from figure 3.11(a) that the ENZ wavelength gradually shifts from the NIR to the visible regime as ρ increases from 0 to 0.6—an effect that is also verified by the results in figure 3.11(b). The projected reflectance of the Sb_2S_3–Ag bilayer with a thickness of 50 nm clearly shows the borderline (depicted as solid white color), thereby demonstrating the transition from low to high reflectance. This exactly follows the ENZ variation for various ρ-values—the property used to design the bilayer stack to act as a metal, dielectric or HMM.

The aim of this study is to design a perfect absorber in the 1.3–1.55 μm window of the NIR regime. In general, multilayer absorbers have been made of stacks comprising material and/or dielectric layers with different thicknesses to manipulate the optical path of the reflected waves and excite resonant modes along the structure [78–80]. The advantage of using the Sb_2S_3–Ag bilayer is that the behavior of bilayers can be adjusted by setting precise ρ-values. Moreover, the effective permittivity of the stack can also be actively tuned thermally—the property that can be used for fine-tuning the absorption spectra at the post-processing step.

We consider the device comprising nine bilayers of Sb_2S_3–Ag located on the SiO_2 substrate, as figure 3.9 depicts. The challenge is to find the order of stacks with different structural parameters (i.e. using different values of ρ, d_z) to achieve maximum absorption in the 0.2–1.6 μm range. The respective metal filling fraction and the thickness of each Sb_2S_3–Ag bilayer are taken as ρ^n and d_z^n, with n being the number of bilayers ($1 \leqslant n \leqslant 9$) where 35 nm $\leqslant d_z^n \leqslant$ 65 nm, thereby being much lower than the operating wavelength of 1.31 μm, and also higher than that required due to fabrication constraints. As discussed in the previous section, the optimization process demands the examination of a large number of design parameters. As such, we again employ the PSO to decrease the computational costs associated with the number of simulations.

Let us consider the target absorption spectrum at the center wavelength of 1.31 μm and the FWHM value of 0.08 μm. In this study, we consider the efficiency parameter with a value of $\alpha = 0.9$. As such, the cost function (CF) can be formulated as follows:

$$CF = \frac{1}{N_s}\left(\sqrt{\sum_{\lambda_{\min}}^{\lambda_{\max}}\left|A(\lambda) - \alpha \exp\left[-\frac{(\lambda - \lambda_{\text{center}})^2}{2\sigma^2}\right]\right|^2} - \sqrt{\sum_{\lambda_{\min}}^{\lambda_{\max}}|T(\lambda)|^2}\right), \tag{3.9a}$$

with

$$\sigma = \frac{FWHM}{2\sqrt{2\ln 2}}. \tag{3.9b}$$

Herein, N_s denotes the number of sampling wavelengths bounded between $\lambda_{\min}$ and $\lambda_{\max}$, and $A(\lambda)$ and $T(\lambda)$ represent the absorptance and transmittance spectra of each particle, respectively.

The optimization algorithm looks for the optimum combination of the thicknesses and filling fractions in the bilayer structures, thereby minimizing the CF, as equation (3.9a) states. Table 3.2 summarizes the details of the optimized design in the present work wherein the structure comprises nine Sb_2S_3–Ag bilayers after 100 iterations. In this table, ρ-values of 0 and 1 represent the pure dielectric (i.e. Sb_2S_3) and pure metallic (i.e. Ag) layers, respectively. Figure 3.12 illustrates the

Table 3.2. Design parameters for the multilayer structure.

Structural parameters	Optimized parameters for each bilayer from top to bottom order
Metal filling fraction (ρ)	0.12, 0.13, 0.13, 0, 0, 0, 1, 0.9, 0.8
Layer thickness (nm)	50, 60, 60, 55, 60, 36, 55, 50, 50

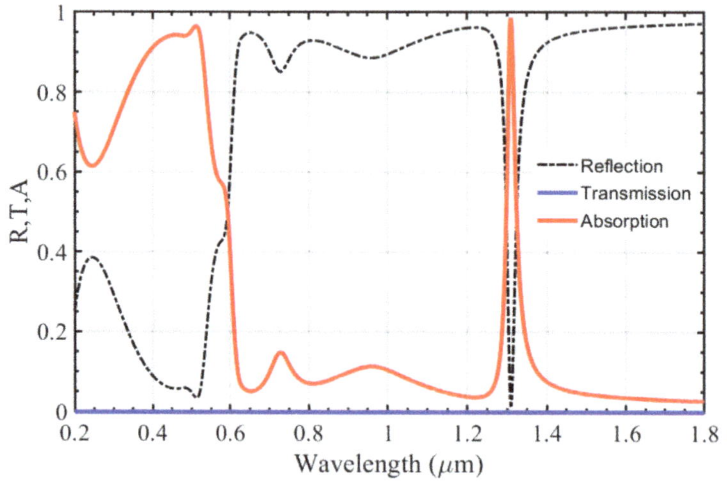

Figure 3.12. Spectral properties of the optimized structure.

spectral response of the resultant structure. We see in this figure that the optimized structure demonstrates perfect absorption at the targeted wavelength of 1.31 μm and, more interestingly, in the visible regime, with a completely suppressed transmission.

It may be of interest to note that the resultant structure is a kind of meta-cavity, which is comprised of three parts—three Sb_2S_3–Ag stacks at the top (which are the HMMs with different filling fractions and thicknesses), three dielectric layers in the middle as spacer, and three metallic layers as the backplane mirror. It is known that HMMs support localized surface plasmon polaritons (LSPPs)—enabling confinement of the electric field in deep nanoscale sizes while allowing the small portion of incident light to penetrate into the spacer [81].

One feature of the above configurations merits some further discussion. The state of the Sb_2S_3 medium can take a wide range of crystallinity ratios; akin to other PCMs. This feature raises the tunability of the whole structure upon altering the ambient temperature. To that end, applying electrical pulses to the embedded graphene sheets beneath the proposed structure will generate sufficient heat to cause changes in the state of Sb_2S_3 in the multilayer configuration (figure 3.9) [61]. We use this feature to numerically simulate the heating effect associated with different crystallinity values (of α- to c-mass) on the absorption characteristic. The

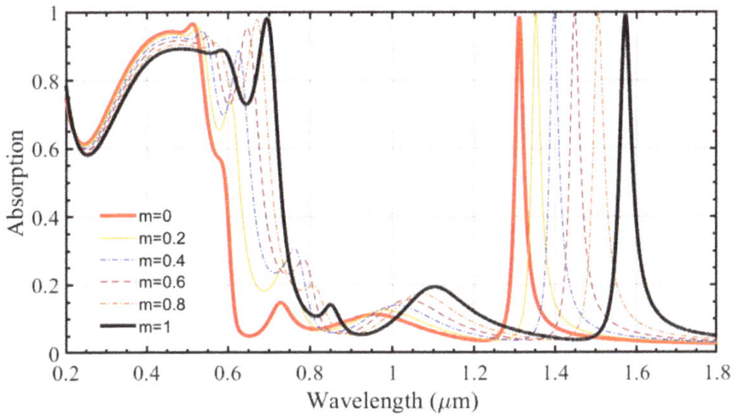

Figure 3.13. Absorption spectra obtained for different crystallinities; the α-state, ($m = 0$), c-state ($m = 1$), and intermediate states ($0 < m < 1$).

absorbance of a structure having the crystallinity ratio of m is explained by the Lorentz–Lorenz formalism [77]

$$\frac{\epsilon_{\text{eff}} - 1}{\epsilon_{\text{eff}} + 2} = m\frac{\epsilon_c - 1}{\epsilon_c + 2} + (1 - m)\frac{\epsilon_a - 1}{\epsilon_a + 2}, \tag{3.10}$$

with m having its values from 0 to 1, corresponding to the α- (i.e. $m = 0$) and c- (i.e. $m = 1$) states, respectively. Moreover, ϵ_{eff}, ϵ_c, and ϵ_a are the respective permittivity values of the Sb_2S_3 medium in the different states due to changing m, completely crystalline c and purely amorphous α.

Figure 3.13 illustrates the absorption spectra obtained for the α-state, ($m = 0$), c-state ($m = 1$), and intermediate states ($0 < m < 1$); the lines are identified in the inset of the figure. It is obvious from this figure that the absorption spectra undergo a band shift in the visible regime from 500 to 770 nm and the absorption peak is slightly suppressed by gradually increasing the crystallinity ratio from $m = 0$ to $m = 1$. The absorption spectrum in the NIR regime also shifts from 1310 to 1570 nm, while the maximum absorption is sustained over 0.99 and the FWHM is slightly expanded from 25 to 37.5 nm. This treatment reveals that adjusting the crystallinity ratio m can precisely tune the absorption response in either the visible or the NIR regime—a feature that is very desirable for sensing and/or spectrometry applications.

In all the situations discussed so far, we have considered that the light beam is impinging onto the top surface of the structure at normal incidence angle. We now see the effect of altering the incidence angle on the absorption characteristics under s- and p-polarized illuminations, and also the α- and c-states of Sb_2S_3. Figure 3.14(a, b) shows the results for the analysis of the absorption response of the structure in the α-state, whereas figure 3.14(c, d) represents the resultant spectra corresponding to the case of the c-state. The analysis shows that the absorption is robust for a wide range of incidence angles; however, the p-polarized waves are less angle-sensitive than the s-polarized waves. A good explanation for this discrepancy can be made by

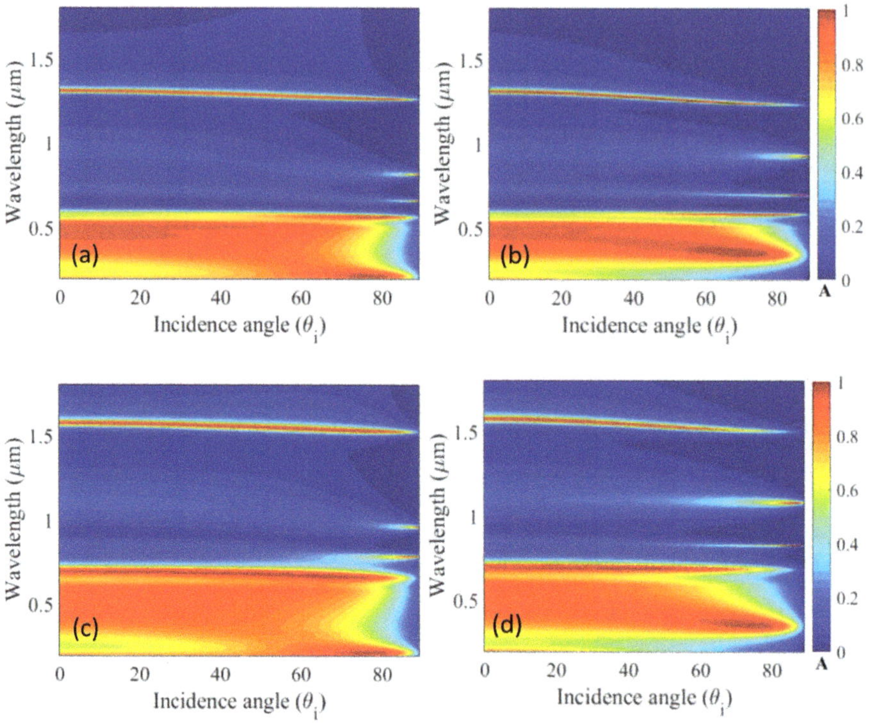

Figure 3.14. Absorption spectra as a function of the incidence angle considering the α- (a, b) and c- (c, d) states of Sb_2S_3: (a, c) the p- and (b, d) the s-polarized incidence radiation.

considering the dependence of the wavenumber of the perpendicular propagation on the incidence angle of the p-polarized beam in the type II HMM region. The out-of-plane component of the wavevector of the p-polarized beam is defined as [61]

$$k_z(\lambda, \theta) = \sqrt{\epsilon_\perp(\lambda)\left(1 + \frac{\sin^2\theta}{|\epsilon_{||}(\lambda)|}\right)}. \tag{3.11}$$

In the type II HMM region, as the parallel component of effective permittivity $|\epsilon_{||}(\lambda)|$ increases, the k_z component gradually becomes independent of the angle of incidence, i.e. $[1 + \frac{\sin^2\theta}{|\epsilon_{||}(\lambda)|}] \approx 1$. A complete analysis of the absorption spectra for both the states of Sb_2S_3 and p- and s-polarized illumination would show the incidence angle dependence of the proposed structure. Figure 3.14(a, c) depicts that the p-polarization state exhibits the maximum absorption in the NIR regime for a wide range of 20°–80° incidence angle, whereupon it undergoes a ~30 nm shift. For the same range of incidence angle (i.e. 20°–80°), the s-polarization exhibits similar behavior in the NIR regime, with a little more shift (of ~35 nm) in the absorption peak, as figure 3.14(b, d) shows.

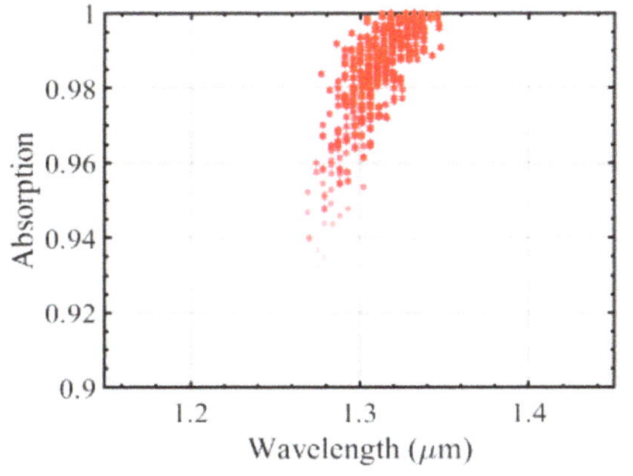

Figure 3.15. Trajectory of the absorption peaks as a function of fabrication errors. The individual layer thicknesses are randomly distributed from –3 to + 3 nm, while the optimized structure is in the α-state of Sb_2S_3 under the illumination of a p-polarized wave at normal incidence.

The fabrication process for thin film multilayer structures is always accompanied by tolerances in manufacturing methods. Within this context, metallic and Sb_2S_3 thin films can be fabricated by PVD techniques. Dong *et al* [82] reported that DC and RF sputtering can be used to successfully realize multilayer structures consisting of Sb_2S_3 films on 3 nm thick metallic layers. Moreover, graphene sheets can be synthesized on ceramic substrate whereupon multilayer films can be transferred [73]. To sustain an appropriate deposition of thin films, the pressure of the deposition chamber and other ambient parameters should be precisely controlled.

However, herein, we investigate the influence of fabrication tolerance on the absorptance of the optimized structure, where the thickness of each bilayer is modified as $\widetilde{d}_z^{\,n} = d_z^{\,n} + \delta_n$, with δ_n being the thickness deviation of the nth stack $(1 \leqslant n \leqslant 9)$. Figure 3.15 illustrates the distribution of the peaks of the resultant absorptance deduced by 200 simulations of the TMM algorithm, while the thickness deviation δ_n has an even distribution in a range of –3 to + 3 nm. These results are obtained for the optimized structure in the α-state and under the normal illumination of the p-polarized light. This study reveals that the structure sustains near-perfect absorption around 1.31 μm with a standard deviation of 15.59 nm.

3.6 Conclusion

This chapter discusses the study of the optical behavior of PCM-embedded HMMs, touching upon the analytical treatment for the optimal design of programmable devices exploiting HMMs' exotic electromagnetic properties. In particular, the optical properties of $GeSe_3$–Al bilayers have been examined for application as the primary color pixels. Stacks of bilayers with equal heights but different structural parameters, namely the thickness and the metallic mass ratios, for the constituent layers have been analysed using the PSO algorithm to optimize the target color reflection. The microheater system has been used to alter the ambient temperature and dynamically modulate the reflected color of pixels which are maintained in a large incidence angle under the *s*- and *p*-polarized excitations. Furthermore, the possibility of realizing tunable absorption in the visible and NIR regimes has been shown, exploiting the HMM behavior of the Sb_2S_3–Ag bilayer stack. Again, the PSO algorithm decides the optimal geometrical parameters that provide a perfect absorption band at 1.31 μm. The same microheater mechanism has been utilized to adjust the absorption band from a 1.310 to a 1.570 μm wavelength. Interestingly, the second absorption band has been found to be in the visible regime, which can be adjusted from 0.5 to 0.77 μm. Both designs offer angle-invariant and polarization-insensitive characteristics.

References

[1] Maier S A 2007 *Palsmonics: Fundamentals and Applications* (New York: Springer)

[2] Lienau C, Noginov M A and Lončar M 2014 ight–matter interactions at the nanoscale *J. Opt.* **16** 110201

[3] Pustovit V N, Urbas A M and Shahbazyan T V 2014 Energy transfer in plasmonic systems *J. Opt.* **16** 114015

[4] Tittl A *et al* 2015 A switchable mid-infrared plasmonic perfect absorber with multispectral thermal imaging capability *Adv. Mater.* **27** 4597–5603

[5] Zhang R, Yu H, Zhang H, Liu X, Lu Q and Wang J 2015 Electronic band-gap modified passive silicon optical modulator at telecommunications wavelengths *Sci. Rep.* **5** 16588

[6] Ghasemi M, Baqir M A and Choudhury P K 2016 On the metasurface based comb filters *IEEE Photon. Technol. Lett.* **28** 1100–3

[7] Krasnok A and Alù A 2020 Active nanophotonics *Proc. IEEE* **108** 628–54

[8] Pourmand M, Choudhury M and P K 2020 Wideband THz filtering by graphene-over-dielectric periodic structures with and without MgF_2 defect layer *IEEE Access* **8** 137385–94

[9] Sheta E M and Choudhury P K 2021 Vanadium nitride-based ultra-wideband nearly perfect metamaterial absorber *J. Nanophoton.* **15** 036007

[10] Sheta E M and Choudhury P K 2021 Nanoengineered hafnium nitride hyperbolic metasurface supported polarization insensitive absorber *IEEE Photon. Technol. Lett.* **33** 1351–4

[11] Choudhury P K 2021 *Metamaterials: Technology and Applications* (Boca Raton, FL: CRC Press)

[12] Pendry J B 2000 Negative refraction makes a perfect lens *Phys. Rev. Lett.* **85** 3966–9

[13] Jacob Z, Alekseyev L V and Narimanov E 2006 Optical hyperlens: far-field imaging beyond the diffraction limit *Opt. Express* **14** 8247–56

[14] Milton G W and Nicorovici N A P 2006 On the cloaking effects associated with anomalous localized resonance *Proc. Royal Soc.* A **462** 3027–59

[15] Schurig D *et al* 2006 Metamaterial electromagnetic cloak at microwave frequencies *Science* **314** 977–80

[16] Kabashin A V *et al* 2009 Plasmonic nanorod metamaterials for biosensing *Nat. Mater.* **8** 867–71

[17] Ghasemi M, Choudhury P K and Dehzangi A 2015 Nanoengineered thin films of copper for the optical monitoring of urine—a comparative study of the helical and columnar nanostructures *J. Electromagn. Waves Appl.* **29** 2321–9

[18] Ghasemi M and Choudhury P K 2018 Complex copper nanostructures for fluid sensing—a comparative study of the performance of helical and columnar thin films *Plasmonics* **13** 131–9

[19] Baqir M A, Choudhury P K and Ibrahim A-B M 2019 A 2019 spectral features of vanadium dioxide-based metasurface for sensing *Proc JSAP-OSA Joint Symp.* 18a-E208-3
[20] Ghasemi M, Roostaei M N, Sohrabi F, Hamidi S M and Choudhury P K 2020 Biosensing application of all-dielectric SiO_2-PDMS meta-stadium grating nanocombs *Opt. Mater. Express* **10** 1018–33
[21] Sheta E M, Choudhury P K and Ibrahim A -B M A 2021 Impact of metasurface deformation on the graphene–$SrTiO_3$ pixelated metamaterial-based sensor *Optik* **242** 167174
[22] Baqir M A and Choudhury P K 2023 Hyperbolic metamaterial-based optical biosensor for detecting cancer cells *IEEE Photon Technol Lett.* **35** 183–6
[23] Sheta E M, Ibrahim A -B M A and Choudhury P K 2023 Metamaterial-based THz polarization-insensitive multi-controllable sensor comprising MgF_2–graphene periodic nano-pillers over InSb thin film *IEEE Trans. Nanotechnol.* **22** 336–41
[24] Govyadinov A A and Podolskiy V A 2006 Metamaterial photonic funnels for subdiffraction light compression and propagation *Phys. Rev.* B **73** 155108
[25] Salgueiro J R and Kivshar Y S 2014 Complex modes in plasmonic nonlinear slot waveguides *J. Opt.* **16** 114007
[26] Poddubny A, Iorsh I, Belov P and Kivshar Y 2013 Hyperbolic metamaterials *Nat. Photon* **7** 948–57
[27] Shekhar P, Atkinson J and Jacob Z 2014 Hyperbolic metamaterials: fundamentals and applications *Nano Conv.* **1** 14
[28] Ferrari L, Wu C, Lepage D, Zhang X and Liu Z 2015 Hyperbolic metamaterials and their applications *Prog. Quantum Electron.* **40** 2015 1–40
[29] Cortes C L, Newman W, Molesky S and Jacob Z 2012 Quantum nanophotonics using hyperbolic metamaterials *J. Opt.* **14** 063001
[30] Guo Y *et al* 2012 Applications of hyperbolic metamaterial substrates *Adv. Optoelectron.* **2012** 1–9
[31] Belov P A, Simovski C R and Ikonen P 2005 Canalization of subwavelength images by electromagnetic crystals *Phys. Rev.* B **71** 193105
[32] Balmain K G, Luttgen A and Kremer P C 2002 Resonance cone formation, reflection, refraction, and focusing in a planar anisotropic metamaterial *IEEE Trans. Antennas Wireless Propagat. Lett.* **1** 146–9
[33] Noginov M, Lapine M, Podolskiy V and Kivshar Y 2013 Focus issue: hyperbolic metamaterials *Opt. Express* **21** 14895–7
[34] Smith D R, Pendry J B and Wiltshire M C K 2004 Metamaterials and negative refractive index *Science* **305** 788–92
[35] Lakhtakia A and Mackay T G 2006 Meet the metamaterials *OSA Opt Photon News* **18** 34–9
[36] Valentine J *et al* 2008 Three-dimensional optical metamaterial with a negative refractive index *Nature* **455** 376–9
[37] Ishii S *et al* 2013 Sub-wavelength interference pattern from volume plasmon polaritons in a hyperbolic medium: sub-wavelength interference in a hyperbolic medium *Laser Photon Rev.* **7** 265–71
[38] Liu Z, Lee H, Xiong Y, Sun C and Zhang X 2007 Far-field optical hyperlens magnifying sub-diffraction-limited objects *Science* **315** 1686
[39] Iorsh I, Poddubny A, Orlov A, Belov P and Kivshar Y S 2012 Spontaneous emission enhancement in metal–dielectric metamaterials *Phys. Lett.* A **376** 185–7
[40] Iqbal N, Zhang S, Choudhury P K, Jin Y and Ma Y 2022 Super-Planckian thermal radiation between 2D hBN monolayers *Int. J. Thermal Sci.* **172** 107315

[41] Li , Dang Y *et al* 2024 Observation of heat pumping effect by radiative shuttling *Nat. Commun.* **15** 5465
[42] Ferrari L 2017 Hyperbolic metamaterials for high-speed optical communications *Doctoral Thesis, Univ. of California*
[43] Kidwai O, Zhukovsky S V and Sipe J E 2012 Effective-medium approach to planar multilayer hyperbolic metamaterials: strengths and limitations *Phys. Rev.* A **85** 053842
[44] Alù A, Salandrino A and Engheta N 2007 Epsilon-near-zero metamaterials and electromagnetic sources: tailoring the radiation phase pattern *Phys. Rev.* B **7** 155410
[45] Dyachenko P N *et al* 2016 Controlling thermal emission with refractory epsilon-near-zero metamaterials via topological transitions *Nat. Commun.* **7** 11809
[46] Raoux S and Wuttig M 2009 *Phase Change Materials: Science and Applications* (Berlin: Springer)
[47] Ovshinsky S R 1991 Reversible electrical switching phenomena in disordered structures *In:* D Adler, B B Schwartz and M Silver *Disordered Materials* ((Boston, MA: Springer))
[48] Ríos C *et al* 2015 Integrated all-photonic non-volatile multi-level memory *Nat. Photon* **9** 725–32
[49] Wuttig M, Bhaskaran H and Taubner T 2017 Phase-change materials for non-volatile photonic applications *Nat. Photon* **11** 465–76
[50] Cheng Z *et al* 2017 On-chip photonic synapse *Sci. Adv.* **3** e1700160
[51] Hosseini P, Wright C D and Bhaskaran H 2014 An optoelectronic framework enabled by low-dimensional phase-change films *Nature* **511** 206–11
[52] Ríos C, Hosseini P, Taylor R A and Bhaskaran H 2016 Color depth modulation and resolution in phase-change material nanodisplays *Adv. Mater.* **28** 4720–6
[53] Baqir M A and Choudhury P K 2019 On the VO_2 metasurface-based temperature sensor *J. Opt. Soc. Am.* B **36** F123–30
[54] Sheta E M, Choudhury P K and Ibrahim A -B M A 2021 Pixelated graphene–strontium titanate metamaterial supported tunable dual-band temperature sensor *Opt. Mater.* **117** 111197 Opt Mater 118, 111225
[55] Pourmand M and Choudhury P K 2022 Nanostructured strontium titanate perovskite hyperbolic metamaterial supported tunable broadband THz Brewster modulator *IEEE Trans. Nanotechnol.* **21** 586–91
[56] Delaney M *et al* 2020 A new family of ultralow loss reversible phase-change materials for photonic integrated circuits: Sb_2S_3 and Sb_2Se_3 *Adv. Func. Mater.* **30** 2002447
[57] Morin F J 1959 Oxides which show a metal-to-insulator transition at the Neel temperature *Phys. Rev. Lett.* **3** 34–6
[58] Kim H T *et al* 2010 Electrical oscillations induced by the metal–insulator transition in VO_2 *J. Appl. Phys.* **107** 023702
[59] Wall S *et al* 2012 Ultrafast changes in lattice symmetry probed by coherent phonons *Nat. Commun.* **3** 721
[60] West P R, Ishii S, Naik G V, Emani N K, Shalaev V M and Boltasseva A 2010 Searching for better plasmonic materials *Laser Photon. Rev.* **4** 795–808
[61] Pourmand M and Choudhury P K 2022 Programmable phase-change medium-assisted hyperbolic metamaterial as a dual-band nearly perfect absorber *J. Opt. Soc. Am.* B **39** 1222–8
[62] Taghinejad H *et al* 2021 ITO-based microheaters for reversible multi-stage switching of phase-change materials: towards miniaturized beyond-binary reconfigurable integrated photonics *Opt. Express* **29** 20449–62

[63] Cocker T L *et al* 2012 Phase diagram of the ultrafast photoinduced insulator–metal transition in vanadium dioxide *Phys. Rev.* B **85** 155120

[64] Miller K J, Haglund R F and Weiss S M 2018 Optical phase change materials in integrated silicon photonic devices: review *Opt. Mater. Express* **8** 2415–29

[65] Shportko K, Kremers S, Woda M, Lencer D, Robertson J and Wuttig M 2008 Resonant bonding in crystalline phase-change materials *Nat. Mater.* **7** 653–8

[66] Kana Kana J B, Ndjaka J M, Vignaud G, Gibaud A and Maaza M 2011 Thermally tunable optical constants of vanadium dioxide thin films measured by spectroscopic ellipsometry *Opt. Commun.* **284** 807–12

[67] Abdollahramezani S *et al* 2020 Tunable nanophotonics enabled by chalcogenide phase-change materials *Nanophotonics* **9** 1189–241

[68] Chamoli S K, Verma G, Singh S C and Guo C 2020 Phase change material-based nano-cavity as an efficient optical modulator *Nanotechnology* **32** 095207

[69] Ghazi Sarwat S *et al* 2019 Strong opto-structural coupling in low dimensional $GeSe_3$ films *Nano Lett.* **19** 7377–84

[70] Soref R A and Bennett B R 1992 Electro-optic Fabry–Perot pixels for phase-dominant spatial light modulators *Appl. Opt.* **31** 675–80

[71] Kristensen A *et al* 2016 Plasmonic colour generation *Nat. Rev. Mater.* **2** 16088

[72] Lee T, Jang J, Jeong H and Rho J 2018 Plasmonic- and dielectric-based structural coloring: from fundamentals to practical applications *Nano Conv.* **5**

[73] Franklin D *et al* 2015 Polarization-independent actively tunable colour generation on imprinted plasmonic surfaces *Nat. Commun.* **6** 7337

[74] Miller B H, Liu H and Kolle M 2022 Scalable optical manufacture of dynamic structural colour in stretchable materials *Nat. Mater.* **21** 1014–8

[75] Robinson J and Rahmat-Samii Y 2004 Particle swarm optimization in electromagnetics *IEEE Trans. Antennas Propagat.* **52** 397–407

[76] Ríos C *et al* 2019 Reversible switching of optical phase change materials using graphene microheaters *Proc CLEO* **2019** SF2H.4

[77] Pourmand M, Choudhury P K and Mohamed M A 2021 Tunable absorber embedded with GST media and trilayer graphene strip microheaters *Sci. Rep.* **11** 3603

[78] Takayama O and Lavrinenko A V 2019 Optics with hyperbolic materials *J. Opt. Soc. Am.* B **36** F38–48

[79] Baqir M A and Choudhury P K 2017 Hyperbolic metamaterial-based UV absorber *IEEE Photon. Technol. Lett.* **29** 1548–51

[80] Pourmand M, Choudhury P K and Mohammed M A Porous gold nanolayer coated halide metal perovskite-based broadband metamaterial absorber in the visible and near-IR regime *IEEE Access* **9** 8912–9

[81] Raether H 1988 *Surface Plasmons on Smooth and Rough Surfaces and on Gratings* (Berlin: Springer)

[82] Dong W *et al* 2019 Wide bandgap phase change material tuned visible photonics *Adv. Funct. Mater.* **29** 1806181

IOP Publishing

Ordered and Disordered Metamaterials
Design and applications
Pankaj K Choudhury and Tatjana Gric

Chapter 4

Full-wave subwavelength characterization of single- and two-layer wire metamaterials at microwave frequencies

Oleg Rybin

In this study, the Clausius–Mossotti equation for the effective electromagnetic response of a metal circular wire metamaterial is obtained at microwave frequencies. The metamaterial is a two-dimensional square array of thin infinite circular cylinders located in free space. The equation is based on obtaining the dipole–dipole approximations of the electric and magnetic polarizabilities of individual cylinder. In order to obtain the approximations for polarizabilities, the solution for the problem of scattering of a p-polarized plane wave by a single-wire grating is implemented using a lattice sum theory. The grating consists of a layer of parallel infinite circular metal wires located in free space. The investigations of the scattering properties and the effective electromagnetic response at GHz frequencies are also carried out for the double wire grating consisting of two similar single-layer gratings symmetrically situated from each other by some distance as well as the same two-layer wire backed by a plane conductive surface. The spectra of the effective constitute properties and the total reflective coefficient are compared with the results of numerical simulations in the subwavelength part of microwave frequencies. The application of the considered two-layer wire metamaterial structures for creating high-impedance surfaces and generating microwave surface plasmon-polaritons are discussed in some detail.

4.1 Introduction

During the last couple of decades, many researchers have been focusing on solving the problem of electromagnetic wave scattering by an ensemble of small scatterers with various spatial periodic arrangements [1–5]. This interest is primarily associated with the existence of resonant phenomena in such periodic structures. The resonance phenomena in ensembles of long metallic cylindrical scatterers are of interest to engineers due to their potential applications in light wave technology [6], antenna

doi:10.1088/978-0-7503-5462-2ch4

engineering [7], plasmonics [8] and many other branches of contemporary applied electromagnetics [9]. Particular interest in such problems is caused by the possibility of using the effective medium theory (EMT) [10, 11] in the theoretical characterization of nanostructures.

As stated, many of the aforementioned studies are associated with the consideration of small particles or scatterers in the form of long thin cylinders. Moreover, interest in studying the optical properties of arrays of circular metal cylinders has increased considerably in the last decade due to the development of numerous theoretical models [9, 12–15]. This is because circular wires are common components of many microwave radio-electronics systems [8, 9, 12]. However, these works devoted to the simultaneous consideration of both the effective electric and effective magnetic responses are still rare [11], despite the fact that the phenomenon of artificial magnetism began to be studied quite a long time ago [16, 17]. This is mostly due to the fact that the latter phenomenon in non-magnetic metal–dielectric composites/metamaterials associated with their effective magnetic response is considerably weaker in subwavelengths than any of the other exotic properties associated with the effective electrical response of these artificial materials [18, 19].

In the first part of this study, the EMT equations are obtained for a metamaterial as an infinitely extended three-dimensional (3D) network of thin metal circular wires with a unit cell with a square cross-section. The formulas are the extension of dipole–dipole approximations based on the Clausius–Mossotti equations for the effective electromagnetic response of the metamaterial under consideration. The derivation of dispersion expressions for the electric and magnetic polarizabilities of the metamaterial precedes obtaining the above formulas in the dipole–dipole approximation. In turn, the dispersion expressions are derived in the subwavelength part of the microwave frequency range based on some ideas and results in [20], where the expressions of electromagnetic field scattered by a single-layer wire grating have been obtained in the dipole approximation.

In the second part of the chapter, the dipole approximations for the total reflection coefficient and the complex relative effective permittivity and permeability of the two-layer wire grating are obtained. The grating consists of two single-layer wire gratings symmetrically separated from each other as well as the same two-layer wire grating backed by a plane conductive surface. The approximations are obtained by using a quite famous approach reported by J R Wait and D Hill for studying the reflective properties of two-layer cross-wire structures [21].

In the third part of the study, an implementation of the two-layer grating as a periodic structure is considered for generating the microwave surface plasmon polaritons. The dispersion equation is obtained in this part. It is shown that the two-layer grating backed by a plane conductive surface can be used for designing high-impedance surfaces operating in the microwave frequency range.

The accuracy of all the approximations obtained in this study is tested by means of performing appropriate comparative simulations using commercial finite-difference time-domain (FDTD) electromagnetic software.

4.2 Electromagnetic response of single-layer composite

4.2.1 The case of *s*-polarization

Consider a metamaterial as a square infinite set of infinitely long thin metal cylinders of circular cross-section periodically arranged in free space, as depicted in figure 4.1, where r_0 is the radius of cylinders, σ is the conductivity of cylinders and a is the dimension of the unit cell (the distance between axes of two adjacent cylinders). It is assumed throughout the study that $r_0 < a < < \lambda$, where λ is the wavelength in the space surrounding cylinders, which, for simplicity, is taken as air in the study to avoid losses.

To obtain the approximations for the complex effective relative permittivity and permeability of the metamaterial under consideration at microwave frequencies, we first consider a single-layer grating of the same cylinders, as depicted in figure 4.2.

Let each cylinder of the single-layer grating be so thin that it can be identified as an elementary dipole. Then the grating can be considered as an infinite ensemble of dipoles arranged symmetrically along the Oy axis. Consider the electric field vector scattered from the grating, which contains only the component E_z. In this case, the (total) electric polarization vector $\overrightarrow{P}$ of the ensemble of the cylinders can be written in the form

$$\overrightarrow{P} = \alpha_E \sum_{n=-\infty}^{n=\infty} \overrightarrow{E_n}\delta(y - na), \tag{4.1}$$

where n is the number of dipoles, α_E is the electric polarizability of the nth electric dipole $\overrightarrow{p_n}$ created by the electric field vector $\overrightarrow{E_n}$

$$\overrightarrow{p_n} = \alpha_E \overrightarrow{E_n}. \tag{4.2}$$

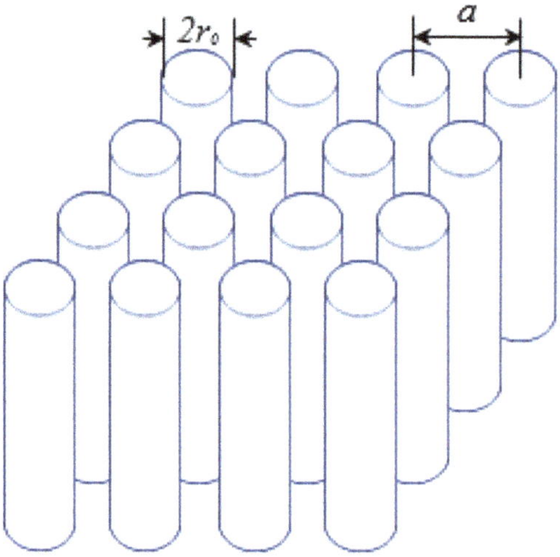

Figure 4.1. The structure of the metamaterial. © [1999] IEEE. Adapted, with permission, from [16].

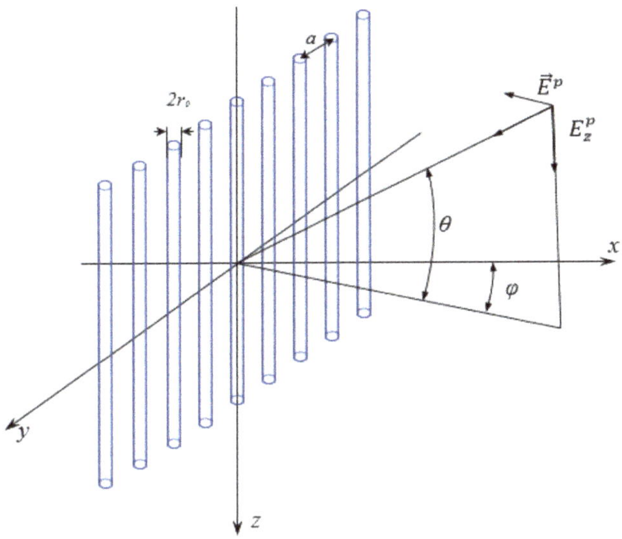

Figure 4.2. Single-layer wire grating with a plane wave at arbitrary incidence. Adapted from [10], copyright (2007), with permission from Elsevier.

Throughout the study, it is assumed that the ensemble of dipoles is excited by its eigenfunctions (eigenwaves), which are determined by the formula

$$\overrightarrow{E_n} = \overrightarrow{E_0} e^{-ihan}, \tag{4.3}$$

where h is the propagation constant to be found later in the study.

Taking into account the harmonic time dependence of all field functions as well as equations (4.1)–(4.3), one can write the expressions for the electric current density $\overrightarrow{j_E}$ as

$$\overrightarrow{j_E} = i\omega\overrightarrow{P} = i\omega\alpha_E\overrightarrow{E_0}\sum_{n=-\infty}^{n=\infty} e^{-ihan}\delta(y - na). \tag{4.4}$$

By making use of Poisson's formula, one can rewrite equation (4.4) in the form

$$\overrightarrow{j_E} = \frac{i\omega}{a}\alpha_E\overrightarrow{E_0}\sum_{n=-\infty}^{n=\infty} \exp(-i(h - 2\pi n/a)y). \tag{4.5}$$

As known from [20], if the incident plane wave is of the form

$$\overrightarrow{E}^{inc}(x, y, z) = \exp\{ik(x\cos\varphi\cos\theta + y\sin\varphi\cos\theta - z\sin\theta)\}, \tag{4.6}$$

where $k = \omega/c$ is the wavenumber in free space and c is the speed of light in vacuum, the electric field component E_z scattered by a single wire is given by

$$E_z = \frac{\mu_0 I\omega}{4}\cos^2\theta e^{-ikz\sin\theta} H_0^{(2)}(k\rho\cos\theta), \tag{4.7}$$

where μ_0 is the permeability of vacuum, $H_0^{(2)}(\cdot)$ is a Hankel function of the second kind of order zero and $\rho = \sqrt{x^2 + y^2}$, I is the amplitude of the electric current excited along the wire by $\overrightarrow{E}^{\text{inc}}$[1] and given by [20]

$$I = -\frac{a}{\frac{Z_0 \cos\theta}{2\cos\varphi} + \frac{i\mu_0\omega a}{2\pi}\cos^2\theta \cdot \ln\left(\frac{a}{2\pi r_0}\right) + aZ_i\cos^2\theta}, \tag{4.8}$$

where $Z_0 = \sqrt{\mu_0/\varepsilon_0}$ is the impedance of free space, ε_0 is the permittivity of vacuum and Z_i is the internal impedance of single wire given by

$$Z_i = \frac{Z_0 I_0(\gamma r_0)}{2\pi I_1(\gamma r_0)}, \tag{4.9}$$

where $\gamma = [i\mu_0\omega/(\sigma + i\omega\varepsilon_0)]^{1/2}$, $I_0(\cdot)$ and $I_1(\cdot)$ are modified Bessel functions of the first kind. Then the expressions for the electric components of electromagnetic field scattered in the direction of the x axis ($\theta = 0$) by the entire ensemble of wires are to be defined by

$$E_z = i\frac{\omega^2 I\mu_0}{4a}\alpha_E E_{0z}\sum_{n=-\infty}^{\infty} H_0^{(2)}(\tilde{\rho}(h - 2\pi n/a)), \tag{4.10}$$

where $\tilde{\rho} = \sqrt{(na - y)^2 + x^2}$.

Another approximation for E_z obtained in [20] using Poisson's formula is given by

$$E_z = \frac{\mu_0\omega I\cos^2\theta}{4}e^{-ikz\sin\theta}\sum_{n=-\infty}^{\infty} e^{ikna\sin\varphi\cos\theta}H_0^{(2)}(k\tilde{\rho}\cos\theta). \tag{4.11}$$

The summation involving Hankel functions in equation (4.11) is slowly convergent and can be transformed to the more convergent series [20]

$$E_z = \frac{i\mu_0\omega I\cos^2\theta}{4\pi}e^{i(ky\sin\varphi\cos\theta - kz\sin\theta)} \sum_{m=-\infty}^{\infty} e^{i2\pi my/a}\frac{\exp\left\{-\frac{2\pi|x|}{a}\sqrt{\left(m + \frac{a\cos\theta\sin\varphi}{\lambda}\right)^2 - \left(\frac{a\cos\theta}{\lambda}\right)^2}\right\}}{\sqrt{\left(m + \frac{a\cos\theta\sin\varphi}{\lambda}\right)^2 - \left(\frac{a\cos\theta}{\lambda}\right)^2}}. \tag{4.12}$$

Note that equation (4.10) takes into account the contributions of all the dipoles, including the central one. Therefore, to calculate the field components E_{0z} and $H_{0\varphi}$ acting on the surface of the reference (central) cylinder ($x = r_0, y = 0, \varphi = 0, \theta = 0$), the field produced by the reference cylinder must be subtracted from the right sides of equation (4.10) and/or equations (4.11) and (4.12). The corresponding

[1] A single wire is considered to be so thin that only a current parallel to its axis is excited on it.

calculations using equations (4.11) and (4.12) finally give, after rather routine algebraic manipulations,

$$E_z = \frac{i\mu_0 \omega I}{4\pi} \sum_{\substack{m=-\infty \\ (m\neq 0)}}^{\infty} \frac{\exp\left\{-\frac{2\pi r_0}{a}\sqrt{m^2 - \left(\frac{a}{\lambda}\right)^2}\right\}}{\sqrt{m^2 - \left(\frac{a}{\lambda}\right)^2}}. \tag{4.13}$$

Substituting equation (4.13) into equation (4.10) finally gives

$$E_z = -\frac{\mu_0^2 I^2 \omega^3}{16\pi a}\alpha_E \left(\sum_{\substack{m=-\infty \\ (m\neq 0)}}^{\infty} \frac{\exp\left\{-\frac{2\pi r_0}{a}\sqrt{m^2 - \left(\frac{a}{\lambda}\right)^2}\right\}}{\sqrt{m^2 - \left(\frac{a}{\lambda}\right)^2}} \right) \sum_{n=-\infty}^{\infty} H_0^{(2)}\left(\tilde{r}\left(h - \frac{2\pi n}{a}\right)\right). \tag{4.14}$$

Equating the right sides of equations (4.14) and (4.11) finally gives the equation for obtaining the unknown α_E as

$$\sum_{n=-\infty}^{\infty} H_0^{(2)}(k\tilde{\rho}) = -\frac{\mu_0 \omega^2 I}{4\pi a}\alpha_E \left(\sum_{\substack{m=-\infty \\ (m\neq 0)}}^{\infty} \frac{\exp\left\{-\frac{2\pi r_0}{a}\sqrt{m^2 - \left(\frac{a}{\lambda}\right)^2}\right\}}{\sqrt{m^2 - \left(\frac{a}{\lambda}\right)^2}} \right) \sum_{n=-\infty}^{\infty} H_0^{(2)}\left(\tilde{\rho}\left(h - \frac{2\pi n}{a}\right)\right). \tag{4.15}$$

Considering $k = h - 2\pi n/a$ for defining the propagation constant h finally gives from equation (4.15)

$$\alpha_E = -\frac{2\pi a}{\mu_0 \omega^2 I} \frac{1}{\sum_{m=1}^{\infty} \frac{\exp\left\{-(2\pi r_0/a)\sqrt{m^2 - (a/\lambda)^2}\right\}}{\sqrt{m^2 - (a/\lambda)^2}}}. \tag{4.16}$$

Note that we switched to semi-infinite summation in the series of equation (4.16) because the summation index is present in the series to a even degree.

Due to a considerable convergence of the series, it is enough to evaluate only a few terms of the series while the rest can be neglected. Indeed, considering $h = h_n = k + 2\pi n/a$ as an equation for calculating the propagation constant of the nth mode exited by the grating, it is possible to obtain a condition for the applicability of EMT [18] ($10a > \lambda_n = 2\pi/h_n \geqslant 5a$). Otherwise the concept of polarizability cannot be inapplicable to the composite under consideration. Thus,

$$\frac{2}{1 + 2n} \geqslant 10. \tag{4.17}$$

Note that the condition in equation (4.17) was also chosen such that the accuracy of all approximations obtained in this work was compatible with the accuracy of the solution obtained in [20].

Let us find the dipole approximation for the magnetic polarizability α_M of the metamaterial under consideration for the current polarization. As shown in [20], the total reflection coefficient of the wire grating is given by

$$R_s = \left(\frac{E_z^{\text{scat}}}{E_z^{\text{inc}}}\right)\Bigg|_{x=0} = -\frac{1}{1 + 2Z_g/K_0}, \tag{4.18}$$

where Z_g is the equivalent shunt impedance of the grating, and it is given by

$$\left.\begin{aligned} \frac{Z_g}{K_0} &= i\frac{a}{\lambda}\cos\theta\sin\varphi \cdot \ln\left(\frac{a}{2\pi r_0}\right) + \frac{\cos\varphi}{\cos\theta}\frac{Z_i}{Z_0}a, \\ K_0 &= \left(\frac{E_z^{\text{inc}}}{H_y^{\text{inc}}}\right)\Bigg|_{x=0} = Z_0\frac{\cos\theta}{\cos\varphi}. \end{aligned}\right\}. \tag{4.19}$$

At the same time, considering the wire grating as an ordered ensemble of dipoles, there is a relationship between the total reflection coefficient R and the polarizabilities α_E and α_M [11]

$$R = \frac{ika}{2}\frac{1}{\varepsilon_0 a^2/\alpha_E - \beta} - \frac{ika}{2}\frac{1}{\mu_0 a^2/\alpha_M - \beta}, \tag{4.20}$$

where β is the so-called self-interaction coefficient, which is, in fact, proportional to a Green function of the grating, i.e. equals a summation over the Green function of free space that is generally dependent on the incidence angle, lattice (grating) constant, grating symmetry and polarization.

The local electric field E_z^{loc} acting on the surface of the reference wire is defined as [21]

$$\left.\begin{aligned} E_z^{\text{loc}} &= E_z^p + \beta_e p, \\ E_z^p &= E_0 e^{-ikx}, \end{aligned}\right\}, \tag{4.21}$$

where $\overrightarrow{p}$ is the electric dipole induced by the array of cylinders as the response of the initial electric field E_z^p and $\beta_e = \beta/\varepsilon_0 a^2$ is the interaction constant for the magnetic dipole. At the same time,

$$p = \alpha_E E_z^{\text{loc}}. \tag{4.22}$$

Solving jointly equations (4.21) and (4.22) with respect to p gives

$$p = \frac{\alpha_E}{1 - \alpha_E\beta_e}E_z^p. \tag{4.23}$$

The main implementation area for the above obtained results is antenna applications [6, 7]. Thus, we consider the observation points in the far zone. In this case, we can hypothetically replace the array/grating by an averaged electric current sheet carrying the electric current: $\overrightarrow{\widehat{J_l}} = i\omega\overrightarrow{p}/a^2$ [22]. Then, the electric field created by this sheet ($x = 0$) is $\overrightarrow{E_{\text{scat}}} = -Z_0\overrightarrow{\widehat{J_l}}/2$. Now, taking into account equation (4.23), the total reflection coefficient of the wire grating is given by

$$R_s = \left(\frac{E_z^{\text{scat}}}{E_z^{\text{inc}}} \right) \Bigg|_{x=0} = -i\omega \frac{Z_0}{2a^2} \frac{\alpha_E}{1 - \alpha_E \beta_e}. \tag{4.24}$$

Taking into account that R_s is already defined by equations (4.18) and (4.19), we obtain from equation (4.25)

$$\beta = \frac{\varepsilon_0 a^2}{\alpha_E} + i\frac{\varepsilon_0 \omega Z_0}{2R_s}. \tag{4.25}$$

Using rather routine mathematical manipulations over equation (4.20) gives an expression that allows us to obtain the approximation for the unknown magnetic polarizability α_M as

$$\frac{\alpha_M}{\mu_0 a^2} = \frac{1 - (\beta - \varepsilon_0 a^2/\alpha_E)(2iR_s/ka)}{\varepsilon_0 a^2/\alpha_E + (\beta + \varepsilon_0 a^2/\alpha_E)\beta(2iR_s/ka)}, \tag{4.26}$$

where the expressions of α_E, R_s and β are defined by equations (4.16) and (4.17), equations (4.18)–(4.20) and equation (4.25), respectively.

The desired expressions for the complex effective relative permittivity ε_{eff} and magnetic permeability μ_{eff} of the metamaterial can be obtained from the Clausius–Mossotti equation

$$\left. \frac{\varepsilon_{\text{eff}} - 1}{\varepsilon_{\text{eff}} + 2} = \frac{\alpha_E}{\varepsilon_0 a^2}, \quad \frac{\mu_{\text{eff}} - 1}{\mu_{\text{eff}} + 2} = \frac{\alpha_M}{\mu_0 a^2} \right\}, \tag{4.27}$$

where the expressions of polarizabilities α_E and α_M are defined by equations (4.16) and equation (4.26).

To validate the obtained expressions of the electromagnetic response of the metamaterial for the case of s-polarization, we compare the simulation results, using equations (4.16)–(4.19), equations (4.24)–(4.26) and the second equality in equation (4.28), with the results of numerical simulations, using the analytical approximations obtained in reference [19] for the case of copper wires with $a = 0.002$ m and $r_0 = 0.0001$ m. The results of comparative modeling for ε_{eff} give the expected result $\text{Re}(\varepsilon_{\text{eff}}) < 0$ if $\omega < \omega_p$. Note that ω_p can be evaluated in terms of the approach proposed in this study. Indeed, one can represent the effective relative permittivity for the considered polarization using the Drude model

$$\varepsilon_{\text{eff}}(\omega) = 1 - \frac{\omega_p^2}{\omega(\omega + i\gamma)}, \tag{4.28}$$

where the approximation of the damping term γ can be defined in the same wave as in [23]

$$\gamma = \varepsilon_0 a^2/\pi {r_0}^2 \sigma \tag{4.29}$$

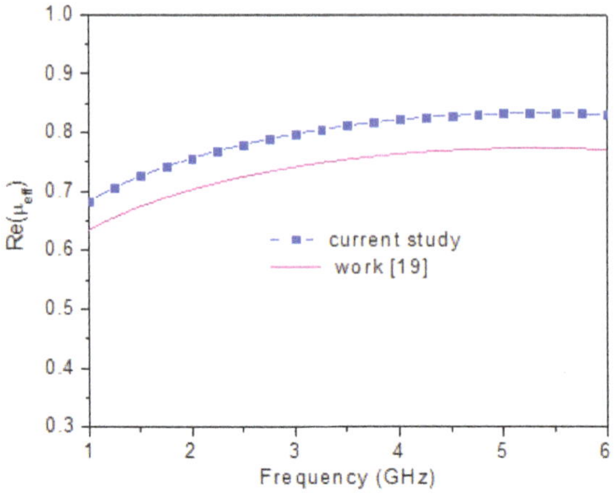

Figure 4.3. Plots of the real part of effective relative permeability for copper wires of radius $r_0 = 0.0001$ m symmetrically located in air with a period of $a = 0.002$ m.

and representing the dissipation of plasmon energy in the composite that occurs whenever $\omega < \omega_p$. Substituting equation (4.27) into equation (4.26) finally gives

$$\omega_p^2 = \frac{\alpha_E - \varepsilon_0 a^2}{3\alpha_E \omega(\omega + i\gamma)}. \tag{4.30}$$

The estimated calculation run for obtaining the plasma frequency ω_p using equation (4.30) is in good agreement with the results of a simulation using the approximations in reference [19], and gives an absolute error of no more than 2%.

The results of the comparative modeling for μ_{eff} are depicted in figure 4.3. Note that the imaginary parts of the effective relative permeability are not presented in this study, since their values did not exceed 10^{-2}. One can see from figure 4.3 that good agreement between the current EMT and the theory of work in [19] is observed. Moreover, the relations of EMT in [19] are noticeably more cumbersome.

4.2.2 The case of *p*-polarization

Let us find the dipole approximation for the electric α_E and magnetic α_M polarizabilities of the metamaterial under consideration in the case of *p*-polarization. It is easy to find the microwave approximation for α_M since the appropriate dipole approximation for the complex effective relative magnetic permeability has been already obtained in [16]:

$$\mu_{\text{eff}} = 1 - \frac{F}{1 + 2i\delta / r_0 \omega \mu_0}, \tag{4.31}$$

where δ is the resistance of the cylinder surface per unit area, which is actually rather a hypothetical amount than a real physical one. However, a relation for calculating δ in terms of real physical amounts was recently obtained in [24]:

$$\delta = 1/\sigma \cdot \Delta, \tag{4.32}$$

where $\Delta = \sqrt{2/\mu_0 \sigma\omega}$ is the skin depth.

Using the second equation in equation (4.27), one can obtain the dipole approximation for the magnetic polarizability α_M for the considered polarization

$$\frac{\alpha_M}{\mu_0 a^2} = \frac{r_0 \omega \mu_0}{(F - 3) r_0 \omega \mu_0 - 6i\delta}. \tag{4.33}$$

According to equation (4.20), in order to find a dipole approximation of the electric polarizability α_E, it is necessary to find the dipole approximations for the appropriate total reflection coefficient R_p and the self-interaction coefficient β.

The p-mode dipole approximation for the total reflection coefficient R_p is already obtained in [25] for the case of normal incidence, and is given by

$$\left.\begin{aligned} R_p &= \frac{ip}{1 - \nu' - ip}, \\ p &= kaF\frac{\eta^2 - 1}{\eta^2 + 1}, \quad v' = \frac{\pi}{3}F\frac{\eta^2 - 1}{\eta^2 + 2} \end{aligned}\right\}, \tag{4.34}$$

where $\eta = \sqrt{\varepsilon_r + i\sigma/\omega\varepsilon_0}$, ε_r is the static relative permittivity of cylinders and $F = \pi r_0^2/a^2$ is the inclusion (cylinder) volume fraction. In order to find a dipole approximation of the self-interaction coefficient β for the considered polarization, we accept the local magnetic field H_z^{loc} acting on the surface of the reference wire in the following form [22]:

$$\left.\begin{aligned} H_z^{\mathrm{loc}} &= H_z^p + \beta_m m, \\ H_z^p &= H_0 e^{-ikx} \end{aligned}\right\}, \tag{4.35}$$

where $\overrightarrow{m}$ is the magnetic dipole induced by the grating as the response of the initial magnetic field H_z^p; $\beta_m = \beta/\mu_0 a^2$ is the interaction constant of the magnetic dipole. At the same time,

$$m = \alpha_M H_z^{\mathrm{loc}}. \tag{4.36}$$

Solving jointly equations (4.35) and (4.36) with respect to m gives

$$m = \frac{\alpha}{1 - \alpha\beta_m} H_z^p. \tag{4.37}$$

For the same reasons as in the case of s-polarization, we will consider the far zone in the case of p-polarization. In this case, we can hypothetically replace the grating by an averaged magnetic current sheet carrying the magnetic current $\overrightarrow{J_m} = i\omega\overrightarrow{m}/a^2$

[22]. Then, the magnetic field created by this sheet ($x = 0$) is $\overleftarrow{H}_{\text{scat}} = -\overrightarrow{J}_m/2Z_0$. Then, taking into account equation (4.37), the total reflection coefficient of the wire grating is given by

$$R_p = \left(\frac{H_z^{\text{scat}}}{H_z^{\text{inc}}}\right)\bigg|_{x=0} = -i\omega\frac{1}{2Z_0a^2}\frac{\alpha_M}{1 - \alpha_M\beta_m}. \tag{4.38}$$

Taking into account that R_p is already defined by equation (4.34), from equation (4.38) we obtain

$$\beta = \frac{\mu_0 a^2}{\alpha_M} + i\frac{\mu_0\omega}{2Z_0R_s}. \tag{4.39}$$

Using rather routine mathematical manipulations over equation (4.20) gives an expression that allows us to calculate the approximation for the unknown magnetic polarizability α_E

$$\frac{\alpha_E}{\varepsilon_0 a^2} = \frac{1 + (\beta - \mu_0 a^2/\alpha_M)(2iR_p/ka)}{\mu_0 a^2/\alpha_M + (\beta - \mu_0 a^2/\alpha_M)\beta(2iR_p/ka)}, \tag{4.40}$$

where the expressions of α_M, R_p and β are defined by equations (4.33), (4.34) and (4.39), respectively.

To validate the obtained expressions for the electromagnetic response of the considered metamaterial for the case of s-polarization, we compare the results of the analytical modeling using equations (4.34), (4.39) and (4.40) and the first equality of equation (4.27) with the results of a numerical simulation using the analytical approximations obtained in [19] for the same parameters as for the case of p-polarization. The results of the comparative simulation for ε_{eff} are depicted in figure 4.4. However, we do not show the plots of the imaginary part of the effective relative permeability, since their values do not exceed 10^{-3}. One can see from this figure that good agreement exists between the current EMT and the theory presented in [19].

It is easy to see from figure 4.4 that the relations of EMT in [19] are noticeably more cumbersome. At the same time, there are no so strong restrictions imposed in [19], as for the case under consideration. Also, most importantly, the Mie resonance is taken into account in [19]. However, in the present study, as will be shown later, it can only be predicted. So, speaking about the terms and definitions of [18], the array has been considered from the point of view of non-resonant composites so far in this study. Therefore, the accuracy of the results in [19] is expectedly higher in a wide frequency range. Nevertheless, the results obtained in this study allow us to modify the theory of lattice sums to consider arrays with any number of identical layers in terms of semi-resonant composites, i.e. composites in which resonance is taken into account.

Indeed, if, for example, we are able to obtain the total reflection coefficient R for the n-layer wire grating, we are, in turn, able to obtain the expression of the self-interaction

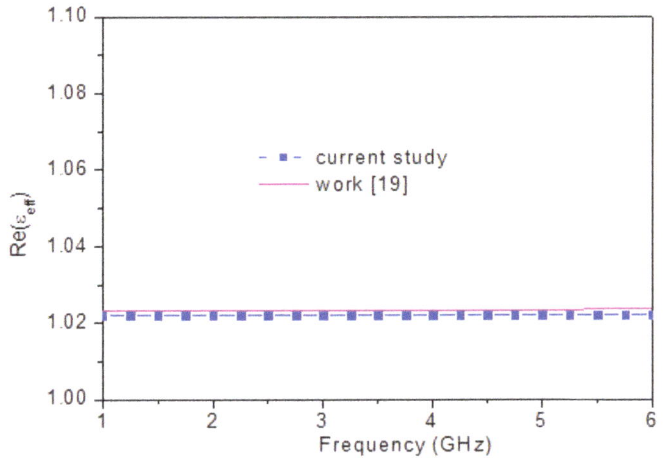

Figure 4.4. Plots of the real part of the effective relative permittivity for copper wires of radius $r_0 = 0.0001$ m symmetrically located in air with period $a = 0.002$ m.

coefficient β from equation (4.20), as long as the expressions of the electric α_E and magnetic α_M polarizabilities are already obtained in [19]. Thus, it is necessary to obtain the total reflection coefficient R of the n-layer wire grating for full-wave characterization of the grating. However, the availability of the coefficient β is rather necessary to obtain any of the parameters R, α_E or α_M, if unknown (for the considered polarization). Thus, for a complete theoretical characterization of a metamaterial as an array of dipoles, it is sufficient to have the expressions for the total reflection coefficient R and the polarizabilities α_E and α_M. To characterize a metamaterial as a scatterer, it is also necessary that subwavelength approximations for the effective permittivity and permeability in given directions are available.

4.3 The electromagnetic response of two-layer composite

4.3.1 The case of *p*-polarization

Let us find the total reflection coefficient R_p of the wire grating consisting of two layers of unloaded infinite metal cylinders with a circular cross-section, as in figure 4.5, where the same designations and restrictions are introduced as for the single-layer grating depicted in figures 4.1 and 4.2. It is also assumed that the first layer grating of the considered two-layer composite has spacing a and is parallel to the axis Oz, and is located in the plane $z = 0$, while the second single-layer grating with the same spacing is parallel to the first one and is located in the plane $z = -a$.

The incident electric field $\overrightarrow{E}^p$ is a plane wave of arbitrary polarization (without a time-harmonic component) given by

$$\overrightarrow{E}^p = \overrightarrow{E_0}e^{ik(x \sin\theta \cos\varphi + y \sin\theta \sin\varphi + z \cos\theta)}, \tag{4.41}$$

where the amplitude vector $\overrightarrow{E_0}$ is the incident field at the origin.

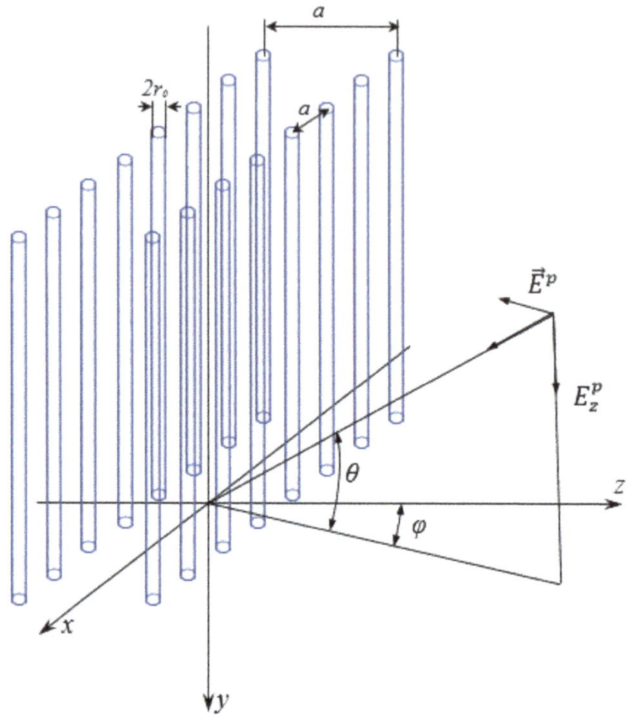

Figure 4.5. Two gratings of unloaded infinitely long wires as finitely long metal cylinders of circular cross-section symmetrically arranged along the y-axis with period a. Adapted from [10], copyright (2007), with permission from Elsevier.

The best-known analytical technique for solving problems similar to the proposed one is the averaged boundary conditions (ABC) approach (described e.g. in [26]). Such approaches employ a technique based on averaging the fields over one unit cell of the mesh containing a finite set of unit cells before applying the boundary conditions. As such, the results would seem to be restricted to perhaps too small spacing of the grating wires in terms of wavelength. Therefore, in this study, we use the well-known and somewhat forgotten the approach proposed by J R Wait and successfully developed by D A Hill. The approach is, for example, presented in [21], where the single-layer wire gratings are perpendicular to each other. Throughout the study, index (1) stands for the upper single-layer grating (located in the plane $z = 0$) while index (2) stands for the second one (located in the plane $z = -a$).

The currents of the x-directed wire gratings can be presented in the following Fourier series form:

$$\left.\begin{aligned} I_{xn}^{(1)} &= \sum_{m=-\infty}^{\infty} A_m \exp\left\{i\left(\frac{2\pi m}{a}y + kx \sin\theta\cos\varphi + na\sin\varphi\right)\right\}, \\ I_{xm}^{(2)} &= \sum_{n=-\infty}^{\infty} B_n \exp\left\{i\left(\frac{2\pi n}{a}y + kx \sin\theta\cos\varphi + ma\sin\varphi\right)\right\} \end{aligned}\right\}, \tag{4.42}$$

where A_m and B_n are to be defined later, m stands for summation over the wires of the upper grating and n stands for summation over the wires of the lower grating.

The electric field scattered by both the wire gratings can be derived from the vector magnetic potential $\vec{\Pi}$, which only has the x-directed components

$$\vec{\Pi} = (\Pi_{x1} + \Pi_{x2})\vec{x_0}. \tag{4.43}$$

Let us obtain the expressions for the components Π_{x1} and Π_{x2}. The potential vector component of the current on the reference wire of the upper grating which coincides with the x-axis is given by [21]

$$\Pi_{x01} = \sum_{n=-\infty}^{\infty} a_m \exp\left\{i\left(kx \sin\theta \cos\varphi + \frac{2\pi m}{a} y\right)\right\} H_0^{(2)}\left(u_m \sqrt{y^2 + z^2}\right). \tag{4.44}$$

Π_{x01} must satisfy the Helmholtz equation

$$(\Delta + k^2)\Pi_{x01} = 0. \tag{4.45}$$

Substituting equation (4.44) in equation(4.45) finally gives

$$u_m = \sqrt{k^2 - \left(\frac{2\pi m}{a}\right)^2 - (k \sin\theta \cos\varphi)^2}. \tag{4.46}$$

Taking into account the definition of the magnetic potential vector $\vec{\Pi}$, one can determine the φ-component of the magnetic field created by the current of the wire

$$H_{\varphi 01} = \frac{k^2}{i\omega\mu_0}\frac{\partial \Pi_{x01}}{\partial \rho} = -\frac{k^2}{i\omega\mu_0}\sum_{n=-\infty}^{\infty} a_m \exp\left\{i\left(kx \sin\theta \cos\varphi + \frac{2\pi m}{a} y\right)\right\} u_m H_1^{(2)}(u_m \rho), \tag{4.47}$$

where $\rho = \sqrt{y^2 + z^2}$.

The current $I_{x0}^{(1)}$ can be obtained from Ampere's law

$$I_{x0}^{(1)} = \lim_{\rho\to r_0}\left(\int_0^{\rho} dr\, H_0^{(2)}(u_m\rho)\right) = \frac{2i\pi r_0 k^2}{\omega\mu_0}\sum_{n=-\infty}^{\infty} a_m \exp\left\{i\left(kx \sin\theta \cos\varphi + \frac{2\pi m}{a} y\right)\right\} u_m H_1^{(2)}(u_m\rho). \tag{4.48}$$

Equating equation (4.48) and the first equality of equation (4.42) finally gives

$$a_m = -\frac{\omega\mu_0}{4k^2} A_m. \tag{4.49}$$

Generalizing equation (4.44) to the case of all wires of the first grating gives

$$\Pi_{x1} = -\frac{\omega\mu_0}{4k^2}\sum_{m=-\infty}^{\infty}\sum_{n=-\infty}^{\infty} A_m \exp\left\{i\left(kx \sin\theta \cos\varphi + \frac{2\pi m}{a} y\right)\right\} H_0^{(2)}\left(u_m \sqrt{(y - na)^2 + z^2}\right) e^{ikna \sin\theta \sin\varphi} \tag{4.50}$$

,

Repeating all the discussions and mathematical manipulations for the case of the second grating finally gives

$$\Pi_{x2} = -\frac{\omega\mu_0}{4k^2} \sum_{m=-\infty}^{\infty}\sum_{n=-\infty}^{\infty} B_n \exp\left\{i\left(kx \sin\theta \cos\varphi + \frac{2\pi n}{a}y\right)\right\} H_0^{(2)}\left(u_n\sqrt{(y-na)^2+(z+h)^2}\right)e^{ikma \sin\theta \sin\varphi}. \tag{4.51}$$

As is well known, the summation involving Hankel functions in equations (4.50) and (4.51) is slowly convergent and can be transformed to more convergent series [20], as was done in equation (4.11). Indeed, straightforward algebra gives

$$\left.\begin{aligned} \Pi_{x1} &= \frac{\omega\mu_0}{4i\pi k^2}\sum_{m=-\infty}^{\infty}\sum_{n=-\infty}^{\infty} A_m \exp\left\{i\left(kx \sin\theta\cos\varphi + \left[2\pi\frac{m+n}{a} + k\sin\theta\sin\varphi\right]y\right)\right\}e^{-\Gamma_{mn}^{(1)}|z|}, \\ \Pi_{x2} &= \frac{\omega\mu_0}{4i\pi k^2}\sum_{m=-\infty}^{\infty}\sum_{n=-\infty}^{\infty} B_n \exp\left\{i\left(kx \sin\theta\cos\varphi + \left[2\pi\frac{n+m}{a} + k\sin\theta\sin\varphi\right]y\right)\right\}e^{-\Gamma_{mn}^{(2)}|z+a|} \end{aligned}\right\}, \tag{4.52}$$

where

$$\Gamma_{mn}^{(1)} = \sqrt{\left(\frac{2\pi n}{a} + k\sin\theta\sin\varphi\right)^2 + \left(\frac{2\pi m}{a}\right)^2 + (k\sin\theta\cos\varphi)^2 - k^2}$$

and

$$\Gamma_{mn}^{(2)} = \sqrt{\left(\frac{2\pi m}{a} + k\sin\theta\sin\varphi\right)^2 + \left(\frac{2\pi n}{a}\right)^2 + (k\sin\theta\cos\varphi)^2 - k^2}.$$

The scattered electric field is defined by

$$\overrightarrow{E}^{sc} = k^2\overrightarrow{\Pi} + \nabla\left(\nabla\cdot\overrightarrow{\Pi}\right). \tag{4.53}$$

Substituting equation (4.52) into equation (4.53) finally gives

$$\left.\begin{aligned} E_x^{sc} &= E_{x1} + E_{x2}, \\ E_{x1} &= \frac{\omega\mu_0}{4i\pi}(1-\sin^2\theta\cos^2\varphi) \sum_{m=-\infty}^{\infty}\sum_{n=-\infty}^{\infty} A_m \exp\left\{i\left(kx\sin\theta\cos\varphi + \left[2\pi\frac{m+n}{a} + k\sin\theta\sin\varphi\right]y\right)\right\}\frac{e^{-\Gamma_{mn}^{(1)}|z|}}{\Gamma_{mn}^{(1)}}, \\ E_{x2} &= \frac{\omega\mu_0}{4i\pi}(1-\sin^2\theta\cos^2\varphi) \sum_{m=-\infty}^{\infty}\sum_{n=-\infty}^{\infty} B_n \exp\left\{i\left(kx\sin\theta\cos\varphi + \left[2\pi\frac{m+n}{a} + k\sin\theta\sin\varphi\right]y\right)\right\}\frac{e^{-\Gamma_{mn}^{(2)}|z+a|}}{\Gamma_{mn}^{(2)}}. \end{aligned}\right\} \tag{4.54}$$

All wire currents must satisfy the following impedance conditions:

$$\left.\begin{aligned}(E_x^p + E_{x1})|_{\substack{y=0\\z=r_0}} &= Z_i I_{xn}^{(1)}|_{\substack{y=0\\z=r_0}}\\ (E_x^p + E_{x2})|_{\substack{y=0\\z=-r_0-a}} &= Z_i I_{xm}^{(2)}|_{\substack{y=0\\z=-r_0-a}}\end{aligned}\right\}, \tag{4.55}$$

where E_x^p is the x-projection of incident field at the origin and Z_i is the internal impedance per unit length of the wires given by equation (4.9). Substituting equations (4.9), (4.41), (4.42) and (4.54) into equation (4.55) finally gives

$$\left.\begin{aligned}&E_{x0}e^{-ikr_0\cos\theta} + \frac{\omega\mu_0}{4i\pi}(1 - \sin^2\theta\cos^2\varphi)\sum_{m=-\infty}^{\infty}\sum_{n=-\infty}^{\infty} A_m \frac{e^{-\Gamma_{mn}^{(1)}r_0}}{\Gamma_{mn}^{(1)}} = Z_i \sum_{m=-\infty}^{\infty} A_m e^{ina\sin\varphi},\\ &E_{x0}e^{-ik(r_0+a)\cos\theta} + \frac{\omega\mu_0}{4i\pi}(1 - \sin^2\theta\cos^2\varphi)\sum_{m=-\infty}^{\infty}\sum_{n=-\infty}^{\infty} B_n \frac{e^{-\Gamma_{mn}^{(1)}(r_0+a)}}{\Gamma_{mn}^{(1)}} = Z_i \sum_{n=-\infty}^{\infty} B_n e^{ima\sin\varphi}\end{aligned}\right\}. \tag{4.56}$$

Taking into account that $e^{-ikr\cos\theta} = \sum_{s=-\infty}^{\infty}(-i)^s J_s(kr)e^{is\theta}$, equation (4.56) can be reduced to the system of equations with regard to A_m and B_n

$$\left.\begin{aligned}&E_{x0}(-i)^m J_m(kr_0)e^{im\theta} + \frac{\omega\mu_0}{4i\pi}(1 - \sin^2\theta\cos^2\varphi)A_m \sum_{n=-\infty}^{\infty}\frac{e^{-\Gamma_{mn}^{(1)}r_0}}{\Gamma_{mn}^{(1)}} = Z_i A_m e^{ina\sin\varphi},\\ &E_{x0}(-i)^n J_n(k(r_0+a))e^{in\theta} + \frac{\omega\mu_0}{4i\pi}(1 - \sin^2\theta\cos^2\varphi)B_n \sum_{m=-\infty}^{\infty}\frac{e^{-\Gamma_{mn}^{(1)}(r_0+a)}}{\Gamma_{mn}^{(1)}} = Z_i B_n e^{ima\sin\varphi}\end{aligned}\right\}, \tag{4.57}$$

where $J_n(\cdot)$ are Bessel functions of the first type. Solving equation (4.57) with respect to A_m and B_n gives

$$\left.\begin{aligned}A_m &= \frac{(-i)^{m+1}J_m(kr_0)\sin\varphi E_0 e^{im\theta}}{Z_i e^{ina\sin\varphi} + \frac{i\omega\mu_0}{4\pi}(1 - \sin^2\theta\cos^2\varphi)\sum_{n=-\infty}^{\infty}\frac{e^{-\Gamma_{mn}^{(1)}r_0}}{\Gamma_{mn}^{(1)}}},\\ B_n &= \frac{(-i)^{n+1}J_n(k(r_0+a))\sin\varphi E_0 e^{in\theta}}{Z_i e^{ima\sin\varphi} + \frac{i\omega\mu_0}{4\pi}(1 - \sin^2\theta\cos^2\varphi)\sum_{m=-\infty}^{\infty}\frac{e^{-\Gamma_{mn}^{(2)}(r_0+a)}}{\Gamma_{mn}^{(2)}}}\end{aligned}\right\}. \tag{4.58}$$

It is possible to avoid infinite summation in equation (4.58). For this reason, one can use the equality proposed in [21]

$$\sum_{n=-\infty}^{\infty}\frac{e^{-\Gamma_{mn}^{(1)}r_0}}{\Gamma_{mn}^{(1)}} \simeq \frac{a}{\pi}\left[\ln\left(\frac{a}{2\pi r_0}\right) + \Delta_m\right] + \frac{1}{\Gamma_{m0}^{(1)}}, \tag{4.59}$$

where $\Delta_m = \frac{1}{2}\sum_{\substack{n=-\infty\\n\neq 0}}^{\infty}\left[\frac{2\pi}{a\Gamma_{mn}^{(1)}} - \frac{1}{|n|}\right]$. Note that the term Δ_m converges as rapidly as it can be considered as a correction term to $\ln(a/2\pi r_0)$, if a/λ is small. Thus, Δ_m can finally be neglected in terms of the EMT [18]. Substituting equation (4.59) into equation (4.58) gives

$$\left.\begin{aligned} A_m &= \frac{(-i)^{m+1}J_m(kr_0)\sin\varphi E_0 e^{im\theta}}{Z_i e^{ima\sin\varphi} + \frac{i\omega\mu_0}{4\pi}(1-\sin^2\theta\cos^2\varphi)\left(\frac{a}{\pi}\ln\left(\frac{a}{2\pi r_0}\right) + \frac{1}{\Gamma^{(1)}_{m0}}\right)}, \\ B_n &= \frac{(-i)^{n+1}J_n(k(r_0+a))\sin\varphi E_0 e^{in\theta}}{Z_i e^{ina\sin\varphi} + \frac{i\omega\mu_0}{4\pi}(1-\sin^2\theta\cos^2\varphi)\left(\frac{a}{\pi}\ln\left(\frac{a}{2\pi r_0}\right) + \frac{1}{\Gamma^{(2)}_{0n}}\right)} \end{aligned}\right\}. \tag{4.60}$$

To find an expression for the reflection coefficient R_p, it is necessary to obtain expressions for all the components (E_x^{sc}, E_y^{sc} and E_z^{sc}) of the scattered field. Taking into account equations (4.52) and (4.53) finally gives

$$\left.\begin{aligned} E_y^{sc} &= E_{y1}^{sc} + E_{y2}^{sc}, \\ E_{y1}^{sc} &= \frac{\omega\mu_0}{2}\left[\frac{m+n}{ka} + \frac{\sin\theta\sin\varphi}{2\pi}\right]\sin\theta\cos\varphi \times \\ &\times \sum_{m=-\infty}^{\infty}\sum_{n=-\infty}^{\infty} A_m \exp\left\{i\left(kx\sin\theta\cos\varphi + \left[2\pi\frac{m+n}{a} + k\sin\theta\sin\varphi\right]y\right)\right\}\frac{e^{-\Gamma^{(1)}_{mn}|z|}}{\Gamma^{(1)}_{mn}}, \\ E_{y1}^{sc} &= \frac{\omega\mu_0}{2}\left[\frac{m+n}{ka} + \frac{\sin\theta\sin\varphi}{2\pi}\right]\sin\theta\cos\varphi \times \\ &\times \sum_{m=-\infty}^{\infty}\sum_{n=-\infty}^{\infty} B_n \exp\left\{i\left(kx\sin\theta\cos\varphi + \left[2\pi\frac{m+n}{a} + k\sin\theta\sin\varphi\right]y\right)\right\}\frac{e^{-\Gamma^{(2)}_{mn}|z+a|}}{\Gamma^{(2)}_{mn}} \end{aligned}\right\} \tag{4.61}$$

and

$$\left.\begin{aligned} E_z^{sc} &= E_{z1}^{sc} + E_{z2}^{sc}, \\ E_{z1}^{sc} &= -\frac{\omega\mu_0}{4\pi k}\sin\theta\cos\varphi\sum_{m=-\infty}^{\infty}\sum_{n=-\infty}^{\infty} A_m \exp\left\{i\left(kx\sin\theta\cos\varphi + \left[2\pi\frac{m+n}{a} + k\sin\theta\sin\varphi\right]y\right)\right\}e^{-\Gamma^{(1)}_{mn}|z|}, \\ E_{z2}^{sc} &= -\frac{\omega\mu_0}{4\pi k}\sin\theta\cos\varphi\sum_{m=-\infty}^{\infty}\sum_{n=-\infty}^{\infty} B_n \exp\left\{i\left(kx\sin\theta\cos\varphi + \left[2\pi\frac{m+n}{a} + k\sin\theta\sin\varphi\right]y\right)\right\}e^{-\Gamma^{(2)}_{mn}|z+a|}. \end{aligned}\right\} \tag{4.62}$$

In the far zone ($|z| > > \lambda$), which is in full agreement with the concept of the effective medium theory ($4\lambda < a$ [18]), only the terms for the constants A_0 and B_0 make the main contribution to the scattered field. In this way, the approximation for the reflection coefficient R_p for the p-polarization is given by

$$R_p = \frac{E_\theta^s}{E_\theta^p} = \frac{E_x^s\cos\theta\cos\varphi + E_y^s\cos\theta\sin\varphi - E_z^s\sin\theta}{E_\theta^p}, \tag{4.63}$$

where E_θ^p is the θ-component of the incident field (4.41), given by

$$\left.\begin{aligned} E_\theta^p &= E_{0\theta}e^{ik(x\sin\theta\cos\varphi + y\sin\theta\sin\varphi + z\cos\theta)}, \\ E_{x0} &= E_{0\theta}\cos\theta\cos\varphi, \\ E_{y0} &= E_{0\theta}\cos\theta\sin\varphi \end{aligned}\right\}. \tag{4.64}$$

In this study, comparative modeling is carried out to evaluate the accuracy of the above obtained theory. Thus, the results of modeling using equations (4.60), (4.63) and (4.64) are compared with the results of numerical simulation using the commercial electromagnetic simulator PLANC FD (v 6.2 created by Information and Mathematical Science Laboratory, Inc.), which is based on the FDTD method. Copper wires with the spacing $a = 0.002$ m and the radius $r_0 = 0.0001$ m are chosen in the comparative modeling. The results of the comparative modeling are depicted in figure 4.6. A good agreement of the results of the analytical modeling with those of the numerical simulations is observed for the case of p-polarization. One can observe from figure 4.6 that the electromagnetic waves in the considered double grating are evanescent in nature and cannot propagate through it ($R_p \approx 1$). Plots of the imaginary part of the total reflection coefficient are not shown here as it turned out to be expectedly low due to air as a medium surrounding the wires. Moreover, the reflection coefficient seems very similar to the one obtained in [27] using another analytical technique. Note that the reflection coefficient is qualitatively and quantitatively similar to the case of a single-layer grating of unloaded wires previously considered in [20, 28] (see figure 4.7). As seen from the figure, the main difference is quantitative—the expected zero point of the reflection coefficient has shifted to a lower frequency due to the influence of the second wire layer, as mentioned in [10]. Apparently, this leads to a redshift of the resonance. Moreover, this shift is greater for a larger number of layers. This effect, as expected, leads to the fact that a structure consisting of several wire layers cannot be transparent at GHz frequencies.

For a more complete description of the optical properties of the wire structure under consideration, one should also study its complex effective electromagnetic response, as was done in [10, 20], for instance. To achieve our goal, the approach proposed in [27] is going to be used in this study independently considering the

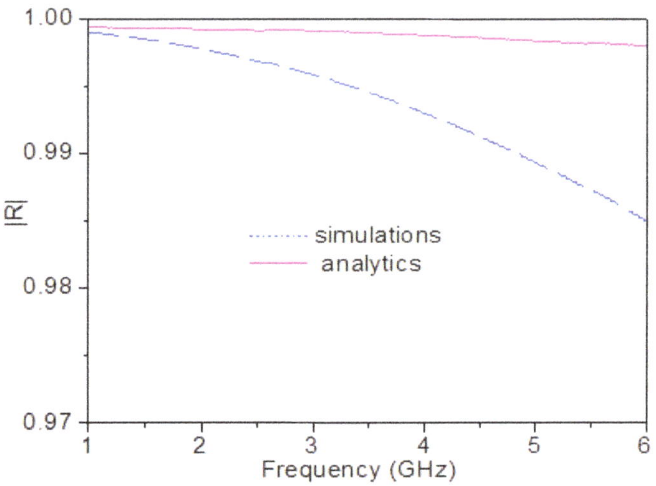

Figure 4.6. Plots of the total reflection coefficient for the two-layer copper wire gratings with $a = 0.002$ m and $r_0 = 0.0001$ m (p-polarization).

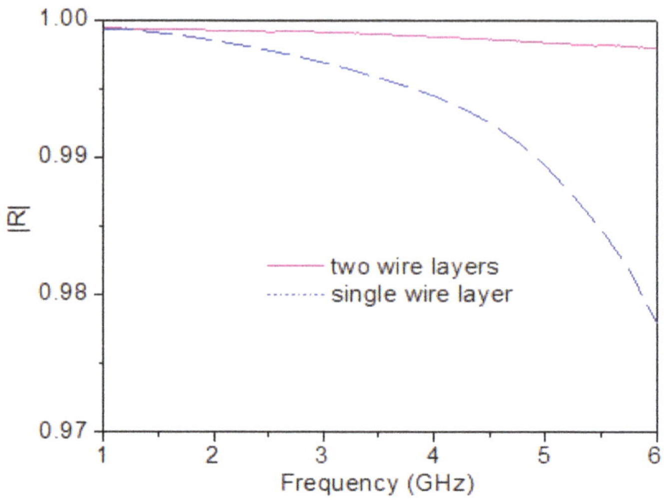

Figure 4.7. Plots of the total reflection coefficient for the single- and two-layer copper wire gratings with $a = 0.002$ m and $r_0 = 0.0001$ m (*p*-polarization).

number of layers in the grating. According to the approach proposed in [26], some restrictions on the total transmission and reflection coefficients must be imposed. These are based on the assumption that the gratings under considerations are flat-like structures. Then, one can assume that the total reflection R and transmission T coefficients satisfy the equalities

$$\left.\begin{array}{l} 1 + R = T, \\ |R|^2 + |T|^2 = 1 \end{array}\right\}. \tag{4.65}$$

Then, one can define the coefficients R and T in the most general form as follows [26]:

$$\left.\begin{array}{l} R = -\cos(\psi)e^{-i\psi}, \\ T = i\sin(\psi)e^{-i\psi} \end{array}\right\}, \tag{4.66}$$

where ψ is the phase, which can be determined relatively arbitrarily. Let $\psi = kn_{\text{eff}}h$, where n_{eff} is the effective refractive index of gratings and h is the thickness of the grating ($h = a$ in the case of a single-layer grating and $h = 2a$ in the case of a two-layer grating, which is being considered). Solving equation (4.66) with regard to n_{eff} finally gives

$$e^{-2ikn_{\text{eff}}a} = -(1 + 2R). \tag{4.67}$$

As shown in [28], solving equation (4.67) gives

$$n_{\text{eff}} = \frac{i}{4k_0a}\ln|1 + 2R| + \frac{2m\pi - \varphi}{4k_0a}, \quad m \in Z, \tag{4.68}$$

where $\varphi = Arg(-1 - 2R)$, and the main branch ($m = 0$) is to be used in equation (4.68) for calculating the complex effective refractive index n_{eff} of the grating under consideration—simply put, the calculation of the refractive index n_{eff} for the main branch is completely consistent with the effective medium approach of work in [29].

In order to obtain microwave approximations for the complex effective relative permittivity $\varepsilon_{\mathrm{eff}}$ and permeability μ_{eff}, one should take into account that in microwave applications with regard to, for example, microwave antenna ground systems, and to various forms of shielding, it is often convenient to express the characteristics (e.g. R and T) of the gratings in terms of a surface impedance [20]

$$R = \frac{-1}{1 + 2Z_{\mathrm{eff}}}, \tag{4.69}$$

where Z_{eff} is the equivalent normalized shunt impedance. At the same time, Z_{eff} is the effective normalized impedance of the gratings, given as $Z_{\mathrm{eff}} = \sqrt{\mu_{\mathrm{eff}}/\varepsilon_{\mathrm{eff}}}$. Substituting this and $n_{\mathrm{eff}} = \sqrt{\mu_{\mathrm{eff}}\varepsilon_{\mathrm{eff}}}$ into equation (4.69) finally gives expressions for the complex effective relative permittivity $\varepsilon_{\mathrm{eff}}$ and permeability μ_{eff} of the grating under consideration:

$$\left.\begin{aligned} \varepsilon_{\mathrm{eff}} &= -\frac{2R}{1 + R} n_{\mathrm{eff}}, \\ \mu_{\mathrm{eff}} &= \frac{n_{\mathrm{eff}}^2}{\varepsilon_{\mathrm{eff}}}. \end{aligned}\right\} \tag{4.70}$$

Note that equations (4.68) and (4.69) can also be used for a grating with any number of layers of parallel circular cylinders. To validate the model defined by equation (4.63) and equations (4.65)–(4.70), one can compare the results of modeling using equations (4.63)–(4.64) and equations (4.66)–(4.70) with the results of FDTD simulations as well as those obtained using the mesoscopic models [10].

Note that the commonly adopted S-parameter retrieval procedure proposed in [30] was implemented in this study, in order to use the results of FDTD simulations. The results of the above numerical comparison are presented in figures 4.8 and 4.9. These show the improved accuracy in the case of the models proposed in this study compared to the models in [10]. The latter difference is quite considerable at GHz frequencies and is caused by the impossibility of taking into account the near field couplings in homogenizing multilayer composites by means of using any average procedure [31].

Figures 4.6–4.9 allow us to conclude that layered metal wire gratings/arrays for the case of s-polarization can be considered as a broadband structure in terms of electromagnetic metamaterials [18]. Moreover, the grating/array can be used to excite and guide microwave surface plasmons [8].

Consider the dependences of the total reflection coefficient and the real parts of the effective parameters on the inclusion (metal) volume fraction $F = \pi(r_0/a)^2$. Such dependences allow us to analyse the accuracy of the approximate analytical solutions in terms of EMT [8, 18, 19]. The corresponding dependences are depicted in figures 4.10–4.12 at 2 GHz frequency. One can see from figure 4.10 that equation (4.63) approximates well the reflectivity of the two-layer grating in a wide frequency

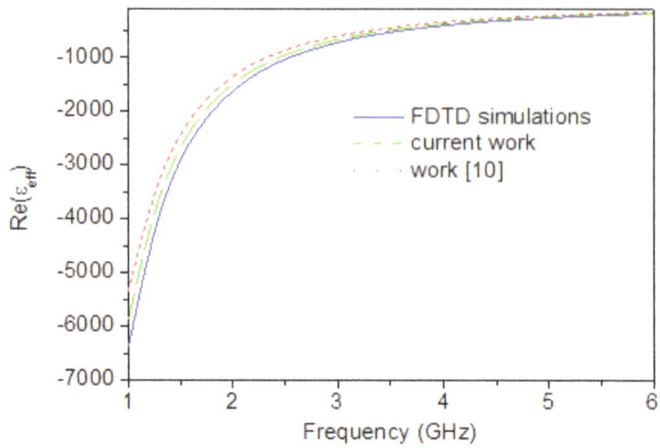

Figure 4.8. Plots of the real part of the effective relative permittivity for two-layer copper wire grating with $a = 0.002$ m and $r_0 = 0.0001$ m (p-polarization).

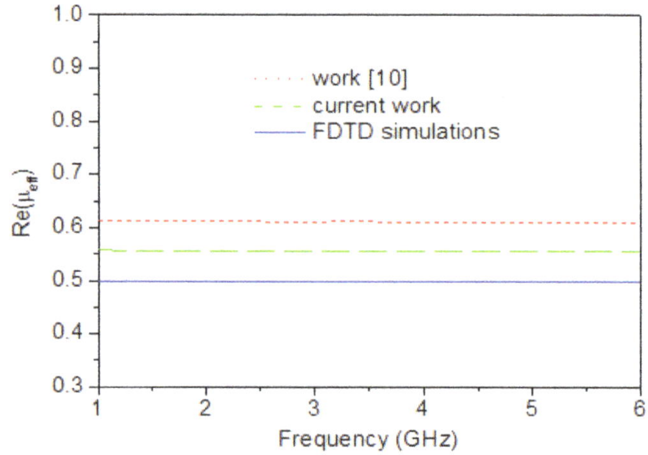

Figure 4.9. Plots of the real part of the effective relative permeability for two-layer copper wire grating with $a = 0.002$ m and $r_0 = 0.0001$ m (p-polarization).

range at the microwave frequency. At the same time, as one can observe from figures 4.11 and 4.12, the accuracy of the approximation for the effective relative permittivity and permeability of the grating may be rather unacceptable for large values of the wire volume fraction: $F > 0.036$ (figures 4.7 and 4.8). This is due to the assumption made when solving the above scattering problem: $|x - d| > >\lambda$.

4.3.2 The case of *s*-polarization

Now, consider the case of s-polarization. The expression for the reflection coefficient R_s is given by [21]

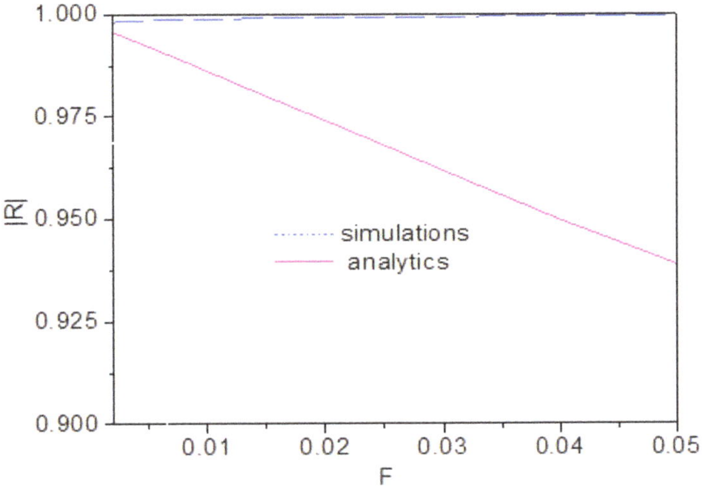

Figure 4.10. Dependence of the total reflectivity on the wire volume fraction F for two-layer copper wire grating at 2 GHz (p-polarization).

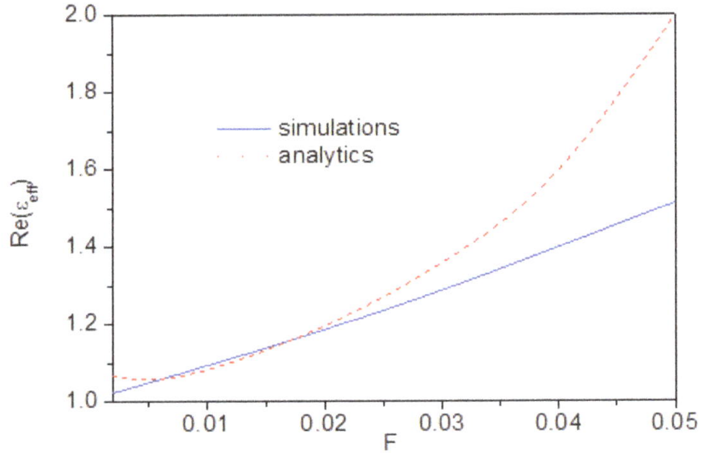

Figure 4.11. Dependence of the real part of effective relative permittivity on the wire volume fraction F for two-layer copper wire grating at 2 GHz (p-polarization).

$$R_s = \frac{E_\varphi^s}{E_\varphi^p} = \frac{-E_x^s \sin\varphi + E_y^s \cos\varphi}{E_\varphi^p}, \tag{4.71}$$

where E_φ^p, the φ-component of the incident field (equation (4.41)) is given by

$$\left.\begin{aligned} E_\varphi^p &= E_{0\varphi} e^{ik(x \sin\theta \cos\varphi + y \sin\theta \sin\varphi + z \cos\theta)}, \\ E_{x0} &= -E_{0\varphi} \sin\varphi, \\ E_{y0} &= E_{0\varphi} \cos\varphi \end{aligned}\right\}. \tag{4.72}$$

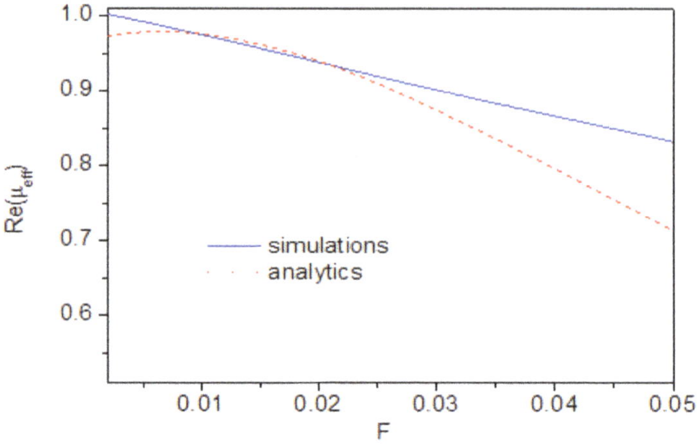

Figure 4.12. Dependence of the real part of effective relative permeability on the wire volume fraction F for two-layer copper wire array at 2 GHz (p-polarization).

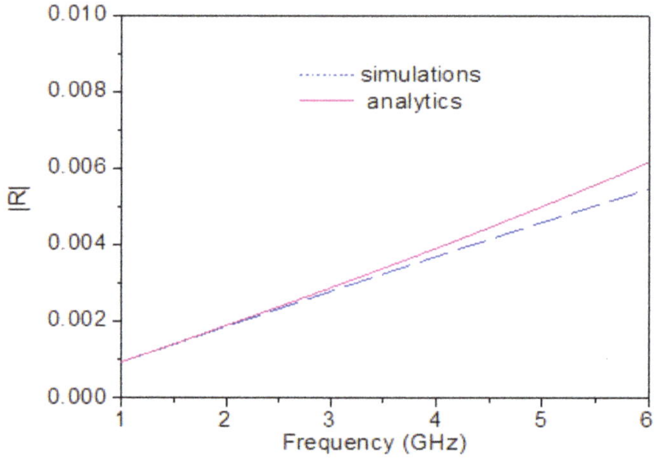

Figure 4.13. Plots of the total reflection coefficient for two-layer copper wire grating with $a = 0.002$ m and $r_0 = 0.0001$ m (s-polarization).

We compare the results of modeling using equations (4.60), (4.71), (4.72) and the results of numerical simulation using the commercial electromagnetic simulator PLANC FD for the total reflection coefficient using the same parameters as used in the case of p-polarization. The results of such a comparison are depicted in figure 4.13, which shows a good agreement of the results of analytical modeling and those of numerical simulations. One can see from this figure that, unlike the case of p-polarization, there is almost full-wave transmission ($R_s \approx 0$) in the case of s-polarization. However, this result is rather a validation for the obtained analytical solution than a principally new result, because that was quite predictable.

Let us learn the complex effective electromagnetic response for the considered polarization, as done for the case p-polarization using equations (4.68)–(4.70). To validate the model defined by equation (4.71), one can compare the results of analytical modeling using equations (4.71), (4.72) and (4.66)–(4.70) with the results of FDTD simulations using the PLANC FD software, as done for the case p-polarization. The results of such a comparison are presented in figures 4.14 and 4.15, which provide the plots of the real part of complex effective parameters with $a = 0.002$ m and $r_0 = 0.0001$ m. The figures show good agreement between the proposed analytical theory and the numerical simulations.

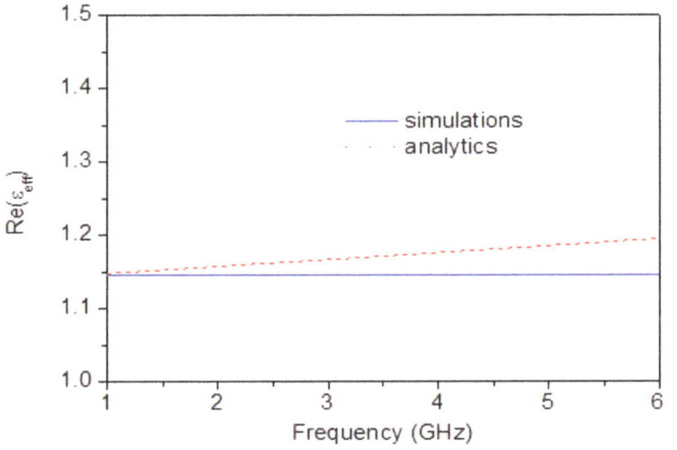

Figure 4.14. Plots of the real part of the effective relative permittivity for two-layer copper wire grating with $a = 0.002$ m and $r_0 = 0.0001$ m (s-polarization).

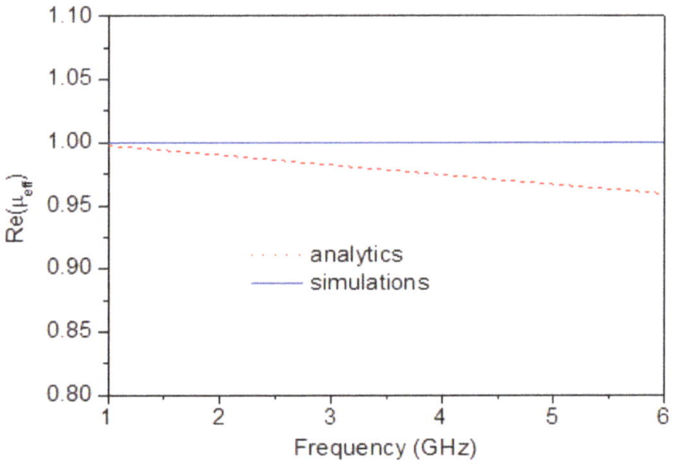

Figure 4.15. Plots of the real part of the effective relative permeability for two-layer copper wire grating with $a = 0.002$ m and $r_0 = 0.0001$ m (s-polarization).

As one can observe from figures 4.14 and 4.15, the grating under consideration behaves as an artificial dielectric in a wide range at microwave frequencies. In fact, the grating can be used to create compact multi-band and multi-directional microwave patch antennas with low-intensity near-fields [6, 7]. Moreover, such antennas should also have improved characteristics, such as power gain and efficiency, compared to patch antennas with dielectric substrates [12]. The foundations of miniaturizations of patch antennas with two-layer wire grating embedded in dielectric substrates are well presented in [6].

Similar to the case of *s*-polarized electromagnetic waves, it is logical to consider the dependences of the total reflection coefficient and the effective parameters of the two-layer wire grating on the wire volume fraction. The corresponding graphs of the dependences are depicted in figures 4.16–4.18 at a frequency of 2 GHz. These figures confirm a good approximation of the reflectivity of the grating at microwave frequencies for the case of *s*-polarization by the analytical models presented by equations (4.71) and (4.72). They also show a decrease in the accuracy of the approximations for the effective electromagnetic response for large values of the wire volume fraction for the case *s*-polarized waves.

Note that the values of $\mathrm{Im}(\mu_{\mathrm{eff}})$ are small for both polarizations, while the values of $\mathrm{Im}(\varepsilon_{\mathrm{eff}})$ are only small for the case of *s*-polarization. However, the values of the tangent of the magnetic loss angle are small for each of the considered polarizations.

The above studies reveal that there are no fundamental differences between the optical and scattering properties of the single- and two-layer wire gratings/arrays. Therefore, the next section will be entirely focused on one of the most widespread applications—the implementation of the aforementioned gratings/arrays in a designing

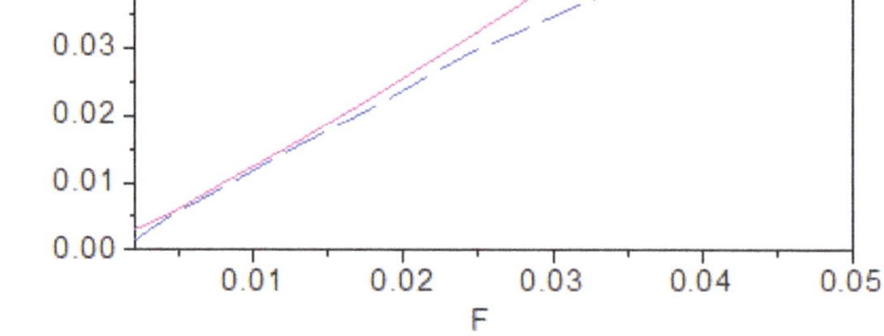

Figure 4.16. Dependence of the total reflectivity on the wire volume fraction *F* for two-layer copper wire array at 2 GHz (*s*-polarization).

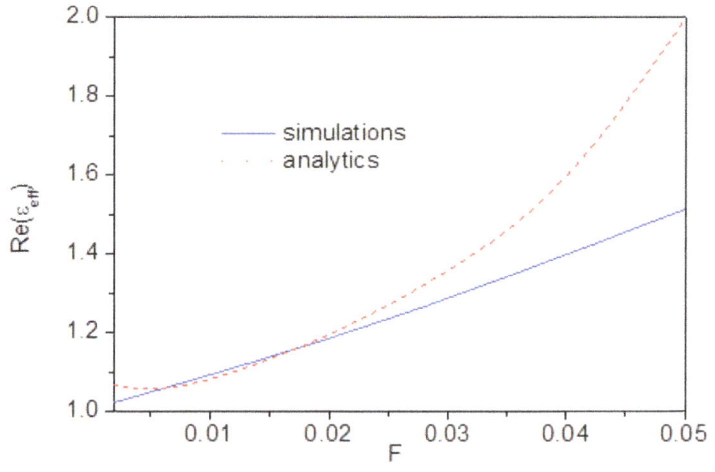

Figure 4.17. Dependence of the real part of effective relative permittivity on the wire volume fraction F for two-layer copper wire array at 2 GHz (s-polarization).

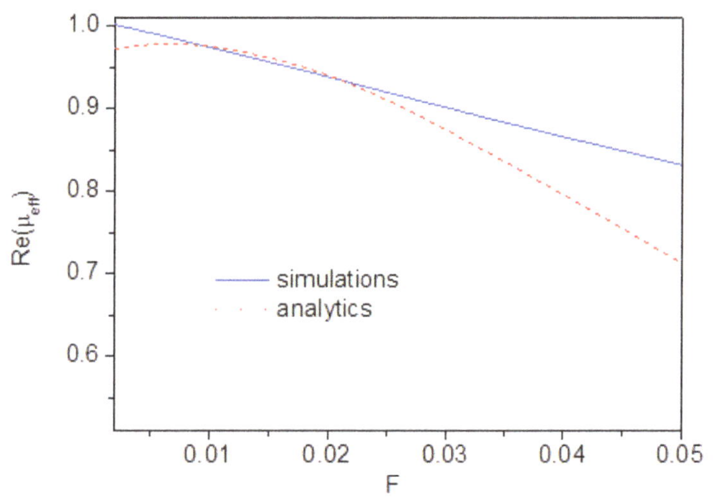

Figure 4.18. Dependence of the real part of effective relative permeability on the wire volume fraction F for two-layer copper wire array at 2 GHz (s-polarization).

high-impedance surfaces. Note that the respective research is conducted in accordance with and within the framework of the basic principles of EMT reported in [18].

4.4 Two-layer wire grating as a high-impedance surface

The single- and two-layer gratings considered above are highly reflective structures. However, the real part of the total reflection coefficient of these gratings is near −1 in a broad range at microwave frequencies. In fact, this is the behavior of an electric wall. It is well known that a grating consisting of identical dipoles arranged near a metallic screen behaves like an electric wall over most of the region of microwave frequency. At the same time, the phase of the reflection coefficient can vary in a resonance manner over a narrow frequency range, depending on the resonance characteristics of the grating elements (or the dipoles) as a result of interaction between the grating and the metallic surface (above which the grating is placed). In that event, the reflection coefficient is equal to $+1$ at the resonance frequency [32]. Substrate-like grating of infinitely long continuous circular wires is an example of high-impedance structure.

Consider the two-layer wire grating as depicted in figure 4.5 but also backed by a conducting wall at the plane $z = -d$. The wall region ($z < -d$) is taken to be inhomogeneous with permittivity ε_g and conductivity σ_g. The first wire layer is located in the plane $z = -h$ and the second one is at the plane $z = 0$ (figure 4.19).

As previously, one can assume that the initial electric field $\overrightarrow{E}^p$ of arbitrary polarization (without a time-harmonic time factor) is defined by equation (4.41) while the currents on the first and second wire layers of the grating are defined by equation (4.42). The fields $\overrightarrow{E}^w$ and $\overrightarrow{H}^w$ due to the above currents and their interactions with the lower half-space ($z > -d$) are defined by [33]

$$\left.\begin{aligned} \overrightarrow{E}^w &= k^2\overrightarrow{\Pi} + \nabla(\nabla\cdot\overrightarrow{\Pi}) - i\mu_0\omega\,\nabla\times\overrightarrow{\Pi}^*, \\ \overrightarrow{H}^w &= k^2\overrightarrow{\Pi}^* + \nabla(\nabla\cdot\overrightarrow{\Pi}^*) + i\varepsilon_0\omega\,\nabla\times\overrightarrow{\Pi} \end{aligned}\right\}, \tag{4.73}$$

where the Hertz potentials $\overrightarrow{\Pi}$ and $\overrightarrow{\Pi}^*$ are defined by equation (4.43) and given by

$$\overrightarrow{\Pi}^* = (\Pi^*_{x1} + \Pi^*_{x2})\overrightarrow{x_0}, \tag{4.74}$$

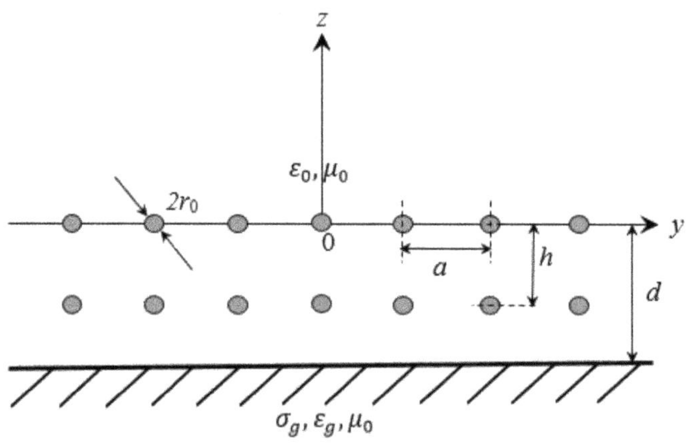

Figure 4.19. Two-layer wire grating located over a homogeneous conducting ground.

where

$$\left.\begin{aligned}
\Pi_{x1} &= \frac{\mu_0\omega}{4ik^2}\sum_{m=-\infty}^{\infty}\sum_{n=-\infty}^{\infty} A_m\left\{e^{-\Gamma^{(1)}_{mn}|z|} + R_{mn}e^{-\Gamma^{(1)}_{mn}|z+2d|}\right\}\frac{e^{i(k_x x + k_y y)}}{\Gamma^{(1)}_{mn}},\\
\Pi_{x2} &= \frac{\mu_0\omega}{4ik^2}\sum_{m=-\infty}^{\infty}\sum_{n=-\infty}^{\infty} B_n\left\{e^{-\Gamma^{(2)}_{mn}|z+h|} + r_{mn}e^{-\Gamma^{(2)}_{mn}|z+2d-h|}\right\}\frac{e^{i(k_x x + k_y y)}}{\Gamma^{(2)}_{mn}},\\
\Pi^{*}_{x1} &= \frac{\mu_0\omega}{4ik^2}\sum_{m=-\infty}^{\infty}\sum_{n=-\infty}^{\infty} A_m S_{mq} e^{-\Gamma^{(1)}_{mn}|z+2d|}\frac{e^{i(k_x x + k_y y)}}{\Gamma^{(1)}_{mn}},\\
\Pi^{*}_{x2} &= \frac{\mu_0\omega}{4ik^2}\sum_{m=-\infty}^{\infty}\sum_{n=-\infty}^{\infty} B_n s_{mq} e^{-\Gamma^{(2)}_{mn}|z+2d-h|}\frac{e^{i(k_x x + k_y y)}}{\Gamma^{(2)}_{mn}}
\end{aligned}\right\}, \tag{4.75}$$

with

$$k_x = \sqrt{\left(\frac{2\pi m}{a}\right)^2 + (k\sin\theta\cos\varphi)^2},\ k_y = \left(\frac{2\pi n}{a}\right) + k\sin\theta\sin\varphi.$$

The coefficients R_{mn}, r_{mn}, S_{mn} and s_{mn} can be obtained by satisfying the continuity condition of the tangential fields E_x^w, H_x^w, E_y^w and H_y^w across the interface at the plane $z = -d$. Making rather routine algebraic manipulations in matching the above boundary conditions finally leads to the following equalities:

$$\left.\begin{aligned}
R_{mn} &= \frac{k^2(\Gamma^{(1)}_{mn} - \Gamma_g K_x)(\Gamma^{(1)}_{mn} - \Gamma_g\varepsilon_\sigma K_x) + (k_x k_y)^2(1 - K_x)^2}{k^2(\Gamma^{(1)}_{mn} - \Gamma_g K_x)(\Gamma^{(1)}_{mn} - \Gamma_g\varepsilon_\sigma K_x) - (k_x k_y)^2(1 - K_x)^2},\\
r_{mn} &= \frac{-2i\varepsilon_0\omega k_x k_y\Gamma^{(1)}_{mn}}{k^2(\Gamma^{(1)}_{mn} - \Gamma_g K_x)(\Gamma^{(1)}_{mn} - \Gamma_g\varepsilon_\sigma K_x) - (k_x k_y)^2(1 - K_x)^2},\\
S_{mn} &= \frac{k^2(\Gamma^{(2)}_{mn} - \Gamma_g K_y)(\Gamma^{(2)}_{mn} - \Gamma_g\varepsilon_\sigma K_y) + (k_x k_y)^2(1 - K_y)^2}{k^2(\Gamma^{(2)}_{mn} - \Gamma_g K_y)(\Gamma^{(2)}_{mn} - \Gamma_g\varepsilon_\sigma K_y) - (k_x k_y)^2(1 - K_y)^2},\\
s_{mn} &= \frac{2i\varepsilon_0\omega k_x k_y\Gamma^{(2)}_{mn}}{k^2(\Gamma^{(2)}_{mn} - \Gamma_g K_y)(\Gamma^{(2)}_{mn} - \Gamma_g\varepsilon_\sigma K_y) - (k_x k_y)^2(1 - K_y)^2},
\end{aligned}\right\} \tag{4.76}$$

where

$$K_x = \frac{k^2 - k_x^2}{k_g^2 - k_x^2},\ K_y = \frac{k^2 - k_y^2}{k_g^2 - k_y^2},\ k_g = \frac{\sqrt{i\mu_0\omega(\sigma_g + i\varepsilon_g\omega)}}{i},\ \varepsilon_\sigma = \frac{\sigma_g + i\varepsilon_g\omega}{i\varepsilon_0\omega},$$
$$\Gamma_g = i\sqrt{k_g^2 - k_x^2 - k_y^2}.$$

To find the coefficients A_m and B_n, it is necessary to impose the boundary conditions for the expressions of the total fields at the wires. This requires finding expressions for the total fields. Let us do this for the case of expression of the total electric field given by

$$\overrightarrow{E}^{\text{tot}} = \overrightarrow{E}^{p}(1 + R_{fr} \cdot \overrightarrow{F}) + \overrightarrow{E}^{w}, \tag{4.77}$$

where R_{fr} is the Fresnel reflection coefficient for the interface $z = -d$ in the absence of wires, while $\overrightarrow{F}$ is some vector function to be found from the boundary conditions. However, in this study, another procedure to find the expression for $\overrightarrow{E}^{\text{tot}}$ (and $\overrightarrow{H}^{\text{tot}}$) is chosen as proposed in [33]. This is because the representation of the full fields in the form of equation (4.77) is more convenient when considering any of the 2D cases. According to the procedure in [33], we should introduce the Hertz potentials for the incident and reflected fields that would exist in the absence of wires. Thus, for the region $z > -d$, these vectors are to be defined in the following form:

$$\left.\begin{aligned}
\pi_{x1} &= \alpha_1\{e^{\Gamma_0 z} + R_{00}e^{-\Gamma_0(z+2d)}\}e^{i(k_x x+k_y y)},\\
\pi_{x2} &= \alpha_2\{e^{\Gamma_0 z} + r_{00}e^{-\Gamma_0(z+2d)}\}e^{i(k_x x+k_y y)},\\
\pi^*_{x1} &= \alpha_1 S_{00}e^{-\Gamma_0(z+2d)}e^{i(k_x x+k_y y)},\\
\pi^*_{x2} &= \alpha_2 s_{00}e^{-\Gamma_0(z+2d)}e^{i(k_x x+k_y y)}
\end{aligned}\right\}, \tag{4.78}$$

where $\Gamma_0 = i\sqrt{k^2 - k_{0x}^2 - k_{0y}^2}$, $k_{0x} = k \sin\theta \cos\varphi$, $k_{0y} = k \cos\theta \sin\varphi$, α_1 and α_2 are the factors prescribed by the incident field. To obtain the expressions for the coefficient α_1 and α_2, we have to substitute equations (4.78), (4.43) and (4.74) into the first equation in equation (4.73). The appropriate algebraic manipulations finally give

$$\left.\begin{aligned}
E_{0x} &= (k^2 - k_{0x}^2)\alpha_1 - k_{0x}k_{0y}\alpha_2,\\
E_{0y} &= \left(k^2 - k_{0y}^2\right)\alpha_2 - k_{0x}k_{0y}\alpha_1
\end{aligned}\right\}, \tag{4.79}$$

where E_{0x} and E_{0y} are just the tangential components of the electric field of the incident field at the origin, as indicated by equation (4.41). Solving equation (4.79) with regard to the variables α_1 and α_2 finally gives:

$$\left.\begin{aligned}
\alpha_1 &= -\frac{\left(k^2 - k_{0y}^2\right)E_{0x} + k_{0x}k_{0y}E_{0y}}{k^2\Gamma_0^2},\\
\alpha_2 &= -\frac{k_{0x}k_{0y}E_{0x} + (k^2 - k_{0x}^2)E_{0y}}{k^2\Gamma_0^2}
\end{aligned}\right\}. \tag{4.80}$$

Substituting equations (4.74), (4.75) and (4.78) into equation (4.73), we obtain

$$\left.\begin{aligned}
&E_x^{\text{tot}} = E_{x1}^{\text{tot}} + E_{x2}^{\text{tot}},\\
&E_{x1}^{\text{tot}} = (k^2 - k_x^2)\left\{\alpha_1(1 + R_{00}e^{-2\Gamma_0 d})e^{-\Gamma_0 r_0}\right.\\
&\left.\quad + \frac{\omega\mu_0}{4i\pi k^2}\sum_{m=-\infty}^{\infty}\sum_{n=-\infty}^{\infty} A_m\left(1 + R_{mn}e^{-2\Gamma_{mn}^{(1)}d}\right)\frac{e^{-\Gamma_{mn}^{(1)}r_0}}{\Gamma_{mn}^{(1)}}\right\}e^{i(k_x x + k_y y)},\\
&E_{x2}^{\text{tot}} = (k^2 - k_x^2)\left\{\alpha_2(1 + r_{00}e^{-2\Gamma_0 d})e^{-\Gamma_0(r_0+h)}\right.\\
&\left.\quad + \frac{\omega\mu_0}{4i\pi k^2}\sum_{m=-\infty}^{\infty}\sum_{n=-\infty}^{\infty} B_n\left(1 + r_{mn}e^{-2\Gamma_{mn}^{(2)}d}\right)\frac{e^{-\Gamma_{mn}^{(2)}(r_0+h)}}{\Gamma_{mn}^{(2)}}\right\}e^{i(k_x x + k_y y)}.
\end{aligned}\right\} \tag{4.81}$$

Now, one can come to the boundary conditions at the wires:

$$\left.\begin{aligned}
E_x^{\text{tot}}\big|_{\substack{y=0\\ z=r_0}} &= Z_i I_{xn}^{(1)}\big|_{\substack{y=0\\ z=r_0}}\\
E_x^{\text{tot}}\big|_{\substack{y=0\\ z=-r_0-a}} &= Z_i I_{xm}^{(2)}\big|_{\substack{y=0\\ z=-r_0-a}}
\end{aligned}\right\}, \tag{4.82}$$

where Z_i is the internal impedance per unit length of the wires given by equation (4.9). Substituting equation (4.81) into equation (4.82) leads to the equations with regard to the variables A_m and B_n in the form

$$\left.\begin{aligned}
&\alpha_1 k^2(1 + R_{00}e^{-2\Gamma_0 d})e^{-\Gamma_0 r_0} + \frac{\omega\mu_0}{4i\pi}\sum_{m=-\infty}^{\infty}\sum_{n=-\infty}^{\infty} A_m\left(1 + R_{mn}e^{-2\Gamma_{mn}^{(1)}d}\right)\frac{e^{-\Gamma_{mn}^{(1)}r_0}}{\Gamma_{mn}^{(1)}}\\
&= \frac{Z_i}{1 - \sin^2\theta\cos^2\varphi}\sum_{m=-\infty}^{\infty} A_m e^{ina\sin\varphi},\\
&\alpha_2 k^2(1 + r_{00}e^{-2\Gamma_0 d})e^{-\Gamma_0(r_0+2h)} + \frac{\omega\mu_0}{4i\pi}\sum_{m=-\infty}^{\infty}\sum_{n=-\infty}^{\infty} B_n\left(1 + r_{mn}e^{-2\Gamma_{mn}^{(2)}d}\right)\frac{e^{-\Gamma_{mn}^{(2)}(r_0+2h)}}{\Gamma_{mn}^{(2)}}\\
&= \frac{Z_i}{1 - \sin^2\theta\cos^2\varphi}\sum_{n=-\infty}^{\infty} B_n e^{ima\sin\varphi}.
\end{aligned}\right\} \tag{4.83}$$

Taking into account that $e^{-ikr} = \sum_{s=-\infty}^{\infty}(-i)^s J_s(kr)$, equation (4.83) can be reduced to the system of equations

$$\left.\begin{aligned}
&\alpha_1 k^2(-i)^m(1 + R_{00}e^{-2\Gamma_0 d})J_m(\Gamma_0 r_0) + \frac{\omega\mu_0}{4i\pi}A_m\sum_{n=-\infty}^{\infty}\left(1 + R_{mn}e^{-2\Gamma_{mn}^{(1)}d}\right)\frac{e^{-\Gamma_{mn}^{(1)}r_0}}{\Gamma_{mn}^{(1)}}\\
&= \frac{Z_i}{1 - \sin^2\theta\cos^2\varphi}A_m e^{ina\sin\varphi},\\
&\alpha_2 k^2(-i)^n(1 + r_{00}e^{-2\Gamma_0 d})J_n(\Gamma_0(r_0 + 2h)) + \frac{\omega\mu_0}{4i\pi}B_n\sum_{m=-\infty}^{\infty}\left(1 + r_{mn}e^{-2\Gamma_{mn}^{(2)}d}\right)\frac{e^{-\Gamma_{mn}^{(2)}(r_0+2h)}}{\Gamma_{mn}^{(2)}}\\
&= \frac{Z_i}{1 - \sin^2\theta\cos^2\varphi}B_n e^{ima\sin\varphi}.
\end{aligned}\right\} \tag{4.84}$$

Solving equation (4.84) with respect to the coefficients A_m and B_n finally gives

$$\left.\begin{aligned} A_m &= \frac{\dfrac{Z_i e^{ima}\sin\varphi}{1-\sin^2\theta\cos^2\varphi} - \dfrac{\omega\mu_0}{4i\pi}\sum\limits_{n=-\infty}^{\infty}\left(1+R_{mn}e^{-2\Gamma_{mn}^{(1)}d}\right)\dfrac{e^{-\Gamma_{mn}^{(1)}r_0}}{\Gamma_{mn}^{(1)}}}{\alpha_1 k^2(-i)^m\left(1+R_{00}e^{-2\Gamma_0 d}\right)J_m(\Gamma_0 r_0)}, \\ B_n &= \frac{\dfrac{Z_i e^{ina}\sin\varphi}{1-\sin^2\theta\cos^2\varphi} - \dfrac{\omega\mu_0}{4i\pi}\sum\limits_{m=-\infty}^{\infty}\left(1+r_{mn}e^{-2\Gamma_{mn}^{(2)}d}\right)\dfrac{e^{-\Gamma_{mn}^{(2)}(r_0+2h)}}{\Gamma_{mn}^{(2)}}}{\alpha_2 k^2(-i)^n\left(1+r_{00}e^{-2\Gamma_0 d}\right)J_n(\Gamma_0(r_0+2h))} \end{aligned}\right\}. \tag{4.85}$$

To be able to calculate the total reflection coefficient for the p- and s-polarizations, it is necessary to obtain the expressions for all three components of the total field $\overrightarrow{E}^{\text{tot}}$. To obtain the expressions for E_y^{tot} and E_z^{tot} (E_x^{tot} was already presented by equation (4.81)), equation (4.73) is to be used. Thus, substituting equations (4.74)–(4.78) into equation (4.73) finally gives

$$\left.\begin{aligned} E_y^{\text{tot}} &= E_{y1}^{\text{tot}} + E_{y2}^{\text{tot}}, \\ E_{y1}^{\text{tot}} &= -\alpha\left(k_x k_y + \left[k_x k_y R_{00} - i\mu_0\omega\Gamma_0 S_{00}\right]e^{-2\Gamma_0 d}\right)e^{-\Gamma_0 z}e^{i(k_x x+k_y y)} - \\ &- \frac{\mu_0\omega}{4ik^2}\left(\sum_{m=-\infty}^{\infty}\sum_{n=-\infty}^{\infty} A_m\left(k_x k_y + \left[S_{mn} + \frac{R_{mn}}{\Gamma_{mn}^{(1)}}\right]e^{-2\Gamma_{mn}^{(1)}d}\right)\right)e^{-\Gamma_{mn}^{(1)}z}e^{i(k_x x+k_y y)}, \\ E_{y2}^{\text{tot}} &= -\beta\left(k_x k_y + \left[k_x k_y r_{00} + \Gamma_0 s_{00}\right]e^{-2\Gamma_0 d}\right)e^{-\Gamma_0(z+h)}e^{i(k_x x+k_y y)} - \\ &- \frac{\mu_0\omega}{4ik^2}\left(\sum_{m=-\infty}^{\infty}\sum_{n=-\infty}^{\infty} B_n\left(k_x k_y - \left[i\mu_0\omega s_{mn} - \frac{r_{mn}}{\Gamma_{mn}^{(2)}}\right]e^{-2\Gamma_{mn}^{(2)}d}\right)\right)e^{-\Gamma_{mn}^{(2)}(z+h)}e^{i(k_x x+k_y y)} \end{aligned}\right\} \tag{4.86}$$

and

$$\left.\begin{aligned} E_z^{\text{tot}} &= E_{z1}^{\text{tot}} + E_{z2}^{\text{tot}}, \\ E_{z1}^{\text{tot}} &= -\alpha_1\Gamma_0\left(k_x + [k_x R_{00} - i\mu_0\omega S_{00}]e^{-2\Gamma_0 d}\right)e^{-\Gamma_0 z}e^{i(k_x x+k_y y)} + \\ &+ \frac{\mu_0\omega}{4k^2}\left(\left(\sum_{m=-\infty}^{\infty}\sum_{n=-\infty}^{\infty} A_m\left(k_x + [k_x R_{mn} - \mu_0\omega S_{mn}]e^{-2\Gamma_{mn}^{(1)}d}\right)\right)e^{-\Gamma_{mn}^{(1)}z}e^{i(k_x x+k_y y)}\right), \\ E_{z2}^{\text{tot}} &= -\alpha_2\Gamma_0\left(k_x + [k_x r_{00} - i\mu_0\omega s_{00}]e^{-2\Gamma_0 d}\right)e^{-\Gamma_0(z+h)}e^{i(k_x x+k_y y)} + \\ &+ \frac{\mu_0\omega}{4k^2}\left(\left(\sum_{m=-\infty}^{\infty}\sum_{n=-\infty}^{\infty} B_n\left(k_x + [k_x r_{mn} - \mu_0\omega s_{mn}]e^{-2\Gamma_{mn}^{(2)}d}\right)\right)e^{-\Gamma_{mn}^{(2)}(z+h)}e^{i(k_x x+k_y y)}\right) \end{aligned}\right\}. \tag{4.87}$$

Similarly, the summations in equations (4.58) and (4.85) converge relatively poorly. To improve the convergence of the above sums, the following approximations of work [33] will be used in this study:

$$\left.\begin{aligned} P_m &= \sum_{n=-\infty}^{\infty}\left(1+R_{mn}e^{-2\Gamma_{mn}^{(1)}d}\right)\frac{e^{-\Gamma_{mn}^{(1)}r_0}}{\Gamma_{mn}^{(1)}} \approx \frac{a}{\pi}\ln\left(\frac{a}{2\pi r_0}\right) + \frac{e^{-\Gamma_{m0}^{(1)}r_0}}{\Gamma_{m0}^{(1)}} + \sum_{n=-\infty}^{\infty}\frac{R_{mn}}{\Gamma_{mn}^{(1)}}e^{-\Gamma_{mn}^{(1)}(r_0+2d)}, \\ Q_n &= \sum_{m=-\infty}^{\infty}\left(1+r_{mn}e^{-2\Gamma_{mn}^{(2)}d}\right)\frac{e^{-\Gamma_{mn}^{(2)}r_0}}{\Gamma_{mn}^{(2)}} \approx \frac{a}{\pi}\ln\left(\frac{a}{2\pi r_0}\right) + \frac{e^{-\Gamma_{0n}^{(2)}r_0}}{\Gamma_{0n}^{(2)}} + \sum_{m=-\infty}^{\infty}\frac{r_{mn}}{\Gamma_{mn}^{(2)}}e^{-\Gamma_{mn}^{(2)}(r_0+2d-2h)} \end{aligned}\right\}. \tag{4.88}$$

The above approximations are in good agreement with the limitations of the EMT, taken into account throughout the study. Indeed, the numerical calculations show that it is sufficient to keep the terms of the sums for $m, n = 0, \pm 1 \pm 2$ in equation (4.86), in order to obtain good accuracy for the proposed mathematical model for $F \leqslant 0.1$.

It must be mentioned that the expressions for the reflection coefficients R_p (the case of p-polarization) and R_s (the case of s-polarization) are defined by equations (4.63), (4.64), (4.71) and (4.72), where $E_x^s = E_x^{\text{tot}} - (k^2 - k_{0x}^2)\alpha_1 + k_{0x}k_{0y}\alpha_2$, $E_y^s = (k^2 - k_{0y}^2)\alpha_2 + k_{0x}k_{0y}\alpha_1$, $E_z^s = E_z^{\text{tot}}$. To assess the reflectance of the wire structure shown in figure 4.19, it is logical to consider the behavior of the magnitude and phase of the total reflection coefficient of this structure as functions of angles for the case of p-polarization. Indeed, as observed in the previous paragraph exclusively for this type of polarization, the two-layer wire gratings structure behaves like a mirror. The corresponding plots of the dependence of magnitude and phase of the reflection coefficient R_p versus the angle θ at fixed values of the angle φ (4.0° and 45°) are shown in figures 4.20 and 4.21. These plots are built for copper wires at a frequency of 100 MHz and for the following parameters:$r_0/a = 0.01$, $h/r_0 = 3.0$, $d/a = 0.1$, $a/\lambda = 0.2$, $\sigma_g = 10^2 S/m$, $\varepsilon_g = 1.8$.

Note that the above parameters set for the numerical simulations are taken from [33] in order to compare the reflective properties of the considered wire structure and that in [33], where the upper single-layer wire grating is rotated with respect to the lower one by 90°. Note that similar structures have been considered experimentally in [34, 35]. Although finite-size samples are used in [34, 35], the results of these works can be used in characterizing the wire structure under consideration and that in [33]. This is due to the fact that $\lambda_m < < L$ ($6 \div 8\lambda_m < L$ in the experiments), where λ_m is the wavelength in the host material of samples and L is their length. Numerical simulations performed for these experiments

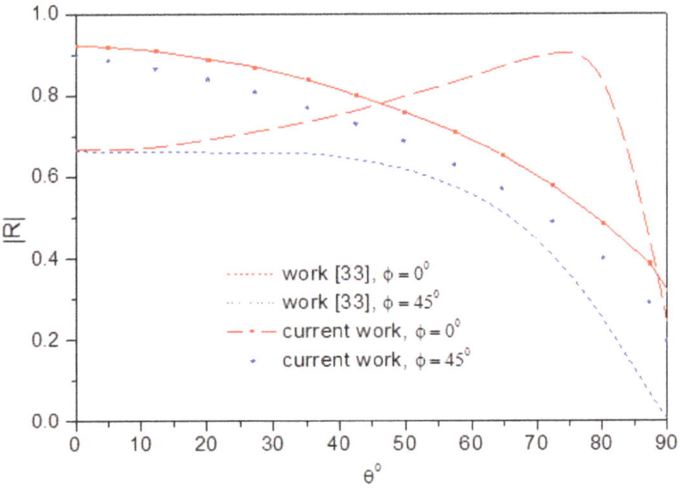

Figure 4.20. Magnitude of R_p plotted against θ at a frequency of 100 Hz for two different values of $\varphi = 0^0, 45^0$ and $r_0/a = 0.01$, $h/r_0 = 3.0$, $d/a = 0.1$, $a/\lambda = 0.2$, $\sigma_g = 10^2 S/m$, $\varepsilon_g = 1.8$.

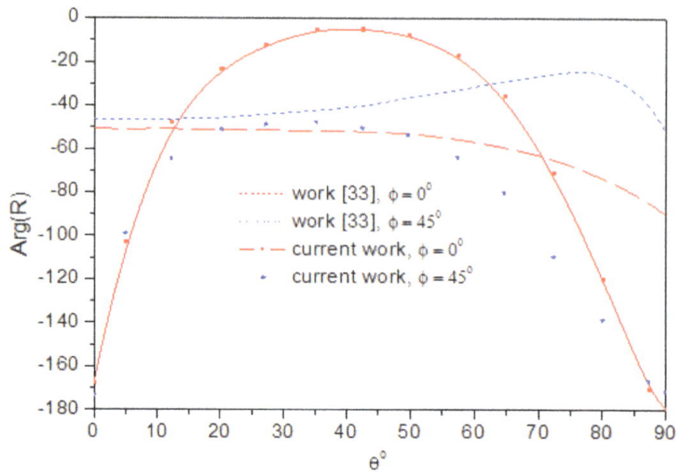

Figure 4.21. Phase of R_p versus of θ at a frequency of 100 Hz for two different values of $\varphi = 0^0$, 45^0 and $r_0/a = 0.01$, $h/r_0 = 3.0$, $d/a = 0.1$, $a/\lambda = 0.2$, $\sigma_g = 10^2 S/m$, $\varepsilon_g = 1.8$.

showed that the finiteness of the dimensions results in dimensional resonances. Moreover, during numerical FDTD simulations, the hypothetical infinitely long flat samples showed no appreciable differences in the *S*-parameters in comparison with the real samples, except for small peaks related to the above dimensional resonances.

As can be seen from figures 4.20 and 4.21, a 90° rotation of one of the single-layer wire gratings leads to completely different scattering properties of the two-layer wire structure. Figure 4.20 enables us to conclude for the predominance of magnitudes in the case of the wire structure under consideration for most angles as compared to the structure in [33]. This result is not obvious in terms of solving a corresponding problem of mathematical physics. Nevertheless, the result is quite predictable in terms of EMT. Indeed, as was shown in [34], the current two-layer wire grating exhibits a greater enhancement in the effective relative permittivity compared to the wire structure considered in [33]. The aforementioned phenomenon of the enhancement in effective permittivity can, in turn, be formulated in terms of the reflectivity considering the electric field distribution between the wires and on the surface of each single-layer wire grating. At the same time, one should mention that the graphs of change in the argument of the total reflection coefficient depicted in figure 21 are difficult to predict for both the wire structures using any approach. However, based on the results of work in [35], one can assume that a small difference in the magnitude of the total reflection coefficient only takes place in the case of non-magnetic wires. This can be explained by the fact that the two-layer wire gratings behave like meta-diamagnets in terms of metamaterials. A noticeable difference in the magnitude dependences is most likely to be expected in the case of ferric wires [35, 36], since one can consider the two-layer wire gratings as artificial ferromagnets or metaferrites in this case [8, 36].

The results of work in [35] also suggest that we should not expect a fundamental change in the reflective properties of the wire structure described in [33] if the single-layer wire gratings of the structure under consideration are swapped. It can be seen in figure 4.21 that, in principle, both structures can be used to design high-impedance surfaces. However, at the same time, figure 4.21 also shows that the most effective high-impedance surface can be designed based on the current wire structure for an angle $\varphi = 0^0$ in a range of θ from 20 to 65°.

4.5 On implementation of the two-layer wire grating for excitation and transmission of microwave plasmon polaritons

As is well known, surface plasmon polaritons can be excited on the metallic surfaces of a planar dielectric–metal–dielectric (DMD) structure in the optical range [37–39]. First of all, this is possible due to the fact that the real part of the dielectric permittivity of metal becomes negative in the optical range. This just occurs in a broad frequency range at microwave frequencies in the case of an s-polarized harmonic EMT wave impinging obliquely on the layer wire gratings considered in the previous sections. However, the practical realization of surface plasmon excitation is not a simple task, since the law of momentum conservation requires wave slowing down for the incident wave on the interface. Using Floquet theorem, the principal possibility for the excitation of microwave surface plasmon polaritons in metamaterials similar to the wire gratings considered in this study is briefly considered in [8] for the case of oblique incidence. However, such excitation was considered in [8] in terms of the effective parameters of the rod gratings. In this study, we will consider the same matter in terms of the surface plasmon resonance on a single wire and EMT as well.

Let us obtain equations for determining the eigenfrequencies of surface plasmons. The appropriate equation directly follows from the denominators of the expressions of equation (4.60) obtained for the coefficients A_m and B_n. Equating the denominators to zero gives just one equality (due to $\Gamma_{s0}^{(1)} = \Gamma_{0s}^{(2)} = \Gamma_s$):

$$\left.\begin{aligned} & Z_i e^{isa\sin\varphi} + \frac{i\omega\mu_0}{4\pi}(1 - \sin^2\theta\cos^2\varphi)\left(\frac{a}{\pi}\ln\left(\frac{a}{2\pi r_0}\right) + \frac{1}{\Gamma_s}\right) = 0, \\ & \Gamma_s = \sqrt{\left(\frac{2\pi s}{a}\right)^2 - k^2\cos^2\theta}, \quad s = 1, 2, 3\ldots \end{aligned}\right\}. \tag{4.89}$$

As can be seen from equation (4.89), all the eigenfrequencies are complex, i.e. $\omega = \omega' + i\omega''$, where ω' stands for the eigen oscillation frequencies while ω'' is associated with plasmon energy damping ($\omega'' > 0$). Equation (4.89) is solved numerically in this study.

Note that the quality factor Q of plasmons can be evaluated by the formula

$$Q = \omega'/2\omega''. \tag{4.90}$$

Let us now use the main principles of EMT to associate the two-layer wire grating under consideration with an effective infinitely long metallic wire of circular cross-section. In this case, one can assume that the axisymmetric waves propagate in the effective cylinder. Then the dispersion equation for the transverse magnetic (TM) waves in the effective cylinder is defined by the equation [38]

$$\frac{k_m}{-\varepsilon_{\text{eff}}}\frac{I_0(k_m r_{\text{eff}})}{I_1(k_m r_{\text{eff}})} = k_d\frac{K_0(k_d r_{\text{eff}})}{K_1(k_d r_{\text{eff}})}, \tag{4.91}$$

where $k_m = \sqrt{k_{\text{sp}}^2 - \varepsilon_{\text{eff}}\mu_{\text{eff}}k^2}$, $k_d = \sqrt{k_{\text{sp}}^2 - k^2}$, k_{sp} is the wavenumber of plasmon wave, ε_{eff} and μ_{eff} are the effective parameters defined by equation (4.70) for the considered polarization, $I_\nu(\cdot)$ and $K_\nu(\cdot)$ are modified Bessel functions of the first and second types, respectively, and $r_{\text{eff}} = \sqrt{2r_0 l(1 - 0.5(r_0/l)^{3/2})}$ is the radius of the effective cylinder [26].

Let us aim to obtain the dispersion relation explicitly from equation (4.90). For this purpose, one can represent the dielectric function ε_{eff} of the effective cylinder by means of the Drude–Lorentz model

$$\varepsilon_{\text{eff}} = 1 - \frac{\omega_p^2}{\omega(\omega + i\gamma)}, \tag{4.92}$$

where typically the damping factor $\gamma = \sigma/\varepsilon_0\omega$ becomes significant in the infrared regime [40]. Thus, we neglect γ in our study (for the microwave frequency range). Substituting equation (4.92) into equation (4.91) finally gives ($\gamma = 0$)

$$\omega = \omega_p\sqrt{\frac{k_d K_0(k_d r_{\text{eff}})I_1(k_m r_{\text{eff}})}{k_d K_0(k_d r_{\text{eff}})I_1(k_m r_{\text{eff}}) + k_m K_1(k_d r_{\text{eff}})I_0(k_m r_{\text{eff}})}}. \tag{4.93}$$

The dispersion curve for copper wires with $r_0 = 0.0001$ m and $a = 0.002$ m is depicted in figure 4.22. One can observe from equation (4.93) and figure 4.22 that, for any radius of the effective cylinder, there is only one plasmon branch compared

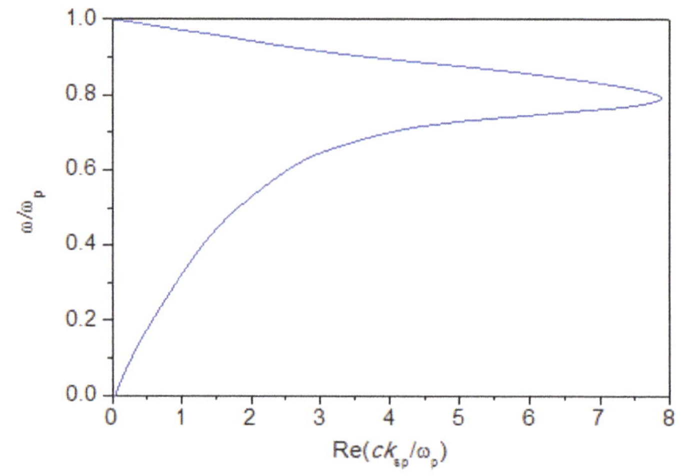

Figure 4.22. Dispersion curve of two-layer copper wire grating located in air.

to any DMD structure, which normally has two branches. This is due to considering a thin effective cylinder. Indeed, if the effective cylinder is thin ($r_{\text{eff}} < < \lambda$), then considering the two-layer wire grating as a thin film, we can assume that the symmetric and antisymmetric plasmons interact so strongly with each other that the difference between them disappears [37]. At the same time, as in the case of a DMD layer, the surface plasmon polariton is only excited by the grating in the case of p-polarization. Note that only the real parts of the solutions of equation (4.89) should be taken as the *working* plasmonic frequencies while using the dispersion curve in figure 4.22.

4.6 Conclusion

The scattering properties and the effective relative permittivity and permeability of single- and two-layer wire gratings are analytically studied in dipole approximation in the microwave frequency range. The single-layer grating is just one layer of metamaterial presented as an infinitely extended 3D network of thin metal circular wires with a unit cell with a square cross-section. The two-layer wire grating consists of two above symmetrical gratings separated by a distance that is equal to the constant of the unit cell. The analytical studies carried out in this study, and confirmed by numerical FDTD simulations, clearly show that the considered layered wire gratings behave almost like a mirror in the case of p-polarization and are almost transparent in the case of s-polarization in a wide range at GHz frequencies. It has been shown that the considered wire gratings can be used to generate surface plasmon polaritons in the microwave frequency range for the case of p-polarization. In particular, the dispersion equation was obtained for the case of a two-layer wire grating. It has also been shown in the study that a two-layer wire grating backed by a plane conductive surface can be used for designing high-impedance surfaces operating over a broad range at microwave frequencies.

Acknowledgments

The author would like to thank his teacher, Dr Georgios Zouganelis, for his support and advice, which led to the results that formed the basis of this study. The author would also like to acknowledge Ms Julia Rybin for her valuable help in preparing the manuscript.

References

[1] Laroche M, Albaladejo S, Gomez-Medina R and Saenz J J 2006 Tuning the optical response of nanocylinder arrays: an analytical study *Phys. Rev.* B **74** 245422

[2] Garcia de Abajo F J 2007 Colloquium: light scattering by particle and hole arrays *Rev. Mod. Phys.* **79** 1267–8

[3] Bulgakov E N, Sadreev A F and Maksimov D N 2015 Light trapping above the light cone in a one-dimensional array of dielectric spheres *Phys. Rev.* A **95** 023816

[4] Kwadrin A, Osorio C I and Koenderink A F 2016 Backaction in metasurface etalons *Phys. Rev.* B **93** 104301

[5] Berkhout A and Koenderink A F 2020 A simple transfer-matrix model for metasurface multilayer systems *Nanophoton* **9** 3985–4007
[6] Choudhury P K (ed) 2021 *Metamaterials: Technology and Applications* (Boca Raton, FL: CRC Press)
[7] Rybin O and Shulga S 2021 Analytical model for miniaturized patch antenna on metaferrites-like substrate *J. Electromagn. Waves Appl.* **35** 2257–68
[8] Rybin O and Shulga S 2021 Broadband applications of a tunable nano-rod metaferrite *J. Quant. Spectrosc. Radiat. Transf.* **263** 107560
[9] Komarov V V 2022 Scattering characteristics of metal circular wire gratings in microwave and terahertz ranges *J. Comput. Electron.* **21** 1166–73
[10] Ikonen P, Saenz E, Gonzalo R, Simovski C and Tretyakov S 2007 Mesoscopic effective material parameters for thin layers modelled as single and double gratings of interacting loaded wires *Metamaterials* **1** 89–105
[11] Albooyeh M, Morits D and Tretyakov S A 2012 Effective electric and magnetic properties of metasurfaces in transition from crystalline to amorphous state *Phys. Rev.* B **85** 205110
[12] Rybin O, Shulga V and Shulga S 2019 The given surface current distribution model of a rectangular patch antenna with metamaterial-like substrate *Results Phys.* **15** 102573
[13] Zalipaev V V and Kosulnikov S Y 2019 Guided electromagnetic waves for periodic arrays of thin metallic wires near an interface between two dielectric media *Proc. R Soc.* A **475** 20180399
[14] Dinia L, Mangini F and Frezza F 2020 Electromagnetic scattering of inhomogeneous plane wave by ensemble of cylinders *J. Telecommun. Inf. Technol.* **3** 86–92
[15] Alvarez-Perez J L 2023 Scattering of a plane wave from a wire grating parallel to the interface between two media *IEEE Trans. Antenn. Propagat.* **71** 1725–35
[16] Pendry J B, Holden A J, Robbins D J and Stewart W J 1999 Magnetism from conductors and enhanced nonlinear phenomena *IEEE Trans. Microw. Theory Technol.* **47** 2075–84
[17] Yannopapas V 2007 Artificial magnetism and negative refractive index in three-dimensional metamaterials of spherical particles at near-infrared and visible frequencies *Appl. Phys.* A **87** 259–64
[18] Rybin O and Shulga S 2020 Generalized broad-band effective medium theory of two-component metamaterials including magnetic ones: a review *J. Electromagn. Waves Appl.* **34** 1513–49
[19] Rybin O and Shulga S 2019 Revised homogenization for two-component metamaterial with non-magnetic metallic cylindrical inclusions *Appl. Phys.* A **125** 125–53
[20] Wait J R 1955 Reflection at arbitrary incidence from a parallel wire grating *Appl. Sci. R* B **4** 393–400
[21] Hill D A and Wait J R 1974 Electromagnetic scattering of an arbitrary plane wave by two nonintersecting perpendicular wire gratings *Can. J. Phys.* **52** 227–37
[22] Tretyakov S 2003 *Analytical Modeling in Applied Electromagnetics* (Boston, MA: Artech House)
[23] Pendry J B, Holden A J, Stewart W J and Youngs I 1996 Extremely low frequency plasmon im metallic mesostructures *Phys. Rev. Lett.* **76** 4773–6
[24] Rybin O and Shulga S 2017 RLC-circuit effective medium approach for two-component non-magnetic metamaterials *Proc. of IEEE 1st Ukrainian Conf. on Electric and Computer Engineering, Kyiv (Ukraine)* pp 127–31

[25] Twersky V 1962 On scattering of wires by the infinite grating of circular cylinders *IRE Trans. Antenn. Propagat.* **10** 737–65

[26] Kontorovich M I, Astrakhan M I, Akimov V P and Fersman G A 1987 *Electrodynamics of Grating Structures* (Moscow: Radio i sviaz) (in Russian)

[27] Rybin O 2023 Microwave properties of a double wire array *J. Comput. Electron.* **22** 1541–8

[28] Decker M T 1959 Transmission and reflection by a parallel wire grating *J. Res. Natl. Bureau Stand-D.* Radio Propag. **63D** 87–90

[29] Rybin O 2013 An advanced optimization technique for layer-specific characterization of slab matematerials *Int. J. Mod. Phys.* C **24** 1350019

[30] Ghodgaonkar D K, Varadan V V and Varadan V K 1990 Free-space measurement of complex permittivity and complex permeability of magnetic materials at microwave frequencies *IEEE Trans. Instrum. Meas.* **39** 387–94

[31] Liu T, Ma S, Yang B, Xiao S and Zhou L 2020 Effective medium theory for multilayer metamaterials: role of near-field correction *Phys. Rev.* B **102** 174208

[32] Mladyonov P L and Prosvirnin S L 2005 Microstrip double-periodic grating of continuous curvilinear metal strips as a high-impedance surface *Telecommun. Radio Eng.* **63** 109–18

[33] Wait J R and Hill D A 1976 Electromagnetic scattering by two perpendicular gratings over a conducting half-space *Radio Sci.* **11** 725–30

[34] Zouganelis G and Rybin O 2006 Two layers magneto-dielectric metamaterial with enhanced dielectric constant as a new ferrite like material *Jpn. J. Appl. Phys.* **45** L1175–8

[35] Rybin O 2009 Enhancement of dielectric constant in metal-dielectric metamaterials *Afr. Rev. Phys.* **3** 49–55

[36] Rybin O 2014 Effective permeability tensor of partially magnetized two-component metaferrites *Mod. Phys. Lett.* B **28** 1450199

[37] Maier S A 2007 *Plasmonics. Fundamental and Applications* (New York: Springer)

[38] Klimov V 2013 *Nanoplasmonics* (Boca Raton, FL: Taylor and Francis)

[39] Barbillon G 2017 *Nanoplasmonics—Fundamental and Applications* (London: InTech Press)

[40] Pendry J B, Holden A J, Robbins D J and Stewart W J 1998 Low frequency plasmons in thin-wire structures *J. Phys. Condens. Matter.* **10** 4785

IOP Publishing

Ordered and Disordered Metamaterials
Design and applications
Pankaj K Choudhury and Tatjana Gric

Chapter 5

Machine learning-based prediction, design, and optimization of optical metamaterials

Jong-ryul Choi, Inseop Byeon and Donghyun Kim

Research and development on optical metamaterials require extensive computational time and resources for the analysis, design, and optimization of their optical properties using conventional calculation tools. To address this challenge, machine learning- and/or artificial intelligence (AI)-based models have been used for the analysis, design, and optimization of various metamaterial properties. The chapter focuses on the latest studies of forward prediction and inverse design of optical metamaterials using machine learning, and their optimization. After conducting a comprehensive examination of these through machine learning and AI, the chapter suggests that the models hold significant potential to drive innovation in the development of various optical elements and associated components.

5.1 Introduction

Metamaterials are artificially designed and fabricated to have properties that do not exist in the natural world [1–4]. They are employed in various applications since they can be used to control or block waves, and are typically made of materials such as metals, ceramics, and dielectrics, often in periodic arrangement with a specific meta-atom structure. Here, we focus on optical metamaterials that consist of an array of nanoscale metallic/dielectric structures on a scale of light wavelength to control the properties of light. By proper design of a meta-atom and/or its periodic arrangement, optical metamaterials can control characteristics of light such as reflection, refraction, diffraction, and interference [5–8].

Optical metamaterials can be designed to achieve transparency in a specific spectral region [9–11], to manipulate the direction of light [12, 13], or to reject and suppress light within a particular waveband [14–16]. These capabilities present opportunities for advancements in industrial and medical technologies. In addition, optical metamaterials have been fabricated to display new phenomena that do not

doi:10.1088/978-0-7503-5462-2ch5

follow existing physical laws [17–19]. Due to their specialized properties, these can be implemented in various applications, such as ultrasensitive optical biosensors [20–26], super-resolution imaging [27–31], highly efficient solar cells [32–36], and display devices [37]. One of the challenges in the development and practical application of optical metamaterials is the efficient design and optimization of metamaterial structures to achieve specific performance under predetermined conditions. Previously, metastructures such as split-ring resonators [38–40] or fishnet structures [41–43] designed at non-optical operating frequency have been applied for optical wavebands with potential limitations in the achievable performance characteristics.

In this chapter, we explore efforts to design metamaterials with machine learning and artificial intelligence (AI), which can optimize metamaterial structures with desired optical properties while the limitations are minimal. To this end, AI can be used to generate various metastructures and to analyse the characteristics of these to determine an optimum. In addition, the use of machine learning- or AI-based algorithms can significantly shorten the time required for metamaterial design and optimization. Existing design methods heavily rely on the experience and knowledge of designers, often resulting in poor accuracy. In contrast, AI learns from large-scale prior data and can enhance design accuracy. Furthermore, utilizing AI in the optimization of metamaterial structures can lead to the discovery of new physical phenomena that demonstrate superior optical performance.

Here, we review some of the recent studies on the design and optimization of optical metamaterials using machine learning and AI with an emphasis on three issues, namely (i) forward prediction of optical parameters in metamaterials by machine learning, (ii) inverse design of optical metamaterials using machine learning, and (iii) machine learning-based optimization of optical metamaterials. Each of these topics is described in detail in the following sections.

5.2 Forward prediction of optical parameters of metamaterials

Evaluation of optical properties of metamaterials is performed using well-known numerical methods, e.g. rigorous coupled-wave analysis (RCWA) [44–48] and the finite-difference time-domain (FDTD) [49–52] technique. Although these methods allow precise calculation of various optical properties of metamaterials, they require extensive computing time and resources. In contrast, while a predictor of the optical properties and light field distribution of metamaterials using machine learning and AI may use significant computing power in the training stage, but this investment in computational resources can yield faster and more efficient results in the prediction stage. Based on this possibility, forward predictors of optical properties have recently been explored; we touch upon these briefly in the following.

Wiecha and Muskens proposed a deep artificial neural network to quickly and accurately predict both the near- and far-field responses [53]. As shown in figure 5.1, the neural network allows the deduction of electromagnetic near- and far-field distributions formed on nanostructures much more quickly than conventional numerical simulation tools. While this method may inherently be less precise

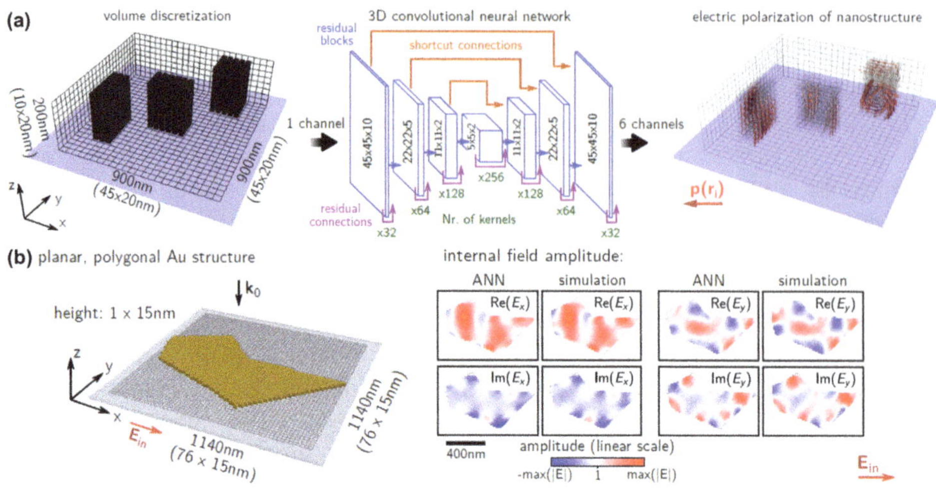

Figure 5.1. (a) Schematic of a deep neural network model for the prediction of electromagnetic fields in nanostructures. (b) Results for electromagnetic field prediction of planar and free-form gold nanostructures. Reprinted with permission from [53]. Copyright (2020) American Chemical Society.

because it is a data-driven approach, it has been confirmed that utilizing a deep neural network for predicting near- and far- field responses results in an error rate of only 5%. This network holds significant potential for applications in studying optical nanostructures, and after a single stage of training can produce optical responses very quickly.

Kaźmierczak *et al* established a machine learning algorithm to predict the properties of optical fibers with integrated nanostructures [54]. In the initial phase, researchers constructed a training dataset of optical properties using the finite element method (FEM) applied to optical fibers. These fibers possessed a core structure arranged in a hexagonal pattern, comprising two distinct types of rods. In this research, which focuses on predicting the zero dispersion wavelength of nanostructured optical fibers, the random forest (RF) algorithm emerged as the most stable and accurate predictor among various machine learning algorithms. Employing the RF algorithm to predict the zero dispersion wavelength across different nanostructured optical fibers resulted in a mean square error of less than 0.03 nm. Furthermore, utilizing RF algorithms for predicting optical properties significantly reduces computational time compared to conventional simulation tools.

In a related study, Kumar *et al* developed a neural network algorithm tailored to predict the optical properties of photonic crystal fibers [55]. The algorithm demonstrates precision in predicting critical information such as effective index, confinement loss, and dispersion across various types of photonic crystal fibers, including those with solid, multiple, and hollow cores. As a result, the network achieves determination of optical parameters in photonic crystal fibers with a computational time 3600 times shorter than conventional methods for calculating optical properties.

On the other hand, Roberts and Hedayati suggested a deep neural network for forward prediction of structural colors produced in optical metamaterials [56]. The computational time required for the deep neural network-based prediction model of structural color in optical metamaterials was reduced by 10^5 times compared to conventional methods based on Maxwell's equations. Additionally, this prediction model achieved an impressive accuracy level of over 96%.

In another study, Gahlmann and Tassin investigated a machine learning algorithm designed to predict the transmission amplitudes and phases of various free-form metasurfaces [57]. Comparing the results obtained from the prediction model based on machine learning algorithms, trained using the optical properties of 36000 samples by FEM, the prediction model demonstrated its robust predictive capabilities with a mean square error of less than 0.003.

5.3 Inverse design of optical metamaterials

Compared to predicting optical properties from a specific nanostructure component, inverse design with predetermined optical properties poses greater challenges. Unlike straightforward calculations of optical properties where parameters such as transmittance and reflectance are determined, inverse design entails considering numerous parameters such as the thickness, size, shape, and material composition (including complex refractive indices) of nanostructures. Consequently, inverse metamaterial design problems exhibit nonlinearities. Generally, solving these challenges involves complex iterative processes, extensive computational time, and significant computing resources.

To address these complexities, several research groups focusing on optical metamaterial design based on nanostructures have presented efficient studies of inverse design of nano-metamaterials with desired optical properties. These studies utilize AI-based algorithms trained on databases containing nanostructures and their corresponding optical properties.

Ma *et al* introduced a deep neural network model consisting of bidirectional neural networks tailored for designing chiral metamaterial nanostructures, which led to exceptionally strong optical responses at predetermined wavelengths [58]. The neural network designed for chiral metamaterials was trained on a dataset comprising 25000 samples of reflection spectra obtained through conventional simulation and validated on 5000 samples. Through this training process, the neural network learned to provide key specifications, such as the thicknesses of dielectric layers, the sizes of gold split-ring resonators, and the angle between these resonators, for chiral metamaterials capable of generating the strongest chiro-optical responses.

Hou *et al* delved into a deep neural network model for split-ring resonator-based metamaterials [59]. This model serves both for predicting reflectance in specific split-ring resonator-based metamaterials and for designing the resonator itself to achieve desired reflectance profiles within a specified optical wavelength range. The neural network, trained using simulation data derived from FEM, exhibited a mean square error of 0.005, indicating high precision.

Similarly, Zhou *et al* developed a machine learning-based model for designing elastic metamaterials [60]. With a pre-trained machine learning model derived from simulation results, this study facilitated the design and production of metamaterial devices capable of generating desired output waveforms.

Liu *et al* contributed to the field by developing and deploying a deep neural network for the inverse design of nanostructured optical components composed of multiple dielectric layers [61]. The neural network was utilized to meticulously design the thickness of each layer within a multi-layer thin film structure that consisted of intersecting SiO_2 and Si_3N_4 layers to achieve desired transmission characteristics within a specific wavelength range.

Malkiel *et al* devised a novel approach for designing nanostructures capable of achieving optical resonance properties [62]. Their method employed a bidirectional deep neural network model with two neural networks as shown in figure 5.2(a). The bidirectional neural network architecture consisted of a geometry-predicting network responsible for generating nanostructure shapes to attain the desired transmission and optical resonance characteristics. Simultaneously, a spectrum-

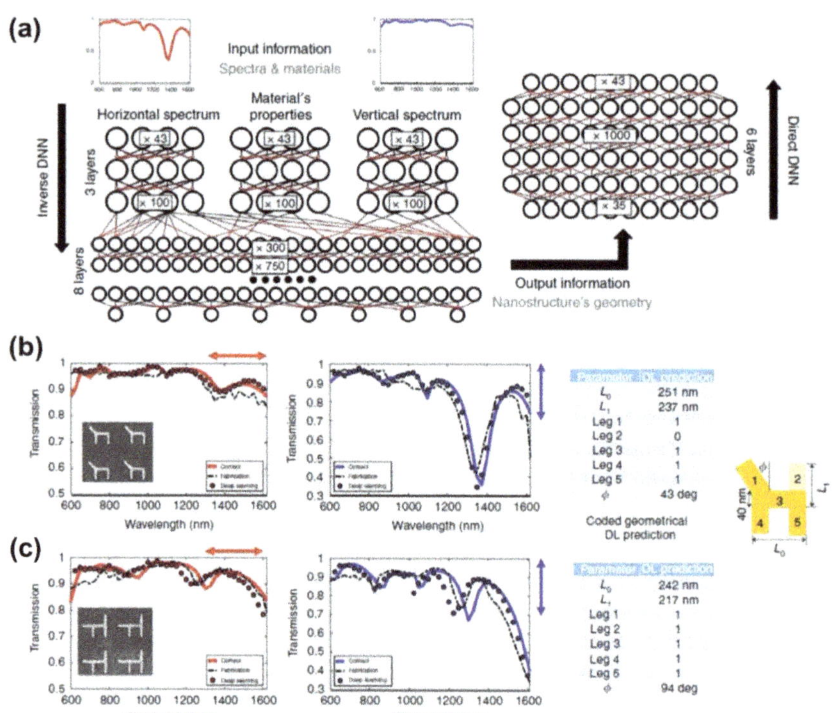

Figure 5.2. (a) Deep neural network architecture to design nanostructures that provide a desired transmittance and optical resonance properties in a specific optical wavelength range. (b–c) Results for two different nanostructured metamaterials and determined parameters. Reproduced from [62] with permission from Springer Nature.

predicting network estimated light transmittance based on the predicted nanostructure geometry. As illustrated in figure 5.2(b), this bidirectional network exhibited significantly enhanced accuracy and speed in metamaterial design compared to conventional methods, where the types, sizes, and materials of nanostructured components are separately calculated and designed.

A related study utilized a generative model: an AI model trained to learn the distribution of input data and generate new data to design metamaterial nanostructures [63]. Unlike conventional simulations of optical properties or machine learning algorithms used for inverse design of nanostructures, the generative model in this study comprised three key components: a generator, which generates nanostructures based on optical properties and noise; a simulator, which resimulated these structures to acquire optical properties; and a critic, which analysed differences between the generated and simulated properties. This approach offers a unique perspective on metamaterial design, leveraging generative modeling to create nanostructures with desired optical properties.

5.4 Machine learning-based optimization of optical metamaterials

In the preceding section, it was highlighted that designing metamaterials with desired optical specifications is more challenging than predicting the optical properties of predetermined optical nanostructured metamaterials. The optimization process for optical metamaterials discussed in this section is slightly more intricate and entails complex parametric calculations compared to the inverse design of optical metamaterials. This complexity arises from the necessity to account for a greater number of variables to derive metamaterials with optimized performance for specific optical properties. In light of these challenges, several studies have been actively exploring the application of machine learning- or AI-based algorithms for optimizing optical metamaterials, as elaborated below.

Asano and Noda utilized a deep learning algorithm to optimize photonic crystal nanocavities for achieving high-quality factors [64]. In particular, the deep learning algorithm recommended a two-dimensional nanocavity configuration with the highest quality factors. The validity of this recommendation was confirmed through rigorous three-dimensional FDTD simulations.

Moon *et al* presented an innovative approach to optimize metamaterials for plasmonic biosensors by employing machine learning algorithms [65]. Based on machine learning-based estimation of optical properties specific to plasmonic biosensors, along with techniques such as data dimension reduction and k-means clustering to classify various types of surface plasmon resonance profiles, the study explored optimized meta-plasmonic structures. These structures were then implemented experimentally for surface plasmon resonance-based optical biosensing of DNA hybridization, as illustrated in figure 5.3. The same group applied convolutional neural network to explore the robustness of surface plasmon imaging [66]. The use of machine learning for plasmonics was also reviewed [67].

Zhang *et al* proposed a machine learning-based approach to optimize transparency in graphene metamaterials [68]. Their research focused on utilizing machine

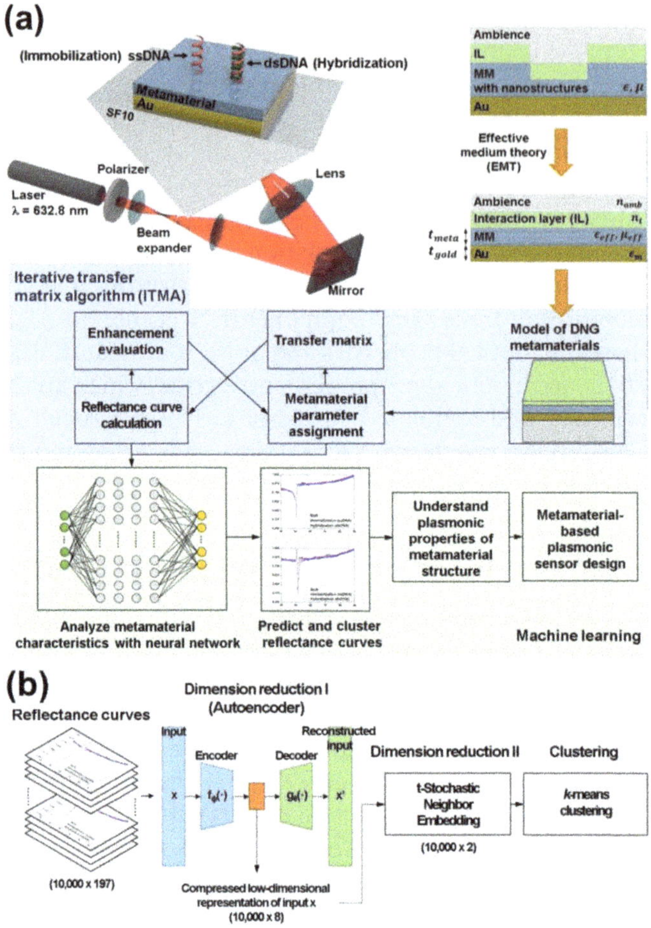

Figure 5.3. (a) Schematics of machine learning-based optical responses prediction and optimization of metamaterials in meta-plasmonic biosensors for DNA (deoxyribonucleic acid). In this research, effective medium theory and iterative transfer matrix algorithm were applied to generate training dataset. (b) Schematics of a reduction of data dimension and k-mean clustering to classify types of surface plasmon resonance profiles for optimization of meta-plamsonic DNA biosensing. Adapted from [65], copyright (2020), with permission from Elsevier.

learning algorithms to predict transmission properties, perform inverse design, and optimize transparency in metamaterial devices comprising two graphene nano-ribbon layers separated by a dielectric layer. Through optimization, they achieved a maximum difference between the optical transmission peaks and transmission depths of up to 0.97. Similarly, Banerji *et al* employed a machine learning-based optimizer to design various types of optimized metamaterial nanoscale devices, including beam splitters and waveguides with 90° and 180° bending capabilities [69].

Generative models, initially employed for the inverse design of metamaterials, have also found application in the optimization of metamaterial nanoscale devices. Jiang *et al* utilized generative adversarial networks (GANs) to design optimized

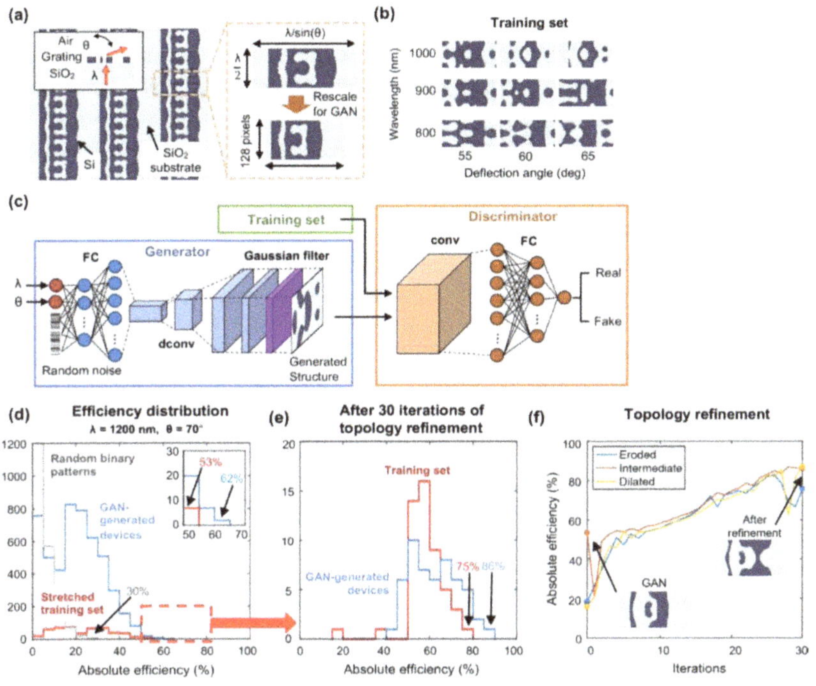

Figure 5.4. (a) Typical optimized metamaterial grating selectively deflects (+1 optical diffraction order). (b) Example of metamaterial gratings as training set. (c) Schematic of GAN to make metamaterial grating with optimized efficiency. (d) Histograms of efficiency distributions in metamaterial grating generated from a generator, training set, and binary patterns randomly produced. (e) Histograms of efficiency distributions in metamaterial gratings generated from a generator, training set, and with a refinement by optimization. (f) Absolute efficiency as a function of iterations via refinement types. Reprinted with permission from [70]. Copyright (2019) American Chemical Society.

free-form diffractive metamaterial gratings [70, 71]. GANs, a prominent type of generative model, operate on a unique principle where two neural networks—the generator and the discriminator—engage in a competitive learning process. The generator generates data that mimic the desired condition to deceive the discriminator, while the discriminator is trained to distinguish between the real and generated data. This adversarial competition fosters rapid and accurate production of data meeting the desired conditions. In their study, the GAN model was effectively employed to design complex metamaterials, as depicted in figure 5.4.

5.5 Conclusion

In this chapter, we reviewed recent studies employing machine learning or AI to analyse, design, and optimize various optical metamaterials. The application of machine learning or AI in these endeavors offers two key advantages. First, it significantly reduces computational time and resource requirements compared to conventional simulations such as RCWA, FDTD, and FEM. Second, it enables the design and optimization of metamaterials with superior optical responses based on novel nanoscale structures, rather than relying solely on existing research-based libraries of nanostructures. It is anticipated that the computational time required for the analysis and design of metamaterials will further decrease with the continued advancement of electronics used in machine learning- and AI-based algorithm operations, including advanced GPUs. Furthermore, advanced AI algorithms such as generative models hold promise for providing novel metamaterial nanostructures with greatly enhanced optical performances.

Simultaneously, accounting for the fabrication conditions of nanostructures as computational parameters is paramount when employing machine learning and AI models for the design and optimization of metamaterials. This consideration is crucial due to the inherent limitations in the resolution of fabricating nanostructures in actual optical metamaterials.

Integration of advancements in nanostructure fabrication technologies with AI for analysis, design, and optimization of metamaterials is expected to yield innovative breakthroughs across various sectors of the optics and photonics arena. These advancements hold the potential to revolutionize the development of optical elements, components, and systems, contributing to the overall progress of the photonics industry.

References

[1] Shelby R A, Smith D R and Schultz S 2001 Experimental verification of a negative index of refraction *Science* **292** 77–9

[2] Zheludev N I and Kivshar Y S 2012 From metamaterials to metadevices *Nat. Mater.* **11** 917–24

[3] Kadic M, Milton G W, van Hecke M and Wegener M 2019 3D metamaterials *Nat. Rev. Phys.* **1** 198–210

[4] Valipour A, Karagozarfard M H, Rakhshi M, Yaghootian A and Sedighi H M 2022 Metamaterials and their applications: an overview *P. I. Mech. Eng. L-J Mat.* **236** 2171–210

[5] Shalaev V M, Cai W S, Chettiar U K, Yuan H K, Sarychev A K, Drachev V P and Kildishev A V 2005 Negative index of refraction in optical metamaterials *Opt. Lett.* **30** 3356–58

[6] Paul T, Menzel C, Rockstuhl C and Lederer F 2010 Advanced optical metamaterials *Adv. Mater.* **22** 2354–57

[7] Soukoulis C M and Wegener M 2010 Optical metamaterials—more bulky and less lossy *Science* **330** 1633–4

[8] Urbas A M *et al* 2016 Roadmap on optical metamaterials *J. Opt.* **18** 093005

[9] Zhang J J, Xiao S S, Jeppesen C, Kristensen A and Mortensen N A 2010 Electromagnetically induced transparency in metamaterials at near-infrared frequency *Opt. Express* **18** 17187–92

[10] Huo P C, Liang Y Z, Zhang S, Lu Y Q and Xu T 2018 Angular optical transparency induced by photonic topological transitions in metamaterials *Laser Photon. Rev.* **12** 1700309

[11] Palmer S J, Xiao X F, Pazos-Perez N, Guerrini L, Correa-Duarte M A, Maier S A, Craster R V, Alvarez-Puebla R A and Giannini V 2019 Extraordinarily transparent compact metallic metamaterials *Nat. Commun.* **10** 2118

[12] Liu X L and Padilla W J 2013 Dynamic manipulation of infrared radiation with MEMS metamaterials *Adv. Opt. Mater.* **1** 559–62

[13] Shi Y Z, Song Q H, Toftul I, Zhu T T, Yu Y F, Zhu W M, Tsai D P, Kivshar Y and Liu A Q 2022 Optical manipulation with metamaterial structures *Appl. Phys. Rev.* **9** 031303

[14] Zhu W R, Zhao X P, Gong B Y, Liu L H and Su B 2011 Optical metamaterial absorber based on leaf-shaped cells *Appl. Phys.* A **102** 147–51

[15] Yu P, Besteiro L V, Huang Y, Wu J, Fu L, Tan H H, Jagadish C, Wiederrecht G P, Govorov A O and Wang Z 2019 Broadband metamaterial absorbers *Adv. Opt. Mater.* **7** 1800995

[16] Soheilifar M R 2019 The wideband optical absorber based on plasmonic metamaterials for optical sensing *Optik* **182** 702–11

[17] Shen N H, Foteinopoulou S, Kafesaki M, Koschny T, Ozbay E, Economou E N and Soukoulis C M 2009 Compact planar far-field superlens based on anisotropic left-handed metamaterials *Phys. Rev.* B **80** 115123

[18] Haxha S, AbdelMalek F, Ouerghi F, Charlton M D B, Aggoun A and Fang X 2018 Metamaterial superlenses operating at visible wavelength for imaging applications *Sci. Rep.* **8** 16119

[19] Chen M K, Wu Y F, Feng L, Fan Q B, Lu M H, Xu T and Tsai D P 2021 Principles, functions, and applications of optical meta-lens *Adv. Opt. Mater.* **9** 2001414

[20] Ma K, Kim D J, Kim K, Moon S and Kim D 2010 Target-localized nanograting-based surface plasmon resonance detection toward label-free molecular biosensing *IEEE J. Sel. Top. Quantum Electron.* **16** 1004–14

[21] Oh Y, Lee W, Kim Y and Kim D 2014 Self-aligned colocalization of 3D plasmonic nanogap arrays for ultra-sensitive surface plasmon resonance detection *Biosens. Bioelectron.* **51** 401–7

[22] Sreekanth K V, Alapan Y, ElKabbash M, Ilker E, Hinczewski M, Gurkan U A, De Luca A and Strangi G 2016 Extreme sensitivity biosensing platform based on hyperbolic metamaterials *Nat. Mater.* **15** 621

[23] Aristov A I, Manousidaki M, Danilov A, Terzaki K, Fotakis C, Farsari M and Kabashin A V 2016 3D plasmonic crystal metamaterials for ultra-sensitive biosensing *Sci. Rep.* **6** 25380

[24] Baqir M A, Farmani A, Fatima T, Raza M R, Shaukat S F and Mir A 2018 Nanoscale, tunable, and highly sensitive biosensor utilizing hyperbolic metamaterials in the near-infrared range *Appl. Opt.* **57** 9447–54

[25] Chen S H, Hu S Q, Wu Y C, Deng D N, Luo Y H and Chen Z 2021 Ultrasensitive biosensor with hyperbolic metamaterials composed of silver and zinc oxide *Nanomater.-Basel* **11** 2220

[26] Yan R Q, Wang T, Yue X Z, Wang H M, Zhang Y H, Xu P, Wang L, Wang Y D and Zhang J Y 2022 Highly sensitive plasmonic nanorod hyperbolic metamaterial biosensor *Photon. Res.* **10** 84–95

[27] Choi J, Kim K, Oh Y, Kim A L, Kim S Y, Shin J S and Kim D 2014 Extraordinary transmission-based plasmonic nanoarrays for axially super-resolved cell imaging *Adv. Opt. Mater.* **2** 48–55

[28] Lee W, Kinosita Y, Oh Y, Mikami N, Yang H, Miyata M, Nishizaka T and Kim D 2015 Three-dimensional superlocalization imaging of gliding mycoplasma mobile by extraordinary light transmission through arrayed nanoholes *ACS Nano* **9** 10896–908

[29] Son T, Lee D, Lee C, Moon G, Ha G E, Lee H, Kwak H, Cheong E and Kim D 2019 Superlocalized three-dimensional live imaging of mitochondrial dynamics in neurons using plasmonic nanohole arrays *ACS Nano* **13** 3063–74

[30] Lee H, Kang K, Mochizuki K, Lee C, Toh K A, Lee S A, Fujita K and Kim D 2020 Surface plasmon localization-based super-resolved Raman microscopy *Nano Lett.* **20** 8951–58

[31] Lee H, Yoo H, Moon G, Toh K A, Mochizuki K, Fujita K and Kim D 2021 Super-resolved Raman microscopy using random structured light illumination: concept and feasibility *J. Chem. Phys.* **155** 144202

[32] Zhang C, Zhou W Z, Sun S, Yi N B, Song Q H and Xiao S M 2015 Absorption enhancement in thin-film organic solar cells through electric and magnetic resonances in optical metamaterial *Opt. Mater. Express* **5** 1954–61

[33] Karami S, Nikoufard M, Shariatmadar S M and Javadi S 2021 Plasmonic hyperbolic metamaterial and nanosphere composite for light trapping as a solar cell: numerical study *Opt. Mater.* **122** 111740

[34] Kumar R, Singh B K and Pandey P C 2022 Broadband metamaterial absorber in the visible region using a petal-shaped resonator for solar cell applications *Phys.* E **142** 115327

[35] El-Hageen H M *et al* 2023 Multilayered gold, MgF_2 and tungsten based ultra wide band infrared absorber for solar cell applications *Mater. Chem. Phys.* **301** 127680

[36] Durmaz H, Li Y Y and Eksiog Y 2023 Broad-band conical-shaped perfect absorber metamaterial for solar cells *Sol. Energy Mat. Sol. Cells* **262** 112569

[37] Kang K, Im S, Lee C, Kim J and Kim D 2021 Nanoslot metasurface design and characterization for enhanced organic light-emitting diodes *Sci. Rep.* **11** 9232

[38] Han J and Lakhtakia A 2009 Semiconductor split-ring resonators for thermally tunable terahertz metamaterials *J. Mod. Opt* **56** 554–7

[39] Chuma E L, Iano Y, Fontgalland G and Roger L L B 2018 Microwave sensor for liquid dielectric characterization based on metamaterial complementary split ring resonator *IEEE Sens. J.* **18** 9978–83

[40] Portosi V, Loconsole A M and Prudenzano F 2020 A split ring resonator-based metamaterial for microwave impedance matching with biological tissue *Appl. Sci.-Basel* **10** 6740

[41] Gómez-Castaño M, Garcia-Pomar J L, Pérez L A, Shanmugathasan S, Ravaine S and Mihi A 2020 Electrodeposited negative index metamaterials with visible and near infrared response *Adv. Opt. Mater.* **8** 2000865

[42] Askari M, Nia Z T and Hosseini M V 2022 Modified fishnet structure with a wide negative refractive index band and a high figure of merit at microwave frequencies *J. Opt. Soc. Am.* B **39** 1282–88

[43] Kalel S and Wang W-C 2023 Realization of broadband negative refractive index in terahertz band by multilayer fishnet metamaterial approach *Prog. Electromag. Res.* C **132** 159–70

[44] Nakagawa W, Tyan R C and Fainman Y 2002 Analysis of enhanced second-harmonic generation in periodic nanostructures using modified rigorous coupled-wave analysis in the undepleted-pump approximation *J. Opt. Soc. Am.* A **19** 1919–28

[45] Byun K M, Kim D and Kim S J 2006 Investigation of the profile effect on the sensitivity enhancement of nanowire-mediated localized surface plasmon resonance biosensors *Sens. Actuators B Chem.* **117** 401–7

[46] Lee Y J, Ruby D S, Peters D W, McKenzie B B and Hsu J W P 2008 ZnO nanostructures as efficient antireflection layers in solar cells *Nano Lett.* **8** 1501–5

[47] Song H, Choi J R, Lee W, Shin D M, Kim D, Lee D and Kim K 2016 Plasmonic signal enhancements using randomly distributed nanoparticles on a stochastic nanostructure substrate *Appl. Spectrosc. Rev.* **51** 646–55

[48] Pham H L, Alcaire T, Soulan S, Le Cunff D and Tortai J H 2022 Efficient rigorous coupled-wave analysis simulation of mueller matrix ellipsometry of three-dimensional multilayer nanostructures *Nanomater.-Basel* **12** 3951

[49] Oubre C and Nordlander P 2004 Optical properties of metallodielectric nanostructures calculated using the finite difference time domain method *J. Phys. Chem.* B **108** 17740–7

[50] Irannejad M, Yavuz M and Cui B 2013 Finite difference time domain study of light transmission through multihole nanostructures in metallic film *Photon. Res.* **1** 154–9

[51] Angelini M, Manobianco E, Pellacani P, Floris F and Marabelli F 2022 Plasmonic modes and fluorescence enhancement coupling mechanism: a case with a nanostructured grating *Nanomater.-Basel* **12** 4339

[52] Pandey P, Seo M K, Shin K H, Lee Y W and Sohn J I 2022 Hierarchically assembled plasmonic metal-dielectric-metal hybrid nano-architectures for high-sensitivity SERS detection *Nanomater.-Basel* **12** 401

[53] Wiecha P R and Muskens O L 2020 Deep learning meets nanophotonics: a generalized accurate predictor for near fields and far fields of arbitrary 3D nanostructures *Nano Lett.* **20** 329–38

[54] Kaźmierczak S, Kasztelanic R, Buczyński R and Mańdziuk J 2024 Predicting optical parameters of nanostructured optical fibers using machine learning algorithms *Eng. Appl. Artif. Intell.* **132** 107921

[55] Kumar H, Jain T, Sharma M and Kishor K 2021 Neural network approach for faster optical properties predictions for different PCF designs *J. Phys. Conf. Ser.* **2070** 012001

[56] Roberts N B and Hedayati M K 2021 A deep learning approach to the forward prediction and inverse design of plasmonic metasurface structural color *Appl. Phys. Lett.* **119** 061101

[57] Gahlmann T and Tassin P 2022 Deep neural networks for the prediction of the optical properties and the free-form inverse design of metamaterials *Phys. Rev.* B **106** 085408

[58] Ma W, Cheng F and Liu Y M 2018 Deep-learning-enabled on-demand design of chiral metamaterials *ACS Nano* **12** 6326–34

[59] Hou Z Y, Tang T T, Shen J, Li C Y and Li F Y 2020 Prediction network of metamaterial with split ring resonator based on deep learning *Nanoscale Res. Lett.* **15** 83

[60] Zhou W J, Wang S Y, Wu Q, Xu X C, Huang X J, Huang G L, Liu Y and Fan Z 2023 An inverse design paradigm of multi-functional elastic metasurface via data-driven machine learning *Mater. Des.* **226** 111560

[61] Liu D J, Tan Y X, Khoram E and Yu Z F 2018 Training deep neural networks for the inverse design of nanophotonic structures *ACS Photon* **5** 1365–9

[62] Malkiel I, Mrejen M, Nagler A, Arieli U, Wolf L and Suchowski H 2018 Plasmonic nanostructure design and characterization via deep learning *Light: Sci. Appl.* **7** 60

[63] Liu Z C, Zhu D Y, Rodrigues S P, Lee K T and Cai W S 2018 Generative model for the inverse design of metasurfaces *Nano Lett.* **18** 6570–6

[64] Asano T and Noda S 2018 Optimization of photonic crystal nanocavities based on deep learning *Opt. Express* **26** 32704–16

[65] Moon G, Choi J R, Lee C, Oh Y, Kim K H and Kim D 2020 Machine learning-based design of meta-plasmonic biosensors with negative index metamaterials *Biosens. Bioelectron.* **164** 112335

[66] Moon G, Son T, Lee H and Kim D 2019 Deep learning approach for enhanced detection of surface plasmon scattering *Anal. Chem.* **91** 9538–45

[67] Moon G, Lee J H, Lee H Y W, Yoo H, Ko K H, Im S and Kim D 2022 Machine learning and its applications for plasmonics in biology *Cell Rep. Phys. Sci.* **3** 101042

[68] Zhang T, Liu Q, Dan Y H, Yu S, Han X, Dai J and Xu K 2020 Machine learning and evolutionary algorithm studies of graphene metamaterials for optimized plasmon-induced transparency *Opt. Express* **28** 18899–916
[69] Banerji S, Majumder A, Hamrick A, Menon R and Sensale-Rodriguez B 2021 Ultra-compact integrated photonic devices enabled by machine learning and digital metamaterials *OSA Contin* **4** 602–7
[70] Jiang J Q, Sell D, Hoyer S, Hickey J, Yang J J and Fan J A 2019 Free-form diffractive metagrating design based on generative adversarial networks *ACS Nano* **13** 8872
[71] Jiang J Q and Fan J A 2019 Global optimization of dielectric metasurfaces using a physics-driven neural network *Nano Lett.* **19** 5366–72

Chapter 6

Terahertz metamaterial-based biosensors

E M Sheta, A B M A Ibrahim and Pankaj K Choudhury

A review is presented focusing on the basics of metamaterial design. Following this, the biosensing application of metamaterials is reviewed and discussed. Emphasis is placed on the polarization-insensitive periodically layered metamaterial configuration composed of blue glass and fluorine-doped tin oxide (FTO), and it is found that the investigated structure is useful for gamma radiation sensing. As such, the investigated metamaterial configuration can be exploited in gamma radiation dosimetry.

6.1 Introduction

Use of the terahertz (THz) frequency range has recently attracted considerable attention on account of its potential applications in a variety of technological domains [1]. It provides a unique spectral window that enables testing of materials' properties. Within this context, metamaterials have surfaced as a lucrative field of study exploring the potential of THz radiation [2, 3]. Through the process of engineering metamaterials to possess particular optical properties, such as a negative refractive index, an unprecedented ability to manipulate THz radiation is achieved [4]. These metamaterials exhibit exceptional spectral properties, most notably their ability to manipulate and direct electromagnetic waves at dimensions considerably smaller than the wavelength. Human-made meta-atoms are capable of performing a multitude of intriguing functions. In particular, metamaterials have been utilized for optical sensing, modulation, filtration, absorption, and polarization conversion at THz frequencies [5–10]. They also have potential in the development of electro-optical and other related devices that function within the THz frequency range [9, 11].

Interestingly, THz radiation has the unique property of penetrating a wide range of materials including but not limited to textiles, papers, and plastics. Consequently, it is utilized in a multitude of crucial applications, such as security, medical imaging, and sensing [6]. Due to their reliance on the interaction between light and the

doi:10.1088/978-0-7503-5462-2ch6

material being detected, conventional optical sensors have restricted measures of sensitivity, specificity, and the overall performance when it comes to sensing, in particular [12, 13]. Engineered metamaterials, however, leverage their unique property of light manipulation at the nanoscale, presenting an innovative framework for applications in optical sensing. Wide-ranging technological sectors, including but not limited to industrial process control, biomedical diagnostics, telecommunications, and environmental monitoring, exercise importance in this matter [14–21].

Tailored arrangement of meta-atoms in the configuration has the capability to induce phase transitions at the interface of metamaterial structures, which can alter the propagation of light wavefronts [22]. THz waves possess substantial benefits and vast potential for biosensing research; sensitivity enhancement remains one of the main potential benefits of THz technology [9]. By substantially enhancing the electric field and the interaction between the THz wave and analyte, metamaterials can exhibit higher sensitivity toward analyte detection [23].

Investigators reported research results focusing on the biosensing and other applications of THz metamaterial devices [24–32]. One work [33] discusses a biosensor design using graphene and metamaterials with a machine learning approach to detect *Mycobacterium tuberculosis* bacteria. The reported label-free biosensor has high sensitivity and prediction accuracy, and can be used in medical diagnostic applications. Another work [30] deals with a metamaterial-based biosensor for liquid sensing in the THz range. The sensor architecture is composed of two quartz substrates imprinted with metallic patterns and sheets. An optimally dense liquid tunnel is produced through the interposition of glass micro-pearls between the quartz substrates. To enhance the sensitivity of the design, it has been crafted to yield relatively large spectral shift per refractive index unit (RIU) by incorporating dual semi-toroidal resonators. The work reports the sensitivity of frequency detection as 52.50 GHz RIU^{-1}. The sensor is utilized experimentally to determine the concentrations of ethanol and sucrose solutions, as well as three other commonly used lubricants. The work demonstrates excellent agreement between the simulation and experimental results.

With the above background, this chapter reviews a few different types of metamaterial-based biosensors, and also incorporates some recent research results on certain layered metamaterial-based THz designs for sensing gamma radiation. However, a brief introduction to the design methodology of metamaterials is presented before venturing to investigate the configurations, designs, and modeling of biosensor devices (in the THz range) comprising nanoengineered metamaterials.

6.2 Metamaterial-based configurations

Metasurfaces are two-dimensional (2D) structures comprising subwavelength elements (or meta-atoms) arranged in a planar configuration, enabling control over the phase, amplitude, and polarization of incidence electromagnetic waves [9]. These meta-atoms are carefully engineered to achieve tailored interactions with the incoming electromagnetic radiation. They can also assume a wide variety of geometries, including the circular, elliptical, split-ring, and many other types of configurations [34–38]. Metasurfaces have remarkable potential in various fields, such as sensing, medical imaging, perfect absorption, optical chirality, superlensing, and planar filters [39–49]. The unit cells (or meta-atoms) are their fundamental components, which primarily govern the electromagnetic response of a metasurface. As such, the functionality of a metasurface is dictated by the dimensions, configuration, and arrangement of these unit cells.

Within this context, a multilayered metamaterial structure is formed by alternating subwavelength-sized layers of different materials, each having unique electromagnetic characteristics and engineered to possess distinct properties so that, when combined, these exhibit exotic electromagnetic responses [50]. By engineering these structures, the electromagnetic wave propagation and/or sustainment can be controlled and manipulated with greater precision.

6.3 Design methodology for metamaterials

6.3.1 Effective medium theory (EMT)

The Bruggeman EMT, and the Maxwell Garnett theory (MGT) are two of the most extensively applied effective medium approaches. However, a minor distinction exists between the two approaches concerning the material properties of each constituent in the mixture and the composite topology upon which they are founded. We consider here Bruggeman's EMT approach.

Let us consider a material with hyperbolic behavior (of the dispersion property) with $\bar{\bar{\varepsilon}}$ and $\bar{\bar{\mu}}$ as the relative permeability and permittivity tensors, respectively. Herein, we take the nonmagnetic nature of media, resulting in the reduction of μ to a unit tensor. Then the form of $\bar{\bar{\varepsilon}}$ is determined by diagonalization as

$$\bar{\bar{\varepsilon}} = \begin{bmatrix} \varepsilon_{xx} & 0 & 0 \\ 0 & \varepsilon_{yy} & 0 \\ 0 & 0 & \varepsilon_{zz} \end{bmatrix}. \tag{6.1}$$

The EMT proposes that the nanoengineered structure can be precisely described as a uniform uniaxial anisotropic material (i.e. $\varepsilon_{xx} = \varepsilon_{yy} = \varepsilon_{\perp}$ and $\varepsilon_{zz} = \varepsilon_{\parallel}$) exhibiting two permittivity directions, namely, one parallel to the propagation plane ($\varepsilon_{\parallel}$) and the other perpendicular to the same ($\varepsilon_{\perp}$). With the propagation of electromagnetic waves parallel to the z-axis orientation, the effective permittivity tensor takes the form

$$\varepsilon_{xx} = \varepsilon_{yy} = \varepsilon_{\perp} = \varepsilon_d \frac{[\varepsilon_m(1 + V) + \varepsilon_d(1 - V)]}{[\varepsilon_m(1 - V) + \varepsilon_d(1 + V)]} \tag{6.2}$$

$$\varepsilon_{zz} = \varepsilon_{\parallel} = V\varepsilon_m + \left(1 - V\right)\varepsilon_d, \tag{6.3}$$

where V is the volume fraction of metal in the configuration. In Bruggeman EMT, the transition between metal and insulator is modeled as percolation. The metal–insulator transition occurs when a metallic domain develops over an insulating domain, creating an inhomogeneous composite with both domains. This EMT assumes a low-volume proportion of conducting materials. This is applicable to metal–dielectric layered systems, where metal is the inclusion and dielectric is the host material.

The behavior of the composite can be modeled as an isotropic ($\varepsilon_{\perp} > 0$; $\varepsilon_{\parallel} > 0$) type I HMM ($\varepsilon_{\perp} < 0$; $\varepsilon_{\parallel} > 0$) or type II HMM ($\varepsilon_{\perp} > 0$; $\varepsilon_{\parallel} < 0$) medium, depending on how we define the composite. In addition to these characteristics, a medium is considered to be metallic if all of the permittivity components have values that are negative, and dielectric if all of the permittivity components are positive-valued. One must note that the type II HMM is highly reflective since it exhibits more metallic properties than the type I HMM [51]. On the other hand, the type I HMMs exhibit low reflection and high transmission. Chapter 3 of this book explains the HMM behavior of metamaterials in greater detail.

6.3.2 Finite integration technique

Computational methods are most frequently employed in the study of engineered structures, and they provide exact solutions to Maxwell's equations without approximation. In electromagnetic simulations, among many others, the finite integration technique (FIT) is an excellent computational method for solving Maxwell's equations [52]. Implemented on the CST Microwave Studio simulation platform, the FIT-based method discretizes the equations in both the time and space domains, and provides a framework for efficient and precise computational solutions. The differential form of Maxwell's equations describes the behavior of the electric and magnetic fields in the presence of sources and materials [53]. The FIT method approximates these equations by integrating them over finite volumes or a simulation domain. This method provides accurate results for electromagnetic simulations by directly and locally treating the equations [54].

The finite-difference time-domain (FDTD) approach [55] and the FIT share some similarities. The computational domain is split into a dual grid doublet designated G and $\widetilde{G}$. The grids are separated from one another in such a way that the center of a cell in one grid corresponds with the corner of a cell in the other grid. The grid electric voltage vector (e), magnetic voltage vector (h), magnetic induction flux (b), and electric displacement flux (d) are the state variables that make up the FIT. While G is used to describe the variables e, b and the electric charge density (q_{ev}), $\hat{G}$ is used to define d, h, and j [56]. Both the edges and the facets of the grid cells are used to establish the state variables that are associated with the $\imath$ th grid cell [57]

$$e_{\imath} = \int_{\mathscr{L}_{\imath}} E.\, ds \tag{6.4}$$

$$h_{\imath} = \int_{\mathscr{L}_{\imath}} H.\, ds \tag{6.5}$$

$$\hat{d}_{\imath} = \int_{\hat{\mathrm{A}}_{\imath}} D.\, d\mathrm{A} \tag{6.6}$$

$$\hat{b}_i = \int_{\mathrm{A}_i} B.\, d\mathrm{A} \tag{6.7}$$

$$\hat{j}_i = \int_{\hat{\mathrm{A}}_i} J.\, d\mathrm{A}, \tag{6.8}$$

when all of these scalar components are gathered in algebraic vectors e, h, $\hat{d}$, $\hat{b}$, and $\hat{j}$. Maxwell's equations can be transformed into discrete matrix–vector equations when evaluated in their integral form for each facet or cell of the mesh [57].

$$\oint_{\partial A} E.\, d\mathrm{S} = -\frac{d}{dt}\int_{\mathrm{A}} \mathrm{B}.\, d\mathrm{A} \xrightarrow{\forall\, \mathrm{A}\in\{A_i\}} Ce = -\frac{d}{dt}\hat{b} \tag{6.9}$$

$$\oint_{\partial A} H.\, d\mathrm{S} = \int_{\mathrm{A}}\left(\frac{d}{dt}D + J\right).\, d\mathrm{A} \xrightarrow{\forall\, \mathrm{A}\in\{\hat{\mathrm{A}}_i\}} \widetilde{C}h = \frac{d}{dt}\hat{D} + \hat{J} \tag{6.10}$$

$$\oint_{\partial V} D.\, d\mathrm{A} = \int_{V} q_{\mathrm{ev}}\, dV \xrightarrow{\forall\, \mathrm{V}\in\{\hat{V}_i\}} \widetilde{S}\hat{D} = q \tag{6.11}$$

$$\oint_{\partial V} B.\, d\mathrm{A} = 0 \xrightarrow{\forall\, \mathrm{V}\in\{V_i\}} S\hat{b} = 0. \tag{6.12}$$

Here (C, S) and $(\hat{C}, \hat{S})$ are support matrix operators that are discrete mappings of the differential operation's 'curl' and 'div' when defined on G and $\widetilde{G}$. Based on the topology of the dual grid doublet, in direct correspondence with the vector analytical properties of continuous operators, the topological matrices S and C, whose elements can only assume the values -1, 1, or 0, satisfy two crucial properties [56]:

$$\text{div curl} = 0 \leftrightarrow \mathrm{SC} = \mathrm{C}\,\hat{S}^T = 0 \tag{6.13}$$

$$\text{curl grad} = 0, \quad \hat{S}\hat{C} = \hat{C}S^T = 0. \tag{6.14}$$

Various consistency and conservation properties of the spatial discretization scheme, such as the stability of time domain schemes, conservation of charge and energy, and orthogonality of eigenmodes, can be derived from these relations and the duality property $\hat{S} = S^T$. Thus, FIT incorporates the discretization approximation via the constitutive material equations. By converting the constitutive relations to discrete constitutive relations, the resulting expressions are [56]

$$\hat{d} = M_{\varepsilon}e + p \tag{6.15}$$

$$\hat{j} = M_{\sigma}e \tag{6.16}$$

$$\hat{b} = M_{\mu}h + m. \tag{6.17}$$

Equations (6.15), (6.16), and (6.17) are the Maxwell grid equations (MGEs), which comprise an all-encompassing system of algebraic equations that are in the same generality as Maxwell's equations. Every matrix operator utilized in the FIT is sparse and intended for structured grid bounded matrices, which enable quick computations and efficient storage. In this situation, M_ε, M_σ, and M_μ are the matrix operators that take into account the average effects of linear polarization of material, electric conductivity, and linear magnetization. Reference [56] provides additional information on the material and geometry averaging techniques that are implemented with the FIT.

6.3.3 Boundary conditions

In the x- and y-directions, a unit cell boundary condition must be applied to analyse a nanoengineered structure [58]. This signifies that the simulation domain is divided into distinct cells, and the electromagnetic properties of the structure within each cell are accounted for in calculations [52]. The unit cell boundary conditions ensure that the electromagnetic fields at each cell boundary are properly treated, allowing for accurate simulation results [58]. In the boundary condition setup, an open z-direction boundary is implemented. This indicates that the electromagnetic waves propagate in the direction of the z-axis. An open boundary condition allows electromagnetic energy to travel freely beyond the simulation domain without being reflected [59]. This boundary condition is suitable for simulating unbounded or semi-infinite systems in which the electromagnetic waves are not limited by the physical boundaries.

6.4 Fabrication techniques

The fabrication of metamaterials is advantageously arranged according to two fundamental strategies—top-down or bottom-up. The top-down approach is exemplified by conventional microelectronic handling innovations, whereas the particulate perspective is closely associated with the rapidly expanding set of alternative nanofabrication strategies [60]. The hierarchical method has its origins in the 1950s theories of Richard Feynman, according to which, larger machines are used to assemble smaller machines, and the smaller machines are used to assemble the smallest machines, etc. Modern microlithography employs an alternative hierarchical method based on techniques such as photolithography and other advanced substances, and subtractive preparation strategies [61]. Electron beam lithography, focused ion beam lithography, colloid monolayer lithography, molecular self-assembly, electrically induced nanopatterning, rapid prototyping, x-ray lithography, chemical vapor deposition, and ion projection lithography are among the nanofabrication techniques used to integrate metamaterial devices [62–64]. However, the two-photon photopolymerization (TPP) and nanoimprint lithography are the two most promising fabrication techniques for multilayer metamaterials.

Fabrication methods based on two-photon photopolymerization (TPP) are considered to be the most promising ones for future manufacture of three-dimensional (3D) metamaterials. Direct single-beam laser writing and the multiple-beam TPP technique offer a sub-diffraction resolution of 100 nm or less. Photopolymerization permits printing with greater resolution than the previously mentioned technology [64]. However, it prints selectively by suitably altering the phase of a resin, leaving behind uncured resin and making multimaterial printing more difficult.

Nanoimprint lithography, the second fabrication method, may also be a successful technique for fabricating 3D metamaterials [62, 65]. This technology platform aims to develop concepts and instruments for the nanopatterning, non-manufacturing, and nanomanipulation of sub-10 nm structures [66, 67]. For instance, scanning probe nanolithography permits the maskless patterning of structures with dimensions as small as a few nanometers.

6.5 Metasurface-based biosensors

Varieties of metasurfaces have been investigated and found to be of great potential in biosensing applications. This section reviews a couple of relevant research reports on biosensors based on metasurface structures in the THz frequency regime.

6.5.1 Hexagonal-shaped gold resonator metasurfaces

Reference [67] reports the design of an engineered metamaterial structure to be implemented as a biosensor with high sensitivity intended for the early detection of cancer. The proposed biosensor consists of two hexagonal-shaped concentric gold resonators formed over a polyamide dielectric layer. The structure exhibits enhanced field confinement and absorption, and the results indicate a superior sensitivity and quality factor, specifically in the detection of diverse forms of cancer cells. The authors obtained the absorption spectra under the variation of the medium refractive index, namely 1.33 and 1.34, using a 1 μm thick analyte layer first. They observed that the change in analyte refractive index n from 1.33 to 1.34 results in a resonance shift of 16 GHz. They also reported a comparison of absorption between the concentric hexagonal and circular configurations of the gold rings, and found that both the configurations attain a high absorption ratio. However, a marginal improvement of 6% was observed in sensitivity when using the circular configuration- (1700 GHz RIU^{-1}) compared to the hexagonal configuration-based sensor (1600 GHz RIU^{-1}). Additionally, the quality factor achieved for the circular ring configuration was 8.38, which is lower than that obtained in the case of hexagonal design of the gold ring (yielding a quality factor of 11.25).

Using different metasurface designs, a range of sensitivity levels can be observed due to the variation in the absorption coefficient with frequency at various gold layer thicknesses. Thus, the authors of [67] reported the absorption spectra employing different values of the analyte thickness, namely 1, 3 and 5 μm, to distinguish between the cancer and normal basal cells. They found that for a 1 μm thick analyte layer the resonance frequency of normal cells is 3184 GHz, and that of the cancer cells is 3152 GHz; these values were used to calculate a shift in resonance frequency of 33 GHz. Their structure showed a sensitivity of 1600 GHz RIU^{-1}. Additionally, they found a sensitivity of 1000 GHz RIU^{-1} employing a 5 μm thick analyte layer [67].

As for the dependence of absorption on the incidence angle in a range of 0°–80°, they reported that, as the incidence angle increases, the resonance frequency undergoes blueshifts, and also the absorption peak decreases. The absorption remains above 98% across the entire incidence angle range of 0°–40°, as indicated by the high quality factor of 10.64. This way, the authors reported a 40° tolerance for the application of the incidence wave [67]. Additionally, the linearity of the proposed biosensor performance in detecting cancer at an early stage was investigated, spanning the n-values from 1.36 to 1.401 with a step size of 0.001 RIU. They found a linear response of the design over the investigated range, thereby claiming the ability of the sensor to detect cancer cells in the earliest stages with an average sensitivity of 1649.8 GHz RIU^{-1}.

In summary, the metamaterial biosensor design discussed above was suggested as a potential device for cancer cell detection, as it demonstrated encouraging outcomes across a range of applications. The design is resistant to fabrication tolerance and possesses a high degree of linearity across the introduced results.

6.5.2 Hexagonal-shaped gold resonator metasurfaces

The study in [68] discussed a THz sensor comprising double-split hexagonal ring metamaterials. In this work, the researchers developed and fabricated a double-open hexagonal ring resonator with a high refractive index sensitivity of 430 GHz RIU^{-1}. Experiments and simulations were employed to illustrate the resonant characteristics of the proposed biosensor. Bovine serum albumin (BSA) and glucose solutions of varying concentrations were utilized to evaluate the sensor performance. The sensitivity of the biosensor to glucose and BSA solutions was determined to be 0.37 GHz mg^{-1} dl^{-1} and 36.44 GHz mg^{-1} ml^{-1}, respectively, with their respective detection limits being 9.867 mg dL^{-1} and 0.16 mg ml^{-1}.

Following the results reported in [68], an important consequence of the redshift in the resonant frequency of the designed biosensor is an increase in the refractive index of the analyte with a thickness of 10 μm. An increase of 0.1 units in the refractive index of the analyte from 1.0 to 1.6 results in a redshift of the resonant frequency from 2.338 to 2.080 THz. Further, a linear fitting was executed to correlate the refractive index of the analyte with the resonant frequency of biosensor, achieving a linearity of 0.999. Additionally, the resonance frequency of biosensor redshifts as the thickness of the analyte increases from 1 to 11 μm, with its refractive index being 1.3 [68]. The authors reported that the computational results fit very well with the nonlinear function. This also determines that the biosensor is sensitive to the thickness of the analyte because it has a logarithmic form of the nonlinear fitting function, and the simulation data fit well [68].

In the same work [68], the transmission spectra of the biosensor differed when it was filled with glucose solutions of 1000, 500, 100, 50, and 10 mg dl^{-1} compared to when it was not loaded with glucose solutions. The resonance frequency of the biosensor undergoes a substantial alteration upon introducing glucose solution. This results in biosensor sensitivity to the surrounding dielectric environment, which the glucose analyte alters on the surface of the biosensor. The resonance frequency of the biosensor experiences an increasing redshift as the concentration of the glucose solution increases [68].

The authors reported the sensing outcomes of the biosensor for various dilutions of BSA solution [68]. In a similar vein, upon addition of BSA, the frequency of the biosensor experiences a redshift. Furthermore, as the dilution of BSA solution increases, the magnitude of the frequency shift diminishes [68]. In their work, the authors denoted the relative concentration of BSA content as 100, as the protein concentration in serum is subject to uncertainty. They reported the corresponding relative concentrations of BSA at dilutions of 10, 20, 50, and 100 to be 20, 10, 5, 2, and 1, respectively. They found a linear relationship between the relative concentration of BSA and the frequency shift; the linearity of this relationship was 0.995.

In summary, biomedical research in the THz regime has experienced substantial expansion, as evidenced by the multitude of investigations examining the potential of THz metamaterials as sensitive biosensors in diverse applications. The prospective impact of these developments on biomedical THz research is revolutionary, as they may facilitate the advancement of biosensors that are both more effective and sensitive, thereby offering a wide range of technological applications.

6.6 Multilayered metamaterial-based biosensors

The estimation of gamma rays is crucial for astrophysical detectors, spectroscopy, medical imaging, and national security [69, 70]. For such applications, detectors currently employ inorganic scintillators or semiconductors with direct conversion [71]. The high energy and resolution of gamma ray images make them difficult to detect. Within the context, it must be emphasized that the metamaterials exhibit potential in applications focusing on absorption, imaging, sensing, and detection [9, 22, 49, 72]. Metamaterial-based sensors and absorbers have been widely discussed in the literature [22, 73, 74]. In sensing applications of metamaterials, the use of phase change media (PCMs) has been attractive. In gamma ray sensing, depending on the dose and composition, gamma radiation can alter the refractive index of glass. In particular, gamma rays affect the optical properties of cobalt glass. Further, blue glass is a secure PCM due to its high radiation sensitivity and reversible phase shift when exposed to ionizing radiation [75].

This section discusses a low-cost multilayer metamaterial configuration for detecting gamma radiation using the blue glass PCM and fluorine-doped tin oxide (FTO). It has been found that the absorption by the structure varies significantly under different parametric and operating conditions. Under numerous gamma radiation treatments, the device exhibits narrow resonance absorption peaks that exceed 90% of the incidence waves. It can detect environmental gamma ray energy, and would have potential for use in astronomy, the detection of nuclear materials, and medicine.

6.6.1 Biosensor comprising multilayered B-co-MP and blue glass

The optical transmittance of FTO films under gamma dose excitation does not differ significantly [76]. The optical parameters of FTO [77] can be calculated using the

Drude–Lorentz model [78], which establishes the relative permittivity to be of the form

$$\epsilon(\omega) = \epsilon_\infty - \frac{\omega_p^2}{\omega^2 + i\gamma\omega} + \sum_k \frac{\omega_{pk}^2}{\omega_{0k}^2 - \omega^2 - i\gamma_k\omega}. \tag{6.18}$$

Table 6.1 shows the optical properties of FTO that we consider in our work [77]. The real and imaginary components of FTO complex permittivity are dispersive in nature across the entire wavelength range; figure 6.1 depicts the relevant plots. We observe that the real component Re(ϵ) is positive across the entire wavelength range of 1000–2500 nm, with the exception of a very narrow range of large wavelengths (~2150–2500 nm) where it is negative. Further, Re(ϵ) decreases gradually with increasing wavelength. The imaginary component Im(ϵ) grows exponentially with wavelength, and is negative in a range of ~1000–1800 nm. Reference [79] describes the effect of gamma radiation doses on the refractive index n; plots in figure 6.2 exhibit the variation in refractive index of B-co-MP (figure 6.2(a)) and blue glass (figure 6.2(b)) in response to gamma doses.

Figure 6.3 presents a schematic of the multilayer metamaterial-based sensor configuration. It consists of N periods of both the blue glass and FTO layers, and the bottom and top layers are thin films of gold (Au) and FTO glass, respectively. We

Table 6.1. Optical parameters of FTO.

ϵ_∞	ω_p (cm^{-1})	γ (cm^{-1})	k	ω_{ok} (cm^{-1})	ω_{pk} (cm^{-1})	γ_k(cm^{-1})
1.8	6466.3	1080.4	1	33 556	4189.3	6678.2
			2	25 869	1545.3	5239.3
			3	19 497	984.22	3624.4

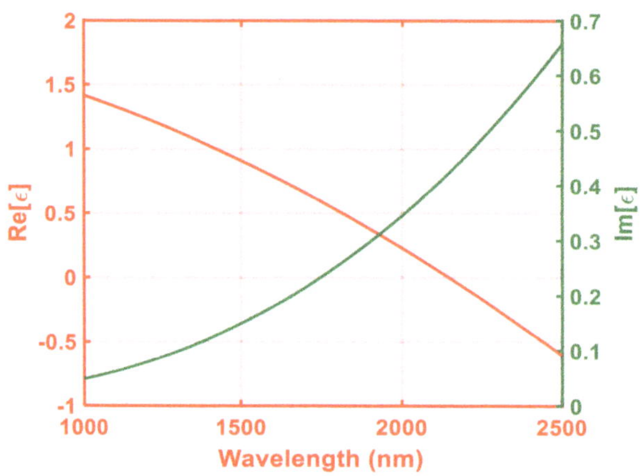

Figure 6.1. The relative permittivity of FTO.

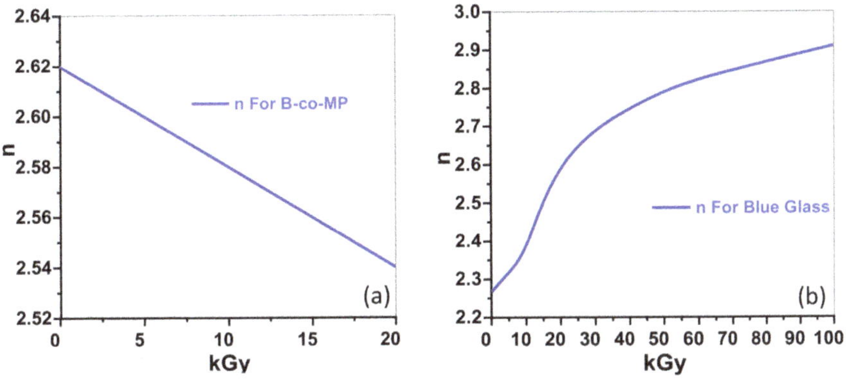

Figure 6.2. Refractive index n for (a) B-co-MP and (b) blue glass under the effect of gamma radiation doses.

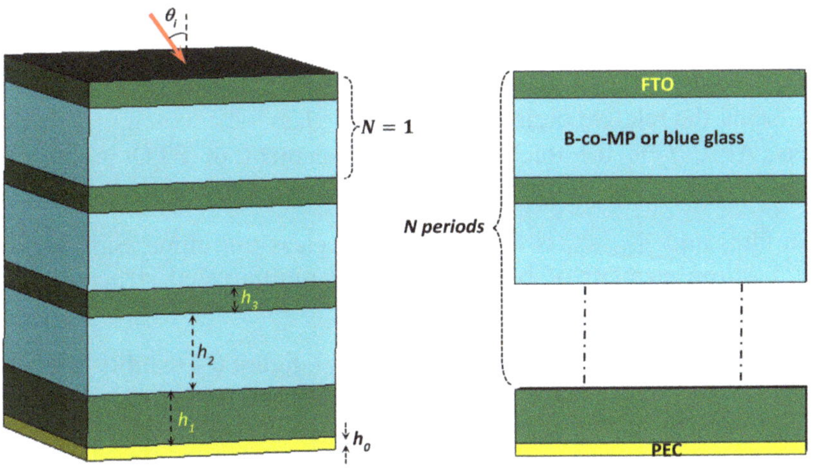

Figure 6.3. (a) 3D schematic of the proposed gamma energy sensor with $N = 3$ and (b) sideview of the structure.

find through numerical simulation that, while some geometrical parameters have fixed effects on the device performance, some others leave significant impact on the characteristic spectra. With reference to figure 6.3, we take the parametric values as $h_0 = 50$ nm, $h_1 = 100$ nm, $h_2 = 500$ nm, and $h_3 = 100$ nm. Under the effect of various gamma radiation doses, we investigate the absorption spectra in relation to changes in the number N of periodic layers.

6.6.2 Analysis

To investigate the spectral properties of the proposed multilayer sensor configuration, we employ the FIT. We impose the unit cell boundary condition on the xy-plane and the open add space condition along the z-axis, which corresponds to the incident wave propagation direction. To ensure minimal transmission, we employ a perfect Au reflector with a thickness greater than the skin depth of the incidence radiation, thereby effectively reducing the transmission coefficients $T(\lambda)$ to a vanishing value. Consequently, the absorption spectra for both the transverse electric (TE) and transverse magnetic (TM) waves can be expressed as $A(\lambda) = 1 - R(\lambda)$, where $R(\lambda)$ is the reflection coefficient. In addition, by considering the scattering S-parameters, the absorption coefficient can be expressed as $A(\lambda) = 1 - |S_{11}|^2$, where S_{11} represents the reflected energy of the metamaterial absorber.

Now, applying the EMT, we evaluate the effective permittivity of the proposed sensor configuration [66]. According to [66, 80], we plot the complex effective permittivity components $\epsilon_{\parallel}$ and $\epsilon_{\perp}$ as function of wavelength; these are given as

$$\epsilon_{\parallel}(\lambda) = f_1\, \epsilon_{\text{FTO}} + f_2\, \epsilon_{\text{blue glass}} \tag{6.19}$$

and

$$\epsilon_{\perp}(\lambda) = \frac{\epsilon_{\text{blue glass}} {}^{*} \epsilon_{\text{FTO}}}{f_2\, \epsilon_{\text{FTO}} + f_1\, \epsilon_{\text{blue glass}}}, \tag{6.20}$$

with ϵ_{FTO} being the relative permittivity of FTO glass, whereas $\epsilon_{\text{blue glass}}$ is that of the blue glass. Also, f_1 is the thickness fraction element of FTO in the multilayer configuration, defined as $f_1 = \frac{T_{\text{FTO}}}{T_{\text{FTO}} + T_{\text{blue glass}}}$, where T_{FTO} is the total thickness of FTO thin films and $T_{\text{blue glass}}$ is that of the blue glass thin films. Similarly, f_2 is the thickness fraction element of blue glass in the multilayered structure, defined as $f_2 = \frac{T_{\text{blue glass}}}{T_{\text{FTO}} + T_{\text{blue glass}}}$.

Using the multilayer configuration in figure 6.3, figure 6.4 exhibits the plots of the effective permittivity (against wavelength) for both the prime and gamma doses of 10 kGy, considering $N = 2$ and the parametric values as $f_1 = 0.33$ and $f_2 = 0.67$. Both of the real components $\Re(\epsilon_{\perp})$ and $\Re(\epsilon_{\parallel})$ show decreasing positive values; however, $\Re(\epsilon_{\perp})$ becomes negative at wavelengths above 2450 nm. In contrast, the

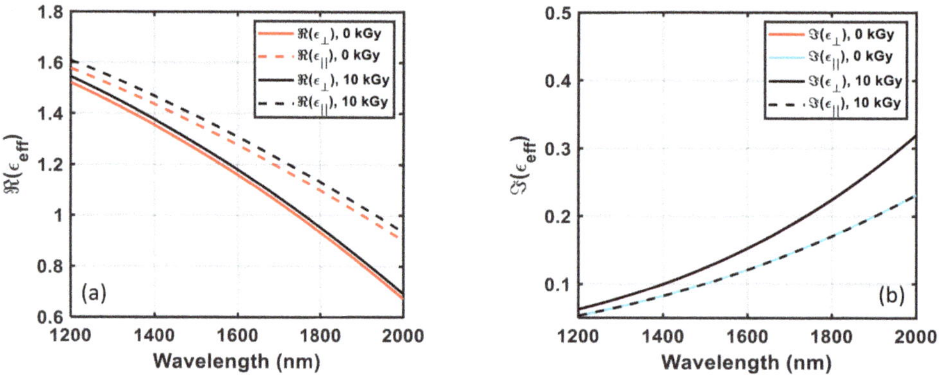

Figure 6.4. Plots of the parallel ($\parallel$) and perpendicular ($\perp$) components of effective permittivity ϵ: (a) real and (b) imaginary parts corresponding to $N = 2$.

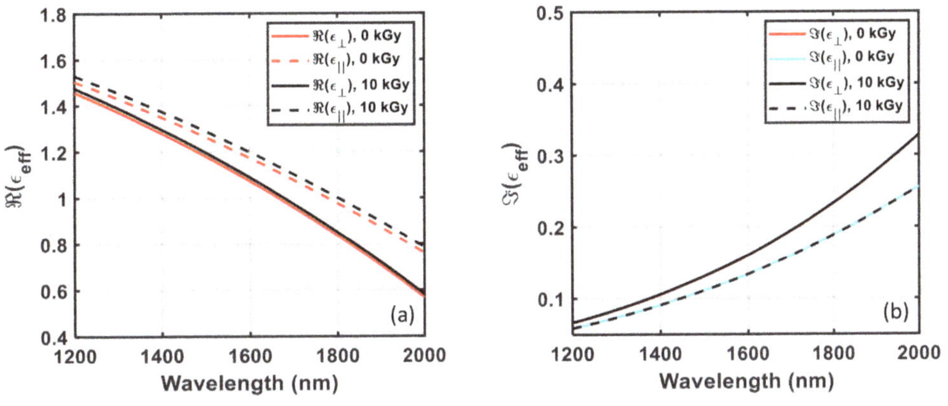

Figure 6.5. Plots of the parallel (‖) and perpendicular (⊥) components of effective permittivity ϵ: (a) real and (b) imaginary parts corresponding to $N = 4$.

imaginary components $\Im(\epsilon_\perp)$ and $\Im(\epsilon_\parallel)$ increase exponentially throughout the entire wavelength regime. Figure 6.5 depicts similar dispersion plots considering another illustrative example of the case where $N = 4$ and all other parameters are unchanged. Clearly, an increase in the number of layers results in a rise in the volume fraction element of blue glass; in our case $f_2 = 0.74$. Comparing the cases of $N = 2$ and $N = 4$, we observe that the imaginary component of effective permittivity $\Im(\varepsilon_{\mathrm{eff}})$ exhibits a similar behavior, with only small differences in their values.

6.6.3 Absorption characteristics

We aim to evaluate device performance when employing B-comp, a PCM known for its gamma ray sensitivity [81]. We investigate the spectral characteristics of the layered metamaterial-based sensor configuration (in figure 6.3) in a wavelength range spanning from 1200 to 2000 nm; figure 6.6 exhibits the absorption response considering a different number N of periods in the layered structure. We clearly observe that the use of B-comp for all cases of N demonstrates low sensitivity.

Due to the limited sensitivity of the B-comp, we now analyse the structure when it comprises blue glass and FTO. In this context, we first focus on the spectral properties of the proposed layered sensor configuration (figure 6.3) comprising blue glass. We consider obtaining the absorption spectra corresponding to the TE- and TM-polarized incidence excitations under the normal and oblique incidence of waves. Increasing the number of stacked layers or the layer thickness can independently tune the presence of surface plasmon resonance (SPR) [82]. That is, this would permit precise SPR localization as well as the construction of multilayer structures that interact strongly with specific wavelength. We keep the layer thickness constant and obtain the spectral response by varying the number of layers.

Figure 6.7 depicts the absorption spectra obtained with the TE-polarized incidence excitation with four different numbers of periods, namely $N = 2$, $N = 3$, $N = 4$, and $N = 5$, and under three different gamma radiation doses, namely, 0 kGy (i.e. in the absence of gamma radiation), 5, and 10 kGy. We clearly

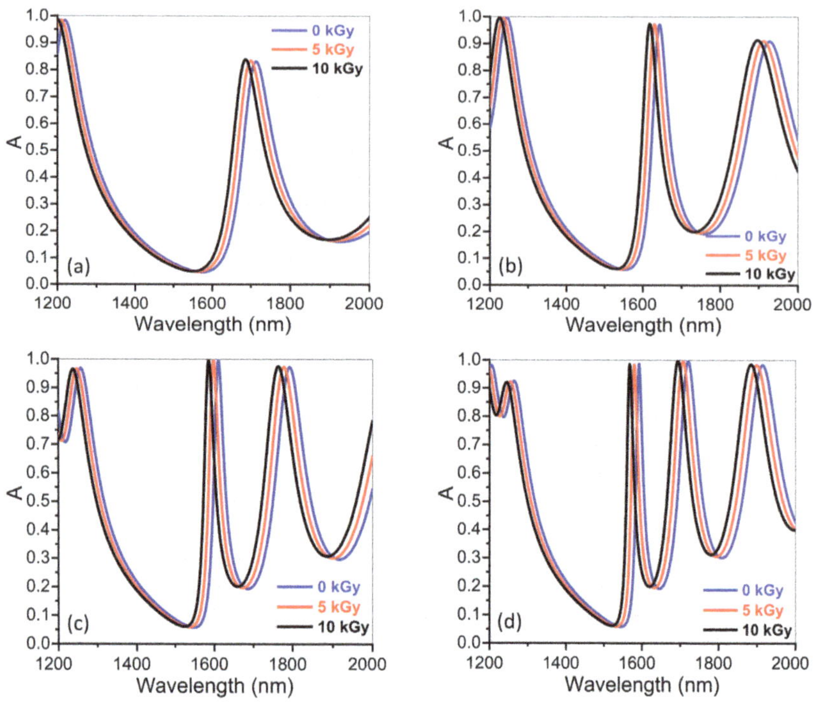

Figure 6.6. Absorption characteristic of the multilayer structure using B-comp, and considering (a) $N = 2$, (b) $N = 3$, (c) $N = 4$, and (d) $N = 5$.

observe the response of the device under varying gamma doses. The increase in the amount of radiation and/or the number of periods redshifts the resonance absorption peaks. In addition, the increase in N results in greater absorption with sufficiently large Q-factor, which makes the device highly sensitive when constructed with a relatively large N-value.

Figure 6.8(a) depicts the absorption spectra under the excitation of the TE and TM waves and varying gamma doses, namely 0–15 kGy with 5 kGy step increases. In these plots, we use $N = 3$. We observe that the absorption plots for both types of incidence excitation are very similar, which determines the configuration to be a polarization-independent device. Figure 6.8(b) depicts the spectral characteristics corresponding to varying polarization angle over a range of 0°–90° at a step of 5°. To plot this figure, we consider the illustrative case of absence of gamma dose (i.e. 0 kGy), also known as the initial state. As the figure depicts, the results demonstrate a stable spectral response throughout the entire range of ϕ. Consequently, the absorption properties remain unaffected due to changes in polarization angle.

6.6.4 Power loss and density under gamma doses

To understand the behavior of the proposed sensing device at the resonance frequency, we choose a resonance wavelength of 1515 nm corresponding to $N = 2$ and 5 kGy of gamma dose (figure 6.7(a)). Figure 6.9(a) depicts the power absorbed

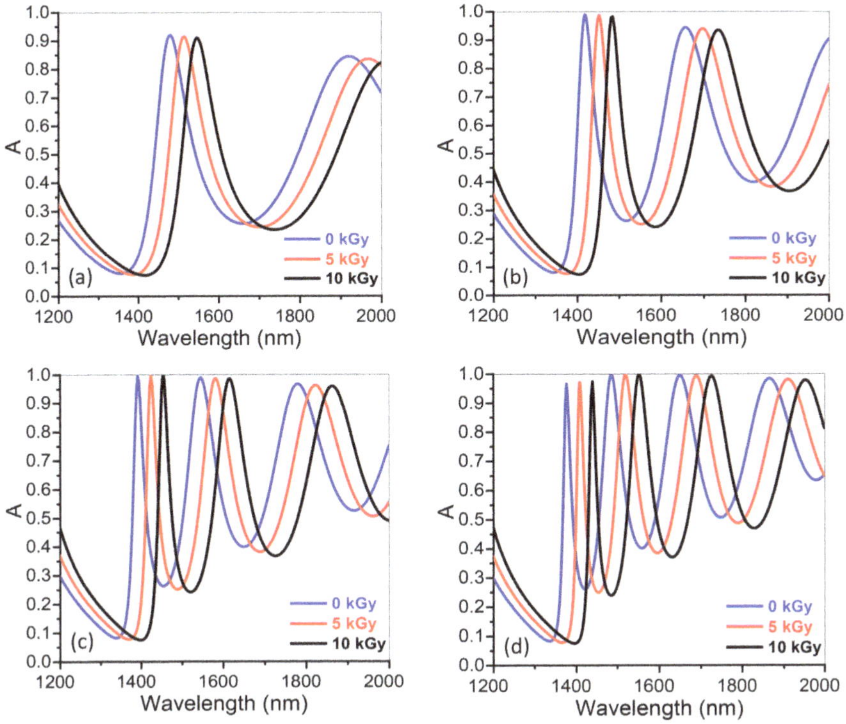

Figure 6.7. Absorptance of the layered structure comprising blue glass, and with (a) $N = 2$, (b) $N = 3$, (c) $N = 4$, and (d) $N = 5$.

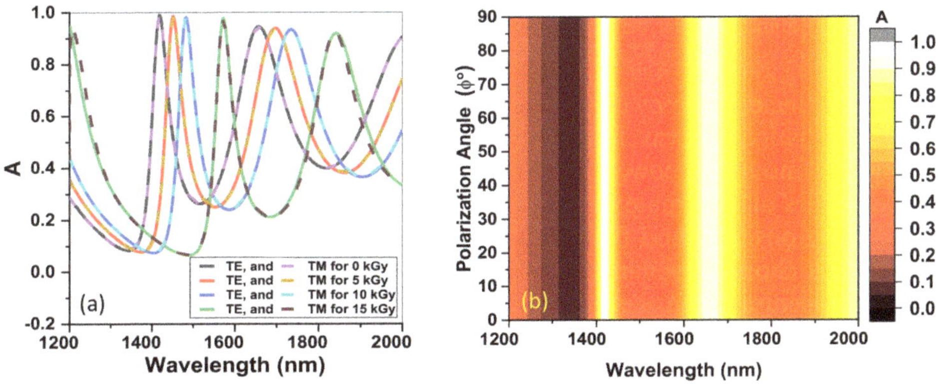

Figure 6.8. (a) Absorption spectra obtained for both the TE and TM waves for $N = 3$, considering different values of gamma radiation doses, and (b) polarization angle (ϕ) vs wavelength at the prime state 0 kGy.

by each medium in the metamaterial-based device comprising nanolayers of blue glass and FTO. Upon being impinged by the incidence radiation, coupling between adjacent layers (of the structure) causes hybridization of plasmonic modes, resulting in the formation of new resonance modes that are localized in distinct regions of the multilayer configuration. It is evident from figure 6.9(a) that FTO absorbs the

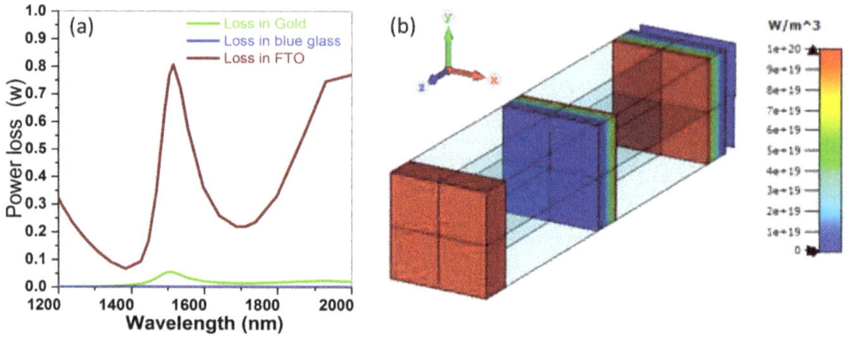

Figure 6.9. Plots of (a) power loss (in each material) against wavelength, and (b) power density at a wavelength of 1515 nm; all under a dose of 5 kGy and $N = 2$.

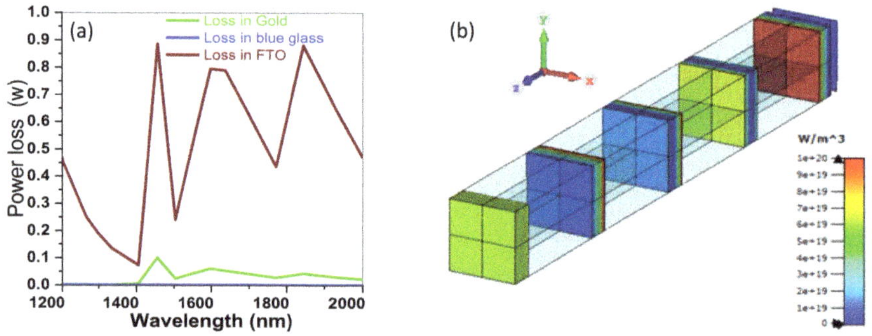

Figure 6.10. Plots of (a) power loss (in each material) against wavelength and (b) power density at a wavelength of 1515 nm; all under the dose of 10 kGy and $N = 4$.

majority of incidence power, whereas blue glass is nearly transparent to the incidence radiation. At wavelengths greater than 1300 nm the thin gold film absorbs a negligible amount of power. Figure 6.9(b) depicts the power density patterns at different layers of metamaterial configuration under the same operating and parametric conditions. We find that the top FTO layer absorbs most of the incidence power at the chosen resonance wavelength of 1515 nm. The remaining absorbed power is concentrated on the upper surface of the bottom FTO layer and the lower surface of the middle FTO layer. This happens because of the power coupling between the middle and bottom FTO layers.

Considering another illustrative case of a 1456 nm resonance absorption wavelength corresponding to $N = 4$ and a gamma dose of 10 kGy (figure 6.7(c)), figure 6.10(a) depicts the power absorbed by each layer (of the metamaterial configuration) and figure 6.10(b) the power density patterns. We see that the FTO layer absorbs the maximum amount of energy. Due to a change in the number of periodic stacks, however, the coupling between the layers is altered, resulting in different power density confinements between the FTO layers, as compared to the situation depicted in figure 6.9(b) (corresponding to $N = 2$).

As stated previously, the optical properties of our configuration change in response to the incidence of gamma radiation intensity, to which the resulting SPR is highly sensitive. Figure 6.11 depicts the initial absorption spectra of blue glass (i.e. in the absence of the gamma radiation). We see that in the operating wavelength regime, the device exhibits narrow absorption peaks suitable for bandpass or bandstop filtration. The number of these (absorption) peaks increases with increasing number of layers in the metamaterial structure, and the positions of these (peaks) shift toward achieving nearly perfect absorption.

Figure 6.12 exhibits the plots of resonance wavelength shifts with the incident gamma radiation strength considering different numbers of layers in the configuration. We observe from this figure that the response is nearly linear and the absorption peaks undergo redshift with increasing gamma radiation exposures. Thus, the radiation of higher doses yields a more pronounced shift in absorption peaks. In an effort to estimate the average sensitivity S, this can be expressed as the shift in resonance wavelength with respect to a change in gamma radiation dose, i.e. $S = \Delta\lambda/\Delta\gamma$. Following this, we find the sensitivity to be $S = 13$ nm kGy^{-1} for $N = 2$, $S = 12.25$ nm kGy^{-1} for $N = 3$, $S = 11.9$ nm kGy^{-1} for $N = 4$, and $S = 11.85$ nm kGy^{-1} for $N = 5$. Clearly, as the number of layers increases, the sensitivity of the proposed layered metamaterial device decreases slowly.

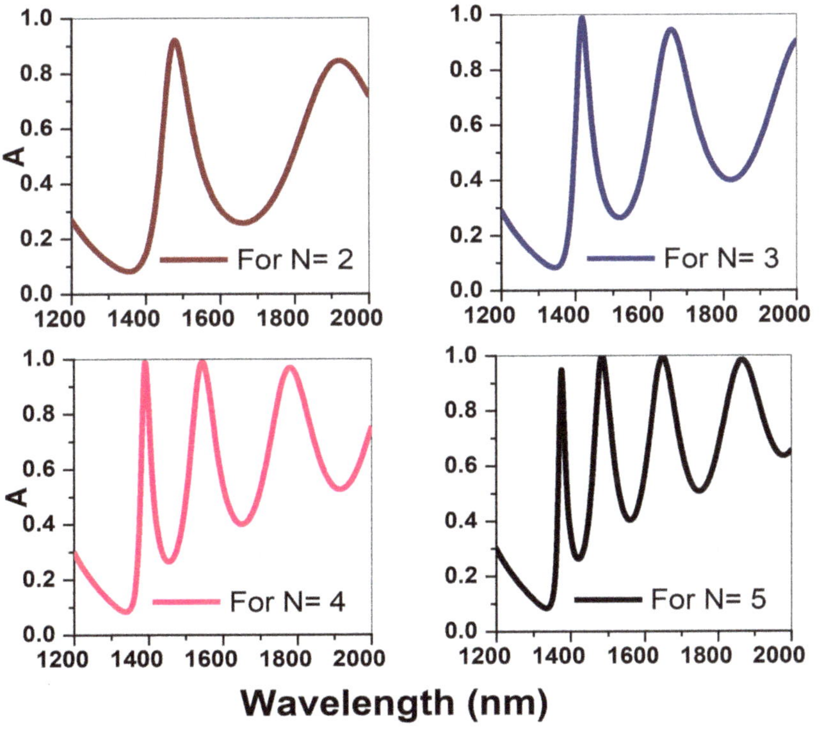

Figure 6.11. Absorptance of the layered metamaterial structure in the prime state corresponding to different values of N.

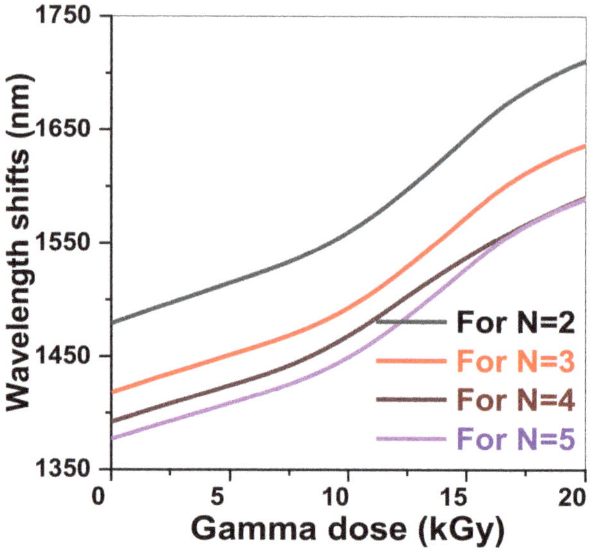

Figure 6.12. Plot of resonance wavelength shifts against the gamma radiation dose corresponding to various N-values.

6.7 Conclusion

In summary, this chapter provides a brief description of biosensing applications for metamaterials. Apart from the various basics related to the design of metamaterials, the chapter focuses on a few varieties of metamaterial configurations to detect biological analytes. There is a major discussion of the polarization-insensitive periodically layered metamaterial configuration comprising blue glass and FTO to be implemented for gamma radiation sensing. Emphasizing the absorptance of the structure, it is reported that the absorption peaks redshift with an increasing number of periods (in the structure). Further, increasing the gamma radiation dose redshifts these peaks. These results indicate that the investigated metamaterial can be exploited in gamma radiation dosimetry using a periodically layered blue glass–FTO configuration.

References

[1] Zhang X -C and Xu J 2010 *Introduction to THz Wave Photonics* (Berlin: Springer)

[2] Chen H *et al* 2008 Electromagnetic metamaterials for terahertz applications *Terahertz Sci.* **1** 42–50

[3] Choudhury P K and El-Nasr M A 2015 Complex metamaterial mediums and THz applications *J. Electromagn. Waves Appl.* **29** 2405–7

[4] Withayachumnankul W and Abbott D 2009 Metamaterials in the terahertz regime *IEEE Photon. J.* **1** 99–118

[5] Chen H -T, Padilla W J, Zide J M O, Gossard A C, Taylor A J and Averitt R D 2006 Active terahertz metamaterial devices *Nature* **444** 597–600

[6] Moghaddas S, Ghasemi M, Choudhury P K and Majlis B Y 2018 Engineered metasurface of gold funnels for terahertz wave filtering *Plasmonics* **13** 1595–601
[7] Pourmand M and Choudhury P K 2020 Wideband THz filtering by graphene-over-dielectric periodic structures with and without MgF_2 defect layer *IEEE Access* **8** 137385–94
[8] Bilal R M H *et al* 2020 On the specially designed fractal metasurface-based dual-polarization converter in the THz regime *Res. Phys.* **19** 103358
[9] Choudhury P K 2022 *Metamaterials—Technology and Applications* (Boca Raton, FL: CRC Press)
[10] Pourmand M and Choudhury P K 2022 Nanostructured strontium titanate perovskite hyperbolic metamaterial supported tunable broadband THz Brewster modulator *IEEE Trans. Nanotechnol.* **21** 586–91
[11] Choudhury S M *et al* 2018 Material platforms for optical metasurfaces *Nanophoton* **7** 959–87
[12] Naresh V and Lee N 2021 A review on biosensors and recent development of nanostructured materials-enabled biosensors *Sensors* **21** 1109
[13] Prabowo B A, Purwidyantri A and Liu K C 2018 Surface plasmon resonance optical sensor: a review on light source technology *Biosensors* **8** 80
[14] Wei J, Ren Z and Lee C 2020 Metamaterial technologies for miniaturized infrared spectroscopy: Light sources, sensors, filters, detectors, and integration *J. Appl. Phys.* **128** 240901
[15] Abdulkarim Y I *et al* 2020 Design and study of a metamaterial based sensor for the application of liquid chemicals detection *J. Mater. Res. Technol.* **9** 10291–304
[16] Tan T *et al* 2019 Renewable energy harvesting and absorbing via multi-scale metamaterial systems for Internet of things *Appl. Energy* **254** 113717
[17] Kadic M, Milton G W, van Hecke M and Wegener M 2019 3D metamaterials *Nat. Rev. Phys.* **1** 198–210
[18] Wang P, Nasir M E, Krasavin A V, Dickson W, Jiang Y and Zayats A V 2019 Plasmonic metamaterials for nanochemistry and sensing *Acc. Chem. Res.* **52** 3018–28
[19] Babicheva V E 2023 Optical processes behind plasmonic applications *Nanomater.* **13** 1270
[20] Góra P and Łopato P 2023 Metamaterials' application in sustainable technologies and anintroduction to their influence on energy harvesting devices *Appl. Sci.* **13** 7742
[21] Shen S *et al* 2022 Recent advances in the development of materials for terahertz metamaterial sensing *Adv. Opt. Mater.* **10** 2101008
[22] Padilla W J and Averitt R D 2022 Imaging with metamaterials *Nat. Rev. Phys.* **4** 85–100
[23] Xiong Z *et al* 2021 Terahertz sensor with resonance enhancement based on square split-ring resonators *IEEE Access* **9** 59211–21
[24] Sarychev A K and Tartakovsky G 2007 Magnetic plasmonic metamaterials in actively pumped host medium and plasmonic nanolaser *Phys. Rev.* B **75** 085436
[25] Chen C, Gao S, Song W, Li H, Zhu S N and Li T 2021 Metasurfaces with planar chiral meta-atoms for spin light manipulation *Nano Lett.* **21** 1815–21
[26] Ullah Z *et al* 2023 Interparticle-coupled metasurface for infrared plasmonic absorption *IEEE Access* **11** 41546–55
[27] Ou H, Lu F, Xu Z and Lin Y 2020 Terahertz metamaterial with multiple resonances for biosensing application *Nanomater.* **10** 1038
[28] Ou H, Lu F, Liao Y, Zhu F and Lin Y 2020 Terahertz metamaterial with multiple resonances for biosensor application *Proc. 2020 Opto-Electron. Commun. Conf. (OECC)* pp 1–3

[29] Yao H, Mei H, Zhang W, Zhong S and Wang X 2022 Theoretical and Experimental Research on Terahertz Metamaterial Sensor With Flexible Substrate *IEEE Photon. J.* **14** 3700109

[30] Deng G, Fang L, Mo H, Yang J, Li Y and Yin Z 2022 A metamaterial-based absorber for liquid sensing in terahertz regime *IEEE Sens. J.* **22** 21659–65

[31] Kumari R, Sharma S, Varshney S K and Lahiri B 2022 Dual-band biosensing with van der Waals assisted optical metasurfaces *IEEE Sens. J.* **22** 15953–60

[32] Sreekanth K V, Luca A D and Strangi G 2015 Hyperbolic metamaterials: design, fabrication, and applications of ultra-anisotropic nanomaterials *NanoSci. Technol.* **100** 447–67

[33] Parmar J, Patel S K, Katkar V and Natesan A 2023 Graphene-based refractive index sensor using machine learning for detection of mycobacterium tuberculosis bacteria *IEEE Trans. Nanobiosci.* **22** 92–8

[34] Baqir M A and Choudhury P K 2012 On the energy flux through a uniaxial chiral metamaterial made circular waveguide under PMC boundary *J. Electromagn. Waves Appl.* **26** 2165–75

[35] Ghasemi M and Choudhury P K 2016 Metamaterial absorber comprised of butt-facing U-shaped nanoengineered gold metasurface *Energies* **9** 451

[36] Bilal R M H, Baqir M A, Choudhury P K, Ali M M and Rahim A A 2020 Tunable and multiple plasmon-induced transparency in a metasurface comprised of silver s-shaped resonator and rectangular strip *IEEE Photon. J.* **12** 4500613

[37] Bilal R M H *et al* 2020 Elliptical metallic rings-shaped fractal metamaterial absorber in the visible regime *Sci. Rep.* **10** 14035

[38] Bilal R M H *et al* 2021 Wideband microwave absorber comprising metallic split-ring resonators surrounded with E-shaped fractal metamaterial *IEEE Access* **9** 5670–7

[39] Baqir M A, Ghasemi M, Choudhury P K and Majlis B Y 2015 Design and analysis of nanostructured subwavelength metamaterial absorber operating in the UV and visible spectral range *J. Electromagn. Waves Appl.* **29** 2408–19

[40] Ghasemi M, Baqir M A and Choudhury P K 2016 On the metasurface based comb filters *IEEE Photon. Technol. Lett.* **28** 1100–3

[41] Baqir M A and Choudhury P K 2017 Hyperbolic metamaterial-based UV absorber *IEEE Photon. Technol. Lett.* **29** 1548–51

[42] Ghasemi M and Choudhury P K 2018 Complex copper nanostructures for fluid sensing—a comparative study of the performance of helical and columnar thin films *Plasmonics* **13** 131–9

[43] Baqir M A and Choudhury P K 2018 Toward filtering aspects of silver nanowire-based hyperbolic metamaterial *Plasmonics* **13** 2015–20

[44] Baqir M A and Choudhury P K 2019 Design of hyperbolic metamaterial-based absorber comprised of Ti nanoshperes *IEEE Photon. Technol. Lett.* **31** 735–8

[45] Baqir M A and Choudhury P K 2019 On the VO2 metasurface-based temperature sensor *J. Opt. Soc. Am.* B **36** F123–30

[46] Sreekanth K V, Mahalakshmi P, Han S, Mani Rajan M S, Choudhury P K and Singh R 2019 Brewster mode-enhanced sensing with hyperbolic metamaterial *Adv. Opt. Mater.* **7** 1900680

[47] Baqir M A, Choudhury P K, Naqvi Q A and Mughal M J 2020 On the scattering and absorption properties of SiO_2–VO_2 core–shell nanoparticle under different thermal conditions *IEEE Access* **8** 84850–7
[48] Ghasemi M, Roostaei N, Sohrabi F, Hamidi S M and Choudhury P K 2020 Biosensing application of all-dielectric SiO_2–PDMS meta-stadium grating nanocombs *Opt. Mater. Express* **10** 1018–33
[49] Pourmand M and Choudhury P K 2023 ght–matter interaction at the sub-wavelength scale: pathways to design nanophotonic devices *Adventures in Contemporary Electromagnetic Theory* ed T G Mackay and A Lakhtakia (Cham: Springer)
Yang Y *et al* 2019 Graphene-based multilayered metamaterials with phototunable architecture for on-chip photonic devices *ACS Photon.* **6** 1033–40
[50] Shekhar P, Atkinson J and Jacob Z 2014 Hyperbolic metamaterials: fundamentals and applications *Nano Converg* **1** 14
[51] Clemens M and Weiland T 2001 Discrete electromagnetism with the finite integration technique *Prog. Electromagn. Res.* **32** 65–87
[52] Clemens M and Weiland T 2001 Discrete electromagnetics: Maxwell's equations tailored to numerical simulations *Int. Compumag. Soc. Newsl.* **8** 13–20
[53] Schuhmann R and Weiland T 2004 Recent Advances in finite integration technique for high frequency applications ed W H A Schilders, E J W Ter Maten and S H M J Houben *Scientific Computing in Electrical Engineering Mathematics in Industry* **4** (Berlin: Springer) pp 46–57
[54] Weiland T 1996 Time domain electromagnetic field computation with finite difference methods *Int. J. Numer. Model. Electron. Networks, Dev. Fields* **9** 295–319
[55] Clemens M, Gjonaj E, Finder P and Weiland T 2000 Numerical simulation of coupled transient thermal and electromagnetic fields with the finite integration method *IEEE Trans. Magn.* **36** 1444–7
[56] Weiland T *et al* 2001 *Ab initio* numerical simulation of left-handed metamaterials: comparison of calculations and experiments *J. Appl. Phys.* **90** 5419–24
[57] Bakır M, Karaaslan M, Dincer F and Sabah C 2017 Metamaterial characterization by applying different boundary conditions on triangular split ring resonator type metamaterials *Int. J. Numer. Model. Electron. Networks, Dev. Fields* **30** e2188
[58] Ye D *et al* 2012 Towards experimental perfectly-matched layers with ultra-thin metamaterial surfaces *IEEE Trans. Antennas Propagat.* **60** 5164–72
[59] Yoon G, Kim I and Rho J 2016 Challenges in fabrication towards realization of practical metamaterials *Microelectron. Eng.* **163** 7–20
[60] Chen Y 2015 Nanofabrication by electron beam lithography and its applications: a review *Microelectron. Eng.* **135** 57–72
[61] Boltasseva A and Shalaev V M 2008 Fabrication of optical negative-index metamaterials: recent advances and outlook *Metamater* **2** 1–17
[62] Chen H-T *et al* 2007 Ultrafast optical switching of terahertz metamaterials fabricated on ErAs/GaAs nanoisland superlattices *Opt. Lett.* **32** 1620–2
[63] Yuan X *et al* 2021 Recent progress in the design and fabrication of multifunctional structures based on metamaterials *Curr. Opin. Solid State Mater. Sci.* **25** 100883
[64] Yoon G, Tanaka T, Zentgraf T and Rho J 2021 Recent progress on metasurfaces: applications and fabrication *J. Phys. D: Appl. Phys.* **54** 383002

[65] Cai W and Shalaev V 2010 *Optical Metamaterials—Fundamentals and Applications* (New York: Springer)
[66] Surdo S, Duocastella M and Diaspro A 2021 Nanopatterning with photonic nanojets: review and perspectives in biomedical research *Micromachines* **12** 256
[67] Azab M Y, Hameed M F O, Nasr A M and Obayya S S A 2021 Highly sensitive metamaterial biosensor for cancer early detection *IEEE Sens. J.* **21** 7748–55
[68] Cai W, Zhu J, Yang Y, Wang X, Qian Z and Fan S 2023 High performance of terahertz sensor based on double-split hexagonal ring metamaterial *IEEE Sens. J.* **23** 22414–20
[69] Wei H and Huang J 2019 Halide lead perovskites for ionizing radiation detection *Nat. Commun.* **10** 1066
[70] Kakavelakis G, Gedda M, Panagiotopoulos A, Kymakis E, Anthopoulos T D and Petridis K 2020 Metal halide perovskites for high-energy radiation detection *Adv. Sci.* **7** 2002098
[71] Chen H-T, Taylor A J and Yu N 2016 A review of metasurfaces: physics and applications *Rep. Prog. Phys.* **79** 076401
[72] Chen F, Cheng Y and Luo H 2020 Temperature tunable narrow-band terahertz metasurface absorber based on InSb micro-cylinder arrays for enhanced sensing application *IEEE Access* **8** 82981–8
[73] Saadeldin A S, Hameed M F O, Elkaramany E M A and Obayya S S A 2019 Highly sensitive terahertz metamaterial sensor *IEEE Sens. J.* **19** 7993–9
[74] Rehren T 2001 Aspects of the production of cobalt-blue glass in Egypt *Archaeometry* **43** 483–9
[75] Oryema B, Jurua E, Madiba I G, Nkosi M, Sackey J and Maaza M 2020 Effects of low-dose γ-irradiation on the structural, morphological, and optical properties of fluorine-doped tin oxide thin films *Radiat. Phys. Chem.* **176** 109077
[76] Miranda H *et al* 2019 Effects of changes on temperature and fluorine concentration in the structural, optical and electrical properties of SnO_2:F thin films *J. Mater. Sci., Mater. Electron.* **30** 15563–81
[77] Kadi M, Smaali A and Outemzabet R 2012 Analysis of optical and related properties of tin oxide thin films determined by Drude–Lorentz model *Surf. Coat. Technol.* **211** 45–9
[78] Abdeldaym A, Sallam O I, Ezz-Eldin F M and Elalaily N A 2021 Influence of gamma irradiation on the optical, thermal and electrical features of blue commercial glass as potential accident dosimetry *J. Phys. Chem. Solids* **157** 110196
[79] Ferrari L, Wu C, Lepage D, Zhang X and Liu Z 2015 Hyperbolic metamaterials and their applications *Prog. Quantum Electron.* **40** 1–40
[80] Sheta E M, Azlan Hamzah A, Azmi U Z M, Mansor I and Choudhury P K 2024 On the γ-radiation dosimetry using a layered metamaterial structure comprising FTO and blue glass *IEEE Trans. Nanotechnol.* **23** 158–63
[81] Prajapati Y K, Pal S and Saini J P 2018 Effect of a metamaterial and silicon layers on performance of surface plasmon resonance biosensor in infrared range *Silicon* **10** 1451–60
[82] Esfandiyari M, Lalbakhsh A, Jarchi S, Ghaffari-Miab M, Mahtaj H N and Simorangkir R B V B 2022 Tunable terahertz filter/antenna-sensor using graphene-based metamaterials *Mater. Des.* **220** 110855

Chapter 7

Ordered and disordered metamaterials—biosensing perspective

Masih Ghasemi and Pankaj K Choudhury

Ordered and disordered structures of metamaterials are briefly reviewed and discussed, considering arrayed forms of configurations with emphasis on their biosensing applications. In particular, the exemplified ordered metamaterial comprises a stadium nanocomb grating structure that can be used in glucose monitoring considering different concentrations of glucose aqueous solution, and the disordered metamaterial configuration assumes the case of BALB/c rat dried DNA thin films of different DNA concentrations deposited over a gold nanolayer. For both kinds of structural configurations, spectral characteristics are evaluated under plasmonic conditions, which determine the investigated approaches to be viable in biosensing.

7.1 Introduction

Metasurfaces are artificially engineered to study the relations of their shapes, geometries, or even the orientations of their constituents along with the electromagnetic waves propagating in them. Modulations of the magnitude, phase, and polarization of electromagnetic waves are the key parameters in evaluating the *on-demand* performance/suitability of metasurface thin films [1–8]. Engineering of the surface of such thin films at nanoscale is practically periodic or non-periodic in nature. Within the context of metasurface-based biosensors, in the case of non-periodic structural designs, an optimization technique is generally implemented to enhance the optical response, whereas in the case of periodic surfaces this response is almost unique and optimized to a certain wavelength region [9–16].

In the context of metamaterials, surface waves exist at the metal–dielectric or dielectric–dielectric interface(s), and the complexity of the roughness at nanoscale and the properties of patterned materials grown over a dielectric platform fundamentally tailor the wave propagation behavior [18–23]. Investigators have reported the generation of plasmonic and Tamm waves, among other surface waves [24, 25].

doi:10.1088/978-0-7503-5462-2ch7

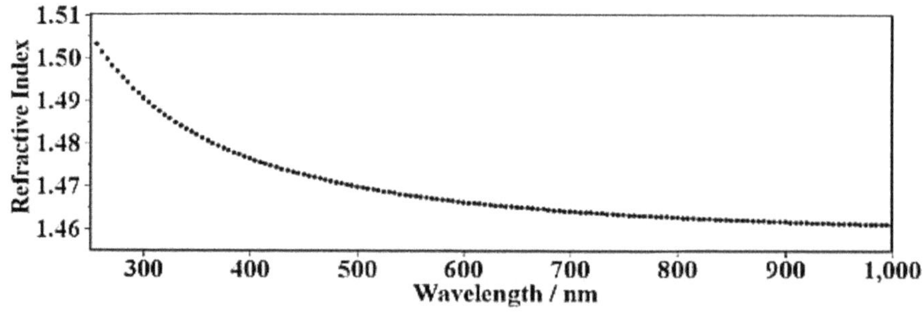

Figure 7.1. Refractive index variation of SiO_2 dielectric film in the visible range. Reproduced from [17].

Tamm surface waves form in circumstances wherein at least one of the two partnering materials is periodically nonhomogeneous normal to the interface guiding the waves [26]. The presence of Tamm waves has been proved experimentally for natural and artificially designed periodic partnering materials [27–29].

The silica (SiO_2) layer as photonic crystals (PCs) with potential localization and confinement of light and small variation in the refractive index (RI) changes in the visible range (figure 7.1) is used extensively in photonic and bio-photonic applications [30, 31]. Creating and/or tailoring of surface wave propagation in complex media is feasible by combining layers of different negative RIs. In this context, the propagation of electrons in sub-wavelength-sized periodic metasurfaces, upon being irradiated by the incident light, is analogous to what electron waves demonstrate in the band gaps of crystalline atomic lattice [32, 33]. The analytical treatment of PCs and solid state crystals in practice is the form of conversion of the solutions of Maxwell's equations into Bloch–Floquet equations. Hence, for instance, if defects such as electronic dopants in solid crystals are present, these behave as point-like cavities, and linear waveguides may lead to the localization of surface waves in artificial defects in PCs [34–36]. Following these explanations, crystals can be considered as waveguides where their physical properties, such as sharp bends and cavities of the wavelength scale dimensions, lead to prefect confinement [37] of energy.

This chapter discusses the propagation of surface waves in two different types of metasurface configurations, namely ordered and disordered metasurfaces. Clearly, the former deals with configurations where the unit cells have certain kinds of defined geometries/shapes, whereas the latter focuses more on completely undefined geometries. We study both types of these structures in the context of biosensing applications considering illustrative measurand examples, and name the designs ordered and disordered meta-biosensors.

7.2 Surface waves in ordered meta-biosensors

We explore Tamm surface waves as a key factor in biosensing applications through spectroscopic ellipsometry at the micron-scale, roughness infiltrated with abundant monosaccharide, also known as glucose. The process for fabrication and spectrometry analysis of the homogeneous and non-homogeneous partnering materials, such as polydimethylsiloxane (PDMS) and engineered symmetric micro-stadium SiO_2 nanocombs, is the key point to discuss. PDMS, with outstanding properties such as chemical stability, optical transparency (from the UV to IR), mechanical compliance, good permeability to gases, thermal stability, and biocompatibility, allows us to select it as the isotropic partnering material [38, 39]. The other participating material is SiO_2 due to its low-cost, long-term mechanical stability, strong adhesion with PDMS, and biocompatibility in sensing applications [40–43]. From the fabrication point of view, further emphasis on the role of SiO_2 is related to the stadium nanocomb structure and the resemblance of this to the defect of the PCs. Metasurfaces in this form play the role of exhibiting a forbidden electron state in a crystal surface [44]. We also analyse the result of an ellipsomtery measurement when the void regions of the ordered meta-biosensors infiltrate different concentrations of glucose in its aqueous solution form. It must be noted that the existence of the Tamm state in a non-metasurface model due to the roughness of an atomic scale is extremely difficult [45].

7.2.1 Fabrication process of micro-stadium SiO_2 nanocombs

The fabrication of periodic structure is a very important step in the mechanism of producing ordered meta-biosensors. Figure 7.2 summarizes the fabrication steps for the micro-projection imprinting of stadium nanocombs on silicon (Si) molds. It is an imprinting lithography process; figure 7.2(a) illustrates an Si mold with micro-sized hollow stadium-shaped grooves, whereas figure 7.2(b) exhibits an array of hollow grooves filled with PDMS [46, 47]. Within this context, figure 7.3 exhibits the unit cell structure of the stadium nanocomb (figure 7.3(a)), and a three-dimensional (3D) view of the overall nanocomb (figure 7.3(b)) comprising the unit cell array of the Si mold.

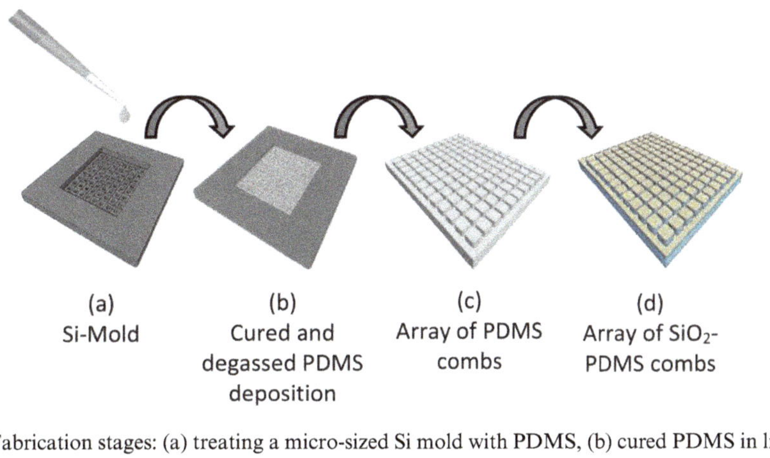

Figure 7.2. Fabrication stages: (a) treating a micro-sized Si mold with PDMS, (b) cured PDMS in liquid phase poured on the Si mold, (c) separation after baking and imprinting processes, and (d) silica-coated micro-sized PDMS array.

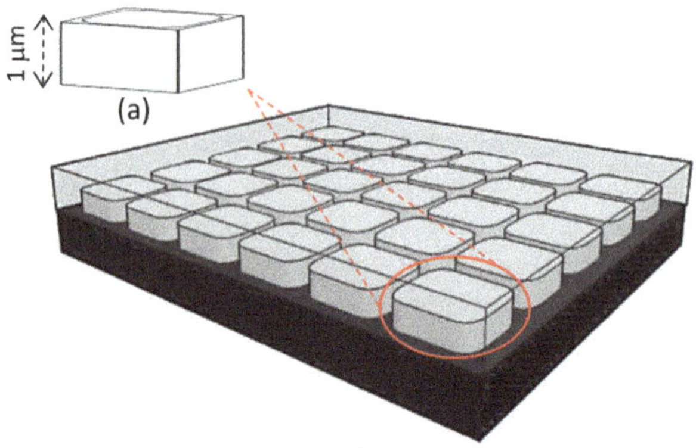

Figure 7.3. (a) Stadium nanocomb unit cell, and (b) schematic 3D view of the unit cell array of the Si mold.

PDMS as an elastomer with a photoresist property is popular in the micro-/nano-fabrication industry. One of the important features of PDMS is its transparency at optical frequencies (240–1100 nM) [48], apart from its biocompatibility feature, which makes it a well-suited material to mimic, for example, blood vessels [49]. PDMS can also be a stable barrier with very low permeability against solvents such as sugar and glucose [50]. Thus, PDMS is promising in applications related to biosensor fabrication.

Among many other PDMS products, Sylgard 184 is widely used in soft lithographic replication of microstructures in microfluidic and microengineering applications [51]. It is resistant to oxidation and hydrolysis, and also thermally stable and modifiable in the microfabrication process [52]. The silicone elastomer kit has two chemical components—the base and curing agents, which can be mixed at different aspect ratios. In practice, however, the standard mixture ratio for fabrication devices is 10:1 [53]. As both agents are thick and viscous in nature, the mixture must be homogenized through a mixer. The homogenized mixture of Sylgard 184 agents exhibit time-dependent flowability, as figure 7.4 shows. Thus, fast pouring of Sylgard 184 mixture over the mold surface is very important in fabrication. The other important consideration during the pouring process is avoiding the formation of bubbles between the mold surface and mixture, which would make the structure less efficient for sensor application or channel quality for micro-channel applications. In this study, we captured the mold with micro-sized features using a CCD sensor module.

The fabrication process for Sylgard 184 mixture involves degassing by keeping it in vacuum, after which the mold enters the curing stage. This includes several steps, such as baking (first step: 30 min at 50 °C, second step: 15 min at 100 °C, and final step: 15 min at 100 °C). The real conversion from Sylgard 184 mixture into PDMS take places after baking in the right time sequences. After proper baking, PDMS with a micro-sized array of stadium combs can be separated from the mold (figure 7.2(c)) upon cooling down (for 24 h) in the normal temperature environment. The last stage of fabrication is popping the silicon layer with 100 nm over the PDMS

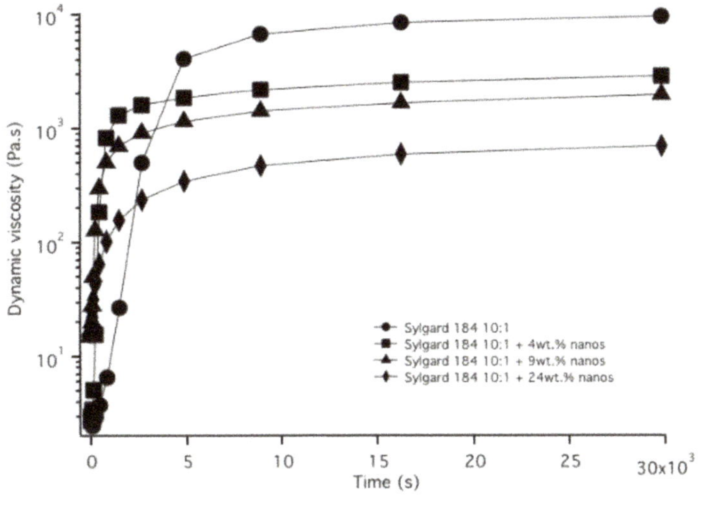

Figure 7.4. Dynamic viscosity dependence of PDMS on time. Adapted from [53], CC BY 4.0.

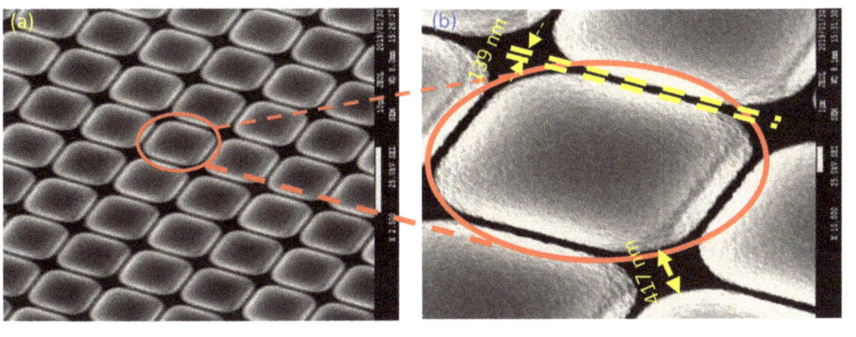

Figure 7.5. (a) Fabricated PDMS stadium combs, and (b) a magnified view.

combs using the sputtering technique (figure 7.2(d)). Figure 7.5(a) shows an SEM image of the finally fabricated grating silica–PDMS nanocombs, the magnified form of which in figure 7.5(b) shows the dimensional features of the same. This structure with nanoscale dimensions can be used as an ordered meta-biosensor.

7.2.2 Ellipsometry method

The ellipsometry technique is an optical measurement method used extensively in characterizing the reflection (or transmission) of light from reflective (or transparent) samples [54–60]. The fundamental feature of ellipsometry is based on measuring alterations in the features of polarized light upon reflection from a sample. The geometrical orientation of oscillations in polarized light carries identical information, which is usable in the optical characterization of samples with micro/nano features [59, 60]. In chemistry and/or biochemistry, ellipsometry of polymer films, self-assembled monolayers, proteins, RNA, DNA, and many other media can be used for optical characterization.

The spectroscopic technique is non-destructive, noncontact, and non-invasive in nature, and only performs with variation in the polarization state of light [61, 62]. The optical properties or physical characteristics (such as thickness, etc) cannot be measured by ellipsometry directly [63]. However, it is useful in the characterization of unknown samples. The fewer the thin film layers, the more efficient the analysis that can be made. The polarization state of light (or the orientation and phase of electric field vector) is the key parameter in elliptical spectrometry. There are many orientations that can be expressed for light polarization, but two special cases of elliptical polarization are circular and linear polarization. When the orthogonal waves are 90° out of phase at equal amplitudes, the polarization is circular. Figure 7.6 exhibits the details of this polarization before and after interacting with thin film. In this specific example, both the linearly polarized waves R_s and R_p are in the same phase, i.e. $\Delta\phi = 0$, while after reflection from complex structured optical thin film, the phase difference ($\Delta\delta \neq 0$) appears between the elliptically polarized waves r_s and r_p.

Assuming a certain optical ellipsometry setup, the polarization state between two orientations can be fixed to any angular value. However, we set this value to 90° in order to emphasize that the polarization is linear and the optical tool, such as a Glan–Taylor prism, can be utilized to produce in-phase polarized waves before interaction with the film surface [64, 65]. This prism has a beam splitter composed of two identical triangular iceland crystals (i.e. crystallized calcium carbonate) with two equal wedge angles, and an air gap on the interface of the left–right crystals. It is evident from figure 7.7 that the optical axis of each triangle is aligned to the plane of reflection. Further, the point where the incident light interacts with the air gap is roughly close to the Brewster angle, and perfectly polarizes the unpolarized wave

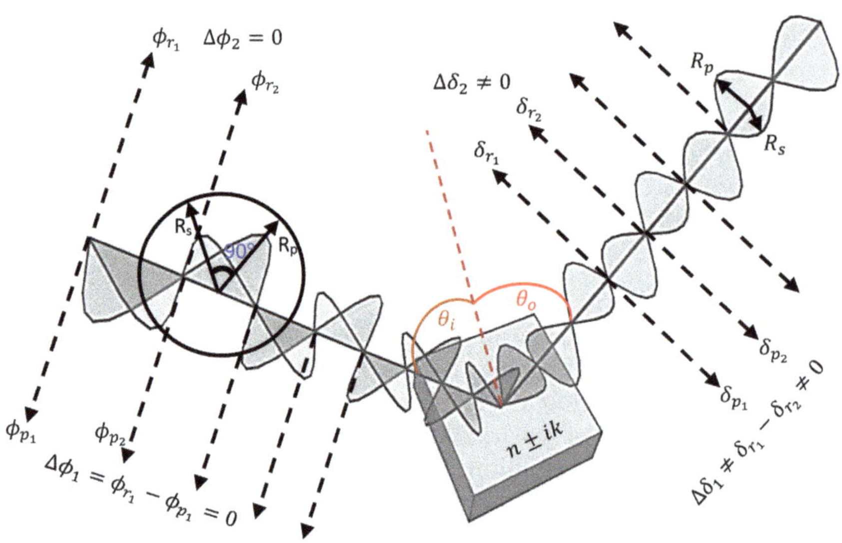

Figure 7.6. Schematic of phase and amplitude changes after reflection of linearly polarized light from the thin film with nanoscale roughness.

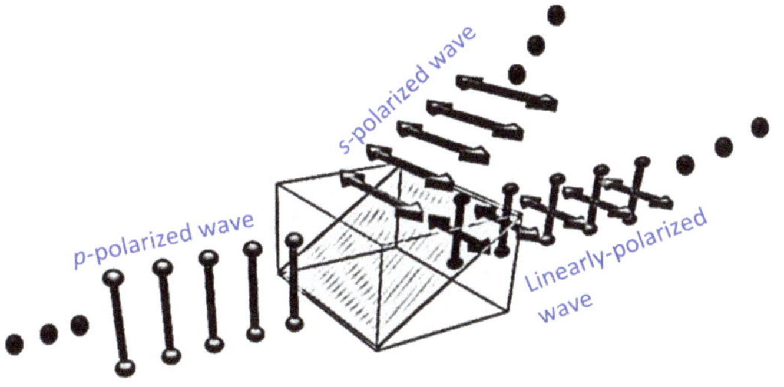

Figure 7.7. Schematic of generating the *p*- or *s*-wave from unpolarized wave using a Glan–Taylor prism.

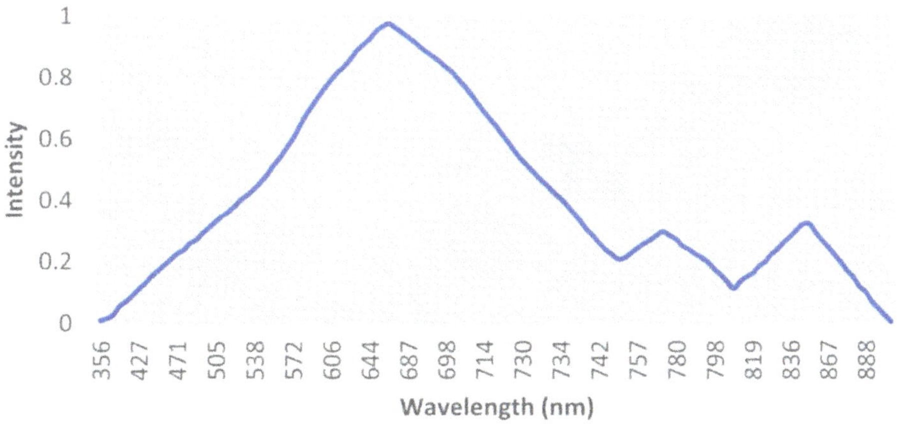

Figure 7.8. Visible broadband spectral response of the halogen lamp.

into the *s*- and *p*-polarizations. These two waves, after passing through the transparent Glan–Taylor prism, are no longer orthogonal to each other; they are in the same phase even when the light source is broadband and non-coherent [66–68].

Practically, any film containing a nanoscaled multilayer structure or two-dimensional (2D) nanostructure over its surface has unknown complex RI, the dependence of which on the light wavelength makes the method of ellipsomtery spectrometry a strong tool to identify surface waves [69–71].

7.2.2.1 Experimental setup and measurement

In experimental measurements, we use a broadband laser source, a laser collimator, a polarizer, converging lenses, a graduated rotary sample, and a broadband spectrometer to compute the parameters Ψ and Δ. The halogen fiber optic illuminator (THORLABS product) is a broadband light source with an output power of 150 W and a spectral wavelength range of 300–900 nm. Figure 7.8 illustrates the spectral response that we consider in our experiment.

Figure 7.9 shows a schematic for the essential elements for treating light before and after interaction with the thin film. Efficient non-coherent broadband light alignment in a certain direction requires a collimator [72]. As figure 7.9 shows, we use a combination of a lens and a pinhole (1 mm diameter) and a G10 Glan–Taylor prism (THORLABS) in the transmission arm. The prism supports a wavelength range of 300 nm to 2.3 μm and has a Brewster's angle of 68°, at which the *p*-wave component is partially separated from the *s*-wave so that it acts as an optical splitter (figure 7.7). The lens used in the reflection arm focuses the beam into the aperture of a photodetector. Figure 7.10 depicts the experimental setup used in the ellipsometry measurement. Herein, the implementation of variable arms practically requires a motorized stage to displace the transmission and reflection arms simultaneously.

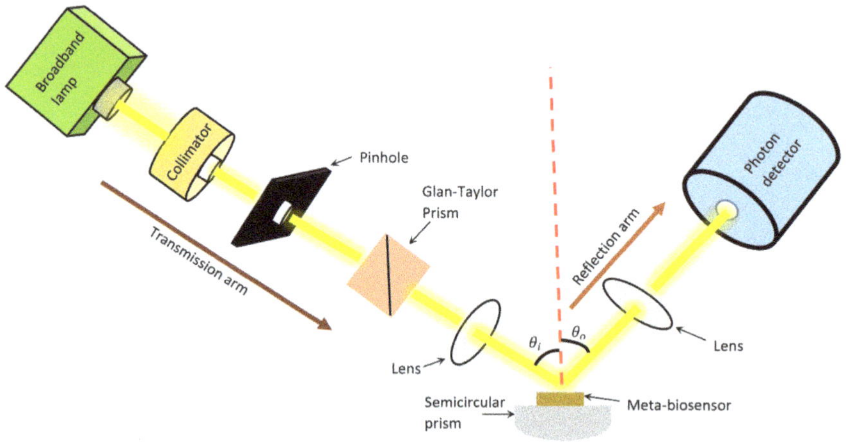

Figure 7.9. Schematic of the optical setup used in ellipsometry.

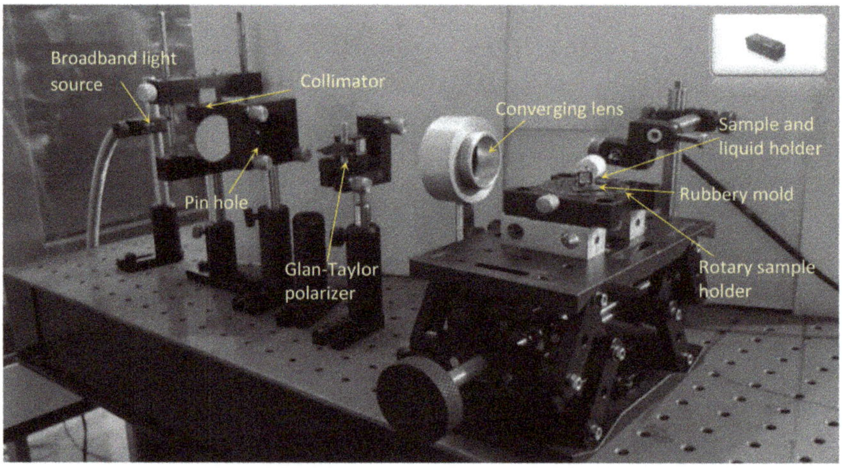

Figure 7.10. Practical measurement setup suitably configured for extraction of ellipsomtery parameters Ψ and Δ.

Furthermore, the critical parameter in the detection and generation of surface waves is the values of the θ_i and θ_o angles. The angle sweeping in the actual setup (figure 7.10) has (fixed) transmission and reflection arms, and the sweeping is only performed by manual rotation of the prism and film mounted over the rotary holder.

In our experiment, we initially keep the broadband spectrum at an angle of 10° by rotary holder. During this process, the photodetector should show the spectral peak, which confirms the condition $\theta_i = \theta_o$; any misalignment of angle results in a vanishing peak. Further, the periphery of the semicircular prism orientation must be aligned with the incident light beam. Thus, the surface of the biosensor must be perpendicular to the plane of rotation. For the fluid measurand (which is glucose in our experiment), to attain optimized interaction with the surface of the thin film and also to avoid leakage from the contact, the surface must be inside the sample and liquid holder. The inset of figure 7.10 presents an image of the sample and liquid holder, the design of which allows the user to replace the sample with a new one after every measurement. Furthermore, we designed a kind of (rectangular) rubbery mold to stick to the semicircular prism and fix the substrate (of the meta-biosensor) to the flat side of the half-circular prism.

7.3 Ordered meta-biosensor analysis and measurement using glucose samples

Figure 7.11 depicts a 3D view of the unit cell for the aforementioned fabricated PDMS-based two-layer meta-biosensor comprising SiO_2 and a PDMS comb structure. The parametric details include the height and width of this unit cell without considering the existing curved edges in the actually fabricated configuration. It also shows the features of one comb design considering the substrate, void region, and fluid measurand. As can be seen, a 100 nm thick SiO_2 layer mounted over the PDMS structure was formed by homogeneous sputtering. Clearly, the SiO_2 layer spreads over the PDMS combs and also the blank space between the neighboring cells during sputtering. We keep the PDMS comb height and width as 250 and 1000 nm, respectively.

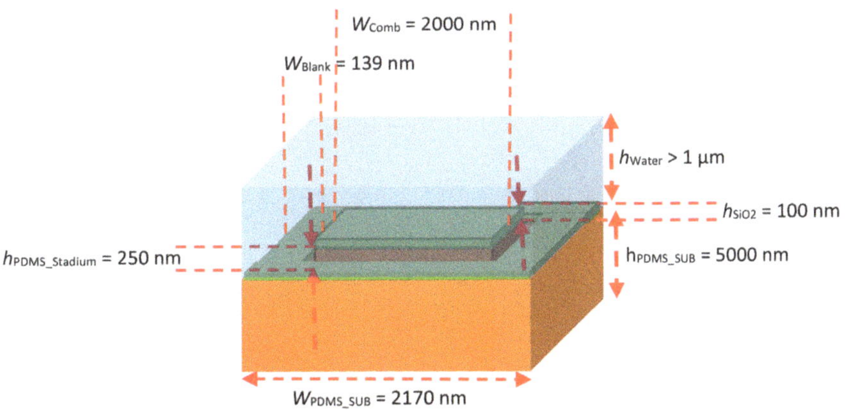

Figure 7.11. Schematic of the unit cell of the biosensor.

Beside the all-dielectric property of the comb metasurface, analysis of the effective permittivity in the visible operating wavelength is essential before moving further to the ellipsomtery results. In computations, since the liquid measurand is an additional layer over the sensor, the fill factor (τ) should be of the form (figure 7.11)

$$\tau = \frac{h_{SiO_2}}{h_{SiO_2} + h_{water/glucose}}, \tag{7.1}$$

which can be used to determine the effective permittivity [73] as

$$\varepsilon_{eff} = \frac{\varepsilon_{SiO_2}(\lambda)(2(1-\tau)\varepsilon_{SiO_2}(\lambda) + (1+2\tau)\varepsilon_{water/glucose}(\lambda)}{(2+\tau)\varepsilon_{SiO_2}(\lambda) + (1-\tau)\varepsilon_{water/glucose}(\lambda)}. \tag{7.2}$$

The RI vs wavelength plots (figure 7.12) of both the SiO_2 and PDMS show a little nonlinearity in the visible range. Meanwhile, the permittivity of SiO_2 in the visible range can be expressed as [74]

$$\varepsilon_{SiO_2} = \frac{0.69617\lambda^2}{\lambda^2 - 0.0684043^2} + \frac{0.4079426\lambda^2}{\lambda^2 - 0.1162414^2} + \frac{0.8974794\lambda^2}{\lambda^2 - 9.896161^2} + 1. \tag{7.3}$$

We now attempt to evaluate the effective permittivity of the developed meta-biosensor considering water and aqueous solutions of different sugar concentrations as measurands with three different thickness values, 25, 50, and 100 nm; figure 7.13 exhibits the results. We see slowly reducing permittivity with increasing wavelength, which is more prominent in the low wavelength range, 200–500 nm. The effective permittivity enhances with increasing sugar concentration and also the thickness of the measurand. This is very much expected owing to the enhanced impedance of the medium in either case. However, its features are almost unaltered upon increasing the measurand thickness from 50 (figure 7.13(b)) to 100 nm (figure 7.13(c)), except the fact that the media become more nondispersive in the latter case. Thus, the role of metasurface becomes more prominent, corresponding to smaller measurand

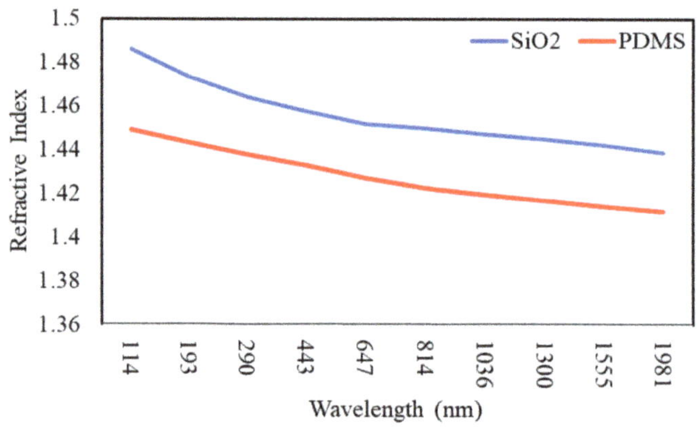

Figure 7.12. Wavelength dependence of refractive index of the two component materials PDMS and SiO_2 used in developing the biosensor.

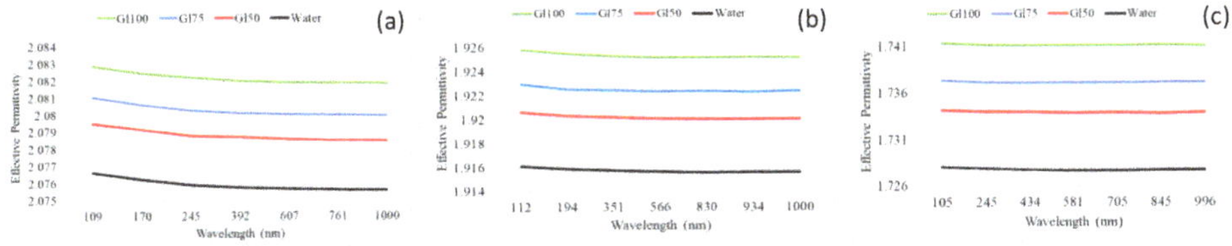

Figure 7.13. Plots of the effective permittivity against wavelength corresponding to the measurand heights (or thicknesses)—(a) 25 nm, (b) 50 nm, and (c) 100 nm—mounted over the ordered meta-biosensor.

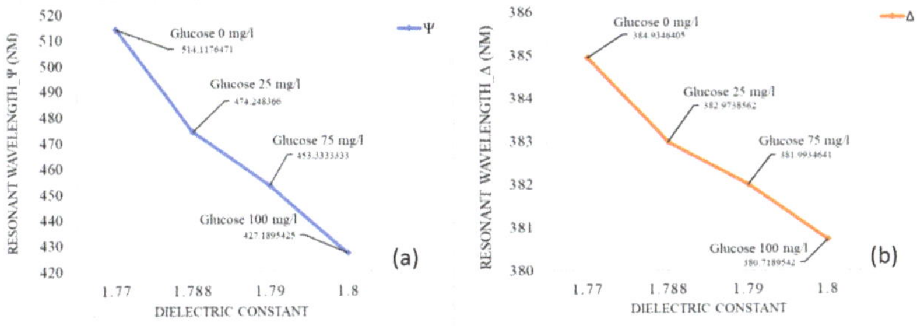

Figure 7.14. Permittivity relation with resonant wavelengths and ellipsomtery parameters Ψ (a) and Δ (b).

thickness. These results also make it clear that the total volume of solution should be maintained constant for each experimental observation as the effective permittivity varies due to this factor as well as the glucose concentration.

Before moving on to the experimental evaluation of the infiltrated ordered meta-biosensor with aqueous solutions, direct ellipsometry measurements of glucose at different wavelengths provide a better understanding of concentration variation. Figure 7.14 shows a measurement analysis of the parameters Ψ and Δ in this regard. This figure shows a trend of permittivity variation of distilled water ($\varepsilon_{\text{Water}}$) and that of the aqueous solutions of different sugar concentrations, as stated above. We see that water and glucose concentrations of 50, 75, and 100 mg l^{-1}, respectively, yield permittivity values of $\varepsilon_{\text{Water}} = 1.776809020$, $\varepsilon_{\text{Glu}_{50}} = 1.7881305841$, $\varepsilon_{\text{Glu}_{75}} = 1.7939387844$, and $\varepsilon_{\text{Glu}_{100}} = 1.80123241$. The resonant wavelength for Ψ corresponding to each sugar concentration is unique, and its value decreases from higher wavelength (i.e. $\lambda = 515$ nm) to lower wavelength (i.e. $\lambda = 420$ nm). The parameter Δ also presents a similar trend of a downward slope with minor differences in the resonant wavelength corresponding to each sugar concentration value. For instance, the resonant wavelength for distilled water in the case of Δ is ~384.65 nm, and that for the aqueous solutions of glucose is ~383 nm (50 mg l^{-1}), ~381.9 (75 mg l^{-1}), and ~380.55 (100 mg l^{-1}), respectively.

7.3.1 Experimental results for glucose monitoring

Figure 7.15 shows the experimentally obtained spectral results while utilizing the ordered meta-biosensor with measurands as stated above. In all plots, we obtain a sharp peak reflectance against wavelength in a range of 200–1050 nm while using

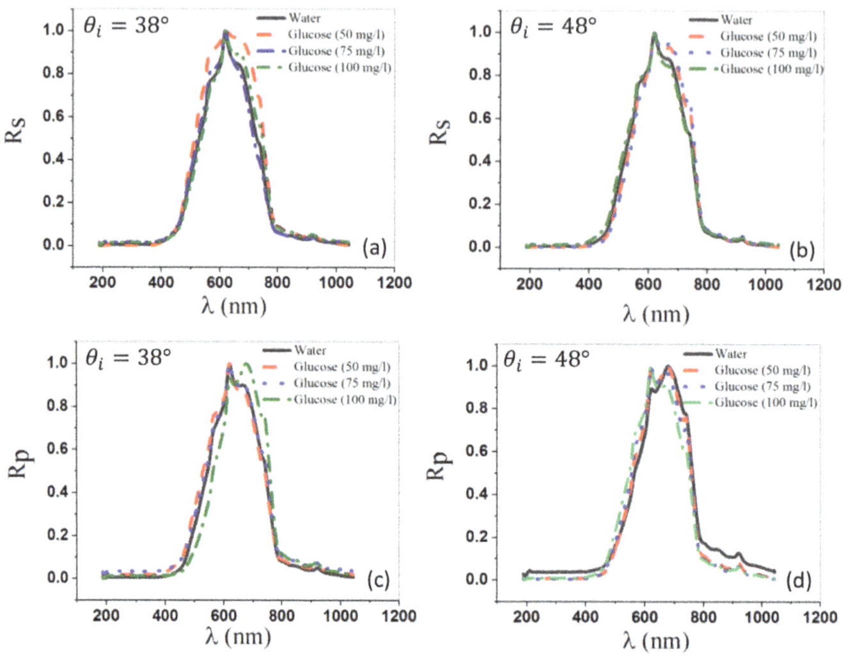

Figure 7.15. Reflectance of the *s*- and *p*-polarized waves at incidence angles of 38° and 48°.

glucose solutions of different concentrations. We see the reflectance peaks exhibit small shifts with altering glucose concentration. This can, however, be eliminated by using other analytical technique such as the kernel cumulative sum algorithm, or experimental techniques such as the lock-in and signal amplification, which can magnify small shifts in resonance condition and improve the spectral quality [75, 76]. In the present work, we implement the method of optical ellipsometry to characterize the reflection of light from the measurand samples. In figure 7.15, R_s and R_p represent the normalized reflectance obtained corresponding to the *s*- and *p*-polarized waves, respectively. We vary the incidence angle over a range of 30°–60° with an interval of 5°, and find reflectance to be the maximum corresponding to 38° and 48°. Table 7.1 shows peak reflectance (i.e. the resonance condition) obtained corresponding to these two angular values while using different measurands.

Figure 7.15 and table 7.1 make it clear that the shifts in reflectance peaks are small, and may not be convenient for using in sensing measurands. Under the circumstances, elliposomtery is a strong tool to expand the scope of the ordered meta-biosensor in sensing with a better view at the shifts. Many experiments and commercialized products implement the relation in RI and measurand concentration [77–80]. Surface waves and complex RI are the two important factors of any optical meta-biosensor [81–83]. The ellipsometry parameters $\Psi(\lambda)$ and $\Delta(\lambda)$ depend on the measurand RI, incidence angle, and wavelength, and at a specific angle these may yield a unique response indicating the resonance behavior. Figure 7.16 illustrates such a resonance wavelength plot of $\Psi(\lambda)$ and $\Delta(\lambda)$ upon varying RI.

Table 7.1. Peak reflectance obtained at different wavelengths corresponding to 38° and 48° oblique incidence angles using different measurands.

Angle of incidence	Measurand (mg l^{-1})	Resonance wavelength (nm)	R_s	R_p
38°	Distilled water	617.7	0.9930	—
38°	Glucose	619.4	1.0	—
38°	Glucose	620.67	0.9979	—
38°	Glucose	621.94	0.9957	—
38°	Distilled water	619.4	—	0.999
38°	Glucose	619.4	—	0.998
38°	Glucose	620.67	—	0.999
38°	Glucose	675.28	—	0.999
48°	Distilled water	619.4	0.999	—
48°	Glucose	621.94	0.998	—
48°	Glucose	621.94	0.998	—
48°	Glucose	621.94	0.995	—
48°	Distilled water	678.20	—	0.999
48°	Glucose	675.28	—	0.999
48°	Glucose	619.4	—	0.997
48°	Glucose	619.4	—	0.998

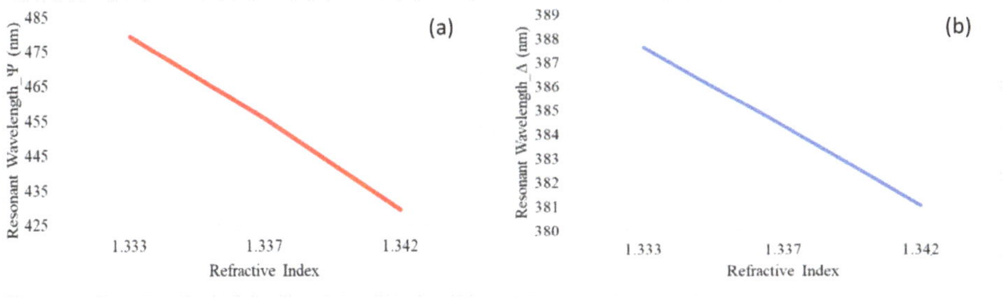

Figure 7.16. Plots of resonance wavelength against RI regarding the Ψ (a) and Δ (b) parameters at an angle of 48°.

Clearly, for an incidence angle of 48°, the resonance wavelength range for Ψ is ~420–500 nm, and that for Δ is ~300–386 nm.

The parameter Ψ can take values between 0° and 90°, whereas Δ can possess two angular ranges, i.e. –180 to 180° and 0 to 360°. As the distilled water has the lowest RI among the measurands in use, we consider it as the reference sample to compute sensitivity S (in ° nm^{-1}) in the form of a coefficient for comparing the performance of Ψ and Δ, using the equations

$$(\Psi_S)_{\text{conc.}} = \left| \frac{\Psi_{\text{water}} - \Psi_{\text{glucose}}}{\lambda_{\text{water}} - \lambda_{\text{glucose}}} \right|_{\text{conc.}} \tag{7.4}$$

$$(\Delta_s)_{\text{conc.}} = \left| \frac{\Delta_{\text{water}} - \Delta_{\text{glucose}}}{\lambda_{\text{water}} - \lambda_{\text{glucose}}} \right|_{\text{conc.}}. \tag{7.5}$$

Figure 7.17 exhibits the ellipsometry responses corresponding to two θ_i-values as 38° and 48°, and tables 7.2, 7.3, 7.4 and 7.5 summarize the obtained results, with tables 7.4 and 7.5 illustrating the obtained sensitivity S-values. As stated previously, we use distilled water as the reference medium, and three different types of aqueous solutions of glucose (having different concentrations).

In ellipsomtery analysis, we start by observing the Ψ-spectra at $\theta_i = 38°$ with increasing glucose concentration, which redshifts the spectral minima from 400 to 500 nm (figure 7.17(a)). We see that the sensitivity values grow nonlinearly at very small slopes, as table 7.4 shows, i.e. $S\_\Psi^{38°}_{GL\_50} > S\_\Psi^{38°}_{GL\_75} > S\_\Psi^{38°}_{GL\_100}$. Increasing the incidence angle to 48° results in shifts in spectral minima of the Ψ-parameter. We see an improvement in sensitivity corresponding to 75 mg rather than 50 and 100 mg. Further, we see from table 7.4 that $S\_\Psi^{48°}_{GL\_75} > S\_\Psi^{48°}_{GL\_50} > S\_\Psi^{48°}_{GL\_100}$. A quick review of the sensitivity values obtained from the Ψ-parameter and also from small nonlinear variations corresponding to 38° and 48° incidence angles in the visible range shows a small improvement in concentration sensing compared to direct measurements of reflection spectra.

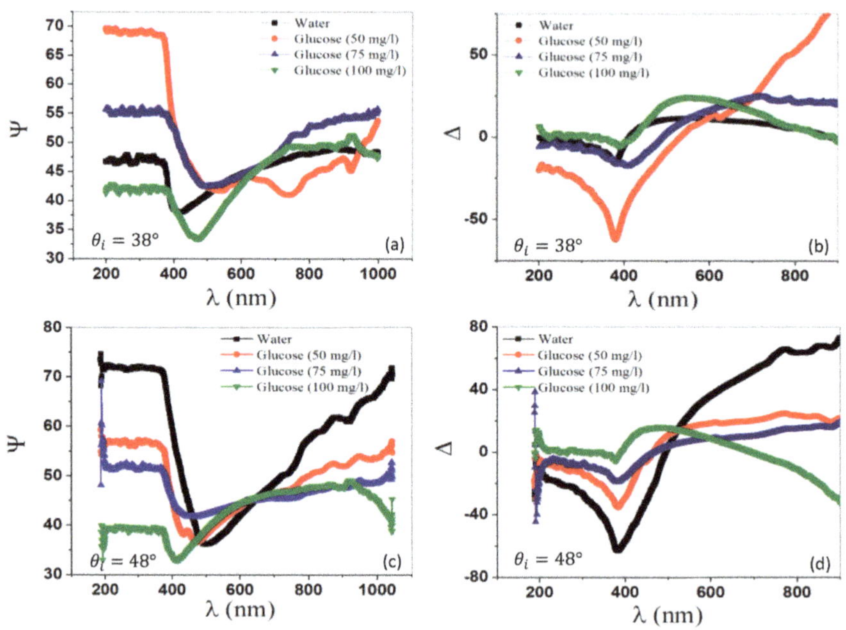

Figure 7.17. Ellipsometry-based spectral analysis of the ordered meta-biosensor infiltrated with water and aqueous solutions of different glucose concentrations corresponding to $\theta_i = 38°$ (a, b) and $\theta_i = 48°$ (c, d) angles of incidence; measurements of the wavelength dependence of Ψ (a, b) and Δ (c, d).

Table 7.2. Ψ and λ corresponding to $\theta_i = 38°$ and $\theta_i = 48°$ (obtained from figure 7.17).

Ψ & λ								
θ	Ψ_{Water}	λ_{Water} (nm)	Ψ_{GL_50}	λ_{GL_50} (nm)	Ψ_{GL_75}	λ_{GL_75} (nm)	Ψ_{GL_100}	λ_{GL_100} (nm)
38°	38°	418	42.1°	525	42.3°	494	33.5°	475
48°	36.2°	503	37.2°	478	42.8°	456	33°	413

Table 7.3. Δ and λ corresponding to $\theta_i = 38°$ and $\theta_i = 48°$ incidence angles (obtained from figure 7.17).

Δ & λ								
θ	Δ_{Water}	λ_{Water}(nm)	Δ_{GL_50}	λ_{GL_50} (nm)	Δ_{GL_75}	λ_{GL_75} (nm)	Δ_{GL_100}	λ_{GL_100} (nm)
38°	−15.2	377	−62.5	379	−17.1	413	−10.9	389
48°	−62.8	384	−35.3	383	−19.1	381	−5	377

Table 7.4. Sensitivity evaluation using the results in table 7.2.

θ	$S\_\Psi_{GL\_50}$ (° nm^{-1})	$S\_\Psi_{GL\_75}$ (° nm^{-1})	$S\_\Psi_{GL\_100}$ (° nm^{-1})
38°	0.038	0.056	0.078
48°	0.04	0.14	0.035

Table 7.5. Sensitivity evaluation using the results in table 7.3.

θ	$S\_\Delta_{GL\_50}$ (° nm^{-1})	$S\_\Delta_{GL\_75}$ (° nm^{-1})	$S\_\Delta_{GL\_100}$ (° nm^{-1})
38°	23.65	0.052	0.358
48°	27.5	14.56	8.25

Figure 7.17 also shows that the amplitude of the Δ-parameter reduces in the order of low to high concentration of glucose. We also note that, besides narrow spectral shifts, the best sensitivity among all the results in figure 7.17, and following the observations shown in table 7.3, corresponds to $\theta_i = 48°$ and the variations in Δ over the visible range. The sensitivity results in table 7.5 determine $S\_\Delta^{48°}_{GL\_50} > S\_\Psi^{48°}_{GL\_75} > S\_\Psi^{48°}_{GL\_100}$. We see that Δ exhibits linear variation with negative slope with increasing glucose concentration used over the top layer of the ordered meta-biosensor.

7.4 Disordered meta-biosensor thin films—DNA sensing

This section reviews the characteristics of the surface plasmon polariton (SPP) upon interacting with the DNA of the BALB/c rat as a biological disordered material utilizing meta-biosensor thin films. However, in the beginning, we attempt to provide a glimpse of the DNA structure followed by the use of DNA and its importance in biological sensing applications.

The importance of DNA as a hereditary material has been studied intensively in the literature [84–91]. In terms of molecular structure, DNA is interesting owing to its dimension being of the nanoscale [92]. Every DNA molecule has two lengthy polynucleotide strands wherein each chain/strand is the combination of four types of nucleotide subunits. Every DNA structure holds two chains of helical structure where the hydrogen bond exists in the base of nucleotides [93]. The backbone of a DNA molecule is in the form of a five-carbon sugar bond with one or more phosphate groups and a nitrogen-containing base. Furthermore, in one block comprising nucleotides in DNA the deoxyribose (sugar), after attaching to the single phosphate family, actually forms DNA. The block base can be one of the nitrogenous bases, nucleobases or nitrogen-containing biological compounds [94]. Four nucleobases are generally shown by adenine (A), cytosine (C), guanine (G), and thymine (T) [95], which are primary or canonical, and regardless of their role in DNA, they individually and directly have important roles in cell metabolism and respiration [96–99].

The geometry of the DNA molecule in 3D space shows nucleotide subunits lined together, and the chemical polarity forms curvature in each strand separately [100]. If we think of each sugar as a block with a protruding knob (the 5′ phosphate) on one side and a hole (the 3′ hydroxyl) on the other, each completed chain, formed by interlocking knobs with holes, will have all of its subunits lined up in the same orientation. Moreover, the two ends of the chain will be easily distinguishable, as one has a hole (the 3′ hydroxyl) and the other a knob (the 5′ phosphate) at its terminus. This polarity in a DNA chain is indicated by referring to one end as the 3′ end and the other as the 5′ end, as in [95].

Another feature of living cells is the existence of a backbone of sugar phosphate within the framework of nucleic acids. Figure 7.18 shows an expressive example of one unit building of the nucleotide related to the adenine base [101]. Sugar is the 3′ end and phosphate is the 5′ end of each nucleotide. The complementary nitrogenous

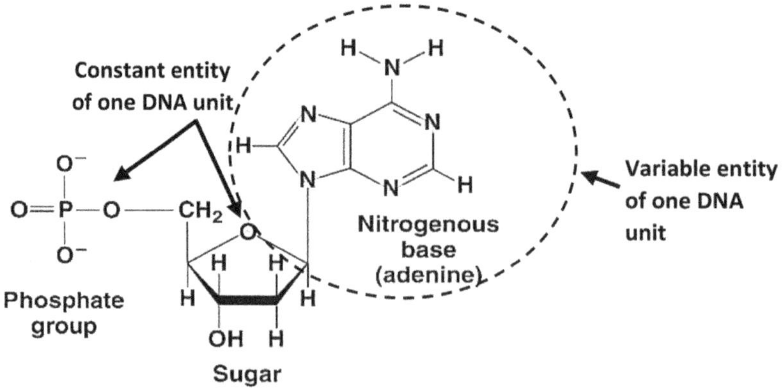

Figure 7.18. Example of one unit of adenine nucleotide in DNA structure.

bases are divided into two groups, the single-ringed pyrimidines and the double-ringed purines. In DNA, the pyrimidines are T and C, and the purines are A and G [102]. Regardless of what member of the pyrimidines or purines groups defines the placement of the nucleotide base, DNA has a strong tendency to dissolve in water through negative charge imposed by bonds created between the phosphorus and oxygen atoms. Furthermore, phosphodiester bonds are accountable for extending or growing in the 5′ to 3′ direction when the molecule is synthesized.

Within this context, the human genome is formed by the combination of 46 DNA molecules or two pairs of chromosomes. In a living cell and inside the nucleus, the basic genome repeating structure contains eight positively charged spherical histones (nucleosomes) and approximately a meter length of DNA packed in a micrometer size in a unit cell [103, 104]. Each base pair (bp) in a DNA strand is $\sim$0.34 nm long, which, along with other billions of bp, can be extracted from the nuclei of blood cells, semen, skin cells, tissue, organs, muscle, brain cells, bone, teeth, hair, saliva, mucus, fingernails, etc.

Different from the ordered structures, which can be engineered artificially, materials such as DNA structures (where they exist in biological media) can be extracted in the form of disordered structures. The extraction of DNA from a single cell is quite complex and expensive, while in the case of many cells, through some chemical techniques, it becomes feasible at low cost [105–107]. Thus, the morphology of DNAs, after extraction from many cells, is relatively large either in the micro- or nanoscale, and is not suitable for broadband plasmonic spectrometry.

In plasmonic spectroscopy, we broadly contend with the one-dimensional (1D), 2D, and 3D nanoparticles [108–111]. The DNA after extraction unwillingly encloses a stack of unknown broken or unbroken DNA strands with dimensions several hundred times greater than the visible wavelength scale. Meanwhile, and similar to the first section on ellipsomtery, the broadband light subject to this section is incoherent in nature, which limits the spectral quality as compared to what one would expect using a single-wavelength laser light source. Thus, it is essential to treat the aqueous form of DNA to enhance its quality in plasmonic spectrometry. The major issue associated with the non-uniform, asymmetrical, and skein morphology of the dissolved DNA piles in water is their non-fitting scale within the scope of plasmonic sensing. Considering this modality of the state of DNA, one possible way to rescale the size appropriately for the dimensions of the plasmonic sensing range is to use the method of dilution. However, estimating the limit of dilution for a certain DNA pile and recognizing how far repetition is required to observe plasmonic phenomenon is challenging. Further, and as an advantageous point, this technique exponentially lessens the target DNA template number in the extracted sample.

We now investigate the plasmonic response of the DNA of a living creature, the BALB/c rat, as a disordered sensing measurand. To achieve this target, we initially apply the progression of serial dilution (for enhancing the interaction of light) from the micro- to nanoscale DNA concentration. It must be recalled that the application of plasmonics as an optical tool is governed by the size, shape, and nature of the material(s) in use, the RI of the measurand, and the thickness of the substrate layer (s) [5, 112–115]. Many plasmonics-related investigations involve a combination array of nanoengineered metallic thin films with small molecular measurands [116–124]. Such explanations determine that DNAs with a large polymerized molecular structure can never be comparable with small size molecules in the usual liquid measurands. Thus, for expanding the surface contact of DNA, a thin homogeneous layer of gold is a notable candidate for plasmonic sensing. In this section, we present the performance of a disordered meta-biosensor by exploiting the plasmonic spectral response of a BALB/c rat's dried DNA (as the disordered large molecular measurand) after deposition over a homogeneous nanolayer thin film of gold.

7.4.1 DNA interaction with a gold nanolayer

Based on the above discussion, every building block of a DNA chain is monopolar in nature. It takes two units of DNA block to form a water molecule. It is very obvious from figure 7.19 that each molecule of a DNA holds negative charge in the aqueous medium [125, 126].

Free electrons distributed over a flat surface of gold are referred to as delocalized electrons. The high electron density property of gold will provide weak electric field with the negative polarity of DNA. A thin surface of gold contains free electrons, and this high electron density property will provide strong binding between the positive polarity of DNA and the negative polarity of the gold layer [127–129]. Figure 7.20 depicts a 3D schematic of two gold thin films, wherein the stockpile of DNA at high concentration (figure 7.20(a)) has non-uniform negative charge distribution over the thin film surface. This also reduces the probability of well-binding to the gold layer compared to the low DNA concentration (figure 7.20(b)). Surface waves quickly vanish in response to thick and dense media, and in broadband plasmonics it is almost impossible to find a trace of them. The operation frequency range of this kind of spectrometry is generally between ~375 and ~750 nm. As such, low DNA concentration, much closer to the morphology of small numbers of DNA strands (or countable units of DNA), essentially improves the compatibility of plasmonic sensing in the visible frequency range.

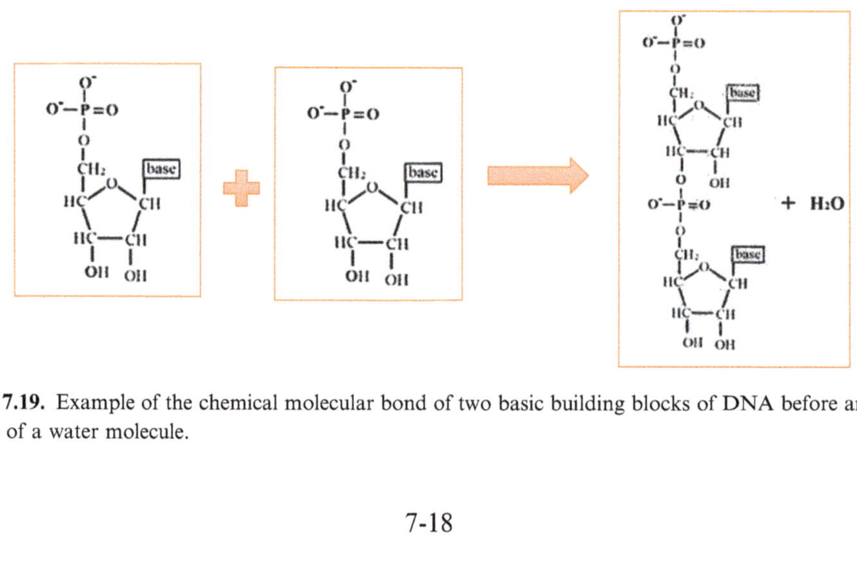

Figure 7.19. Example of the chemical molecular bond of two basic building blocks of DNA before and after release of a water molecule.

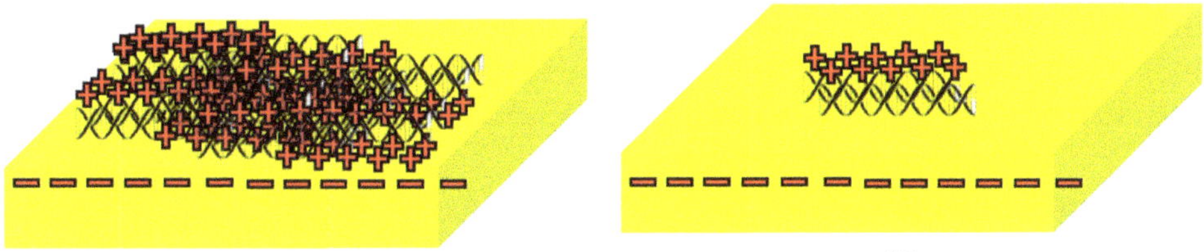

Figure 7.20. Demonstrative schematic of negative charge distributions for (a) high and (b) low concentrations of DNA over a gold nanolayer.

7.4.2 Extracting DNA from a BALB/c rat blood sample

There are many techniques available in the literature discussing DNA extraction from biological samples [130, 131]. In this work, the target DNA selected for extraction from the red blood cells of a BALB/c rat was chosen because of the many genomic DNA similarities of this mammal with humans [133–135]. The extraction process for this experiment begins by taking 700 μl blood from BALB/c rat and putting this in 2 ml of ethylenediaminetetraacetic acid (EDTA)—an amino acid-like molecule, extensively used as a blood clot prevention factor [135–137]. In the human body, its role is to infuse slowly into blood cells and bind with their minerals, such as calcium, iron, and copper, and enhance the conveyance of them to organs such as the kidney [138]. The next step is adding 700 μl TRIzol (i.e. a monophasic solution of phenol and guanidinium isothiocyanate that simultaneously solubilizes biological media and denatures protein) to the blood sample [139, 140]. This solution for efficient chemical interaction is then mixed by vortex mixer for 30 s. Then the pH is stabilized as the DNA is sensitive to pH variation. Hence, Tris or hydroxymethyl aminomethane pH buffer is added to the solution and kept in the shaker for 10 min.

Separating nucleic acids from other cellular substances is the most important step for unimpaired access to the genome substances of cells. Thus, 200 μl cold phenol–chloroform (that lessen enzymes factor) is added to the mixture and centrifuged at a speed of 12 000 rpm at a temperature range of 4 °C–10 °C for 15 min. In the next step, non-genomic substances in the form of low-density liquid phase appear above the denser genome liquid phase medium in the microtube container. This lighter liquid can be removed, and the remaining denser liquid phase can be mixed with cold isopropanol alcohol for the purpose of removing the H_2O hydration shell of phosphate and enhancing the DNA extraction process [141, 142]. The infusion process for isopropanol alcohol is lengthy, and will take around 30 min until the chemical interaction is completed. Further extraction of unwanted blood cellular substances is attained by centrifugation at a speed of 10 000 rpm for 10 min. After separation of the upper liquid phase, 500 μl of 75% ethanol is added to the solution to improve DNA extraction. The final centrifuge at a speed of 8000 rpm for 5 min precipitates the DNA strands and the sediment appears in the bottom of microtube. Finally, the extracted DNA is dissolved in 30 μl DPEC (diethylpyrocarbonate) and sterile-filtered water, and heated by dry heat blocks at 57 °C for 15 min to preserve the solubility of DNA (in DEPC water). The spectrometric analysis is performed using a NanoDrop 8000 UV–vis spectrophotometer, which determines the concentration of the extracted DNA to be 883 ng μl^{-1}; this is sufficient for dilution progress —the part we discuss in the following subsection.

7.4.3 Serial dilution process for BALB/c rat DNA

The step before DNA deposition over the gold nanolayer is dilution, which must be conducted serially, and broadband plasmonic spectrometry is to be applied at every step to see the effect of dilution. This experiment was aimed at reducing the concentration of BALB/c DNA in the liquid phase of deionized water without changing the solution volume. The DNA of different concentrations must be dried after dilution. We evaluate the surface wave behavior in the case of dried DNA detection related to better polarity binding of the bipolar DNA molecules with the gold nanolayer. Thus, subtracting the biological template in the liquid phase helps resize the concentration scale from <100 ng μl^{-1} to <1 ng μl^{-1} in the serial dilution method. In practice, the preliminary soluble BALB/c rat DNA, extracted directly from blood, is tested by a NanoDrop 8000 UV–vis spectrophotometer to measure the initial solution concentration in 1 μl water. The measured concentration before applying the dilution process is 88.3 ng/40 μl (figure 7.21). We apply serial dilution to 1 μl of solution containing 20.825 ng of BALB/c rat DNA. In practice, we change the coefficient ratios (i.e. 0.1, 0.05, and 0.0025 ng μl^{-1}) at every step to reach the scale of <1 ng μl^{-1} in a smaller number of dilution steps (figure 7.21). In our study, we mainly consider DNA concentrations <10 ng μl^{-1}. Thus, for the following steps, we applied 2.825, 1.041 25, and 0.520 625 ng μl^{-1} for DNA deposition over the gold nanolayer thin film.

7.4.4 Dehydrating the BALB/c rat DNA

The dehydration of BALB/c rat DNA is important as the water molecule evaporates and is separated from DNA molecules. It must be recalled that different states of a matter have different permittivity, and therefore, as the wet DNA has its own strong

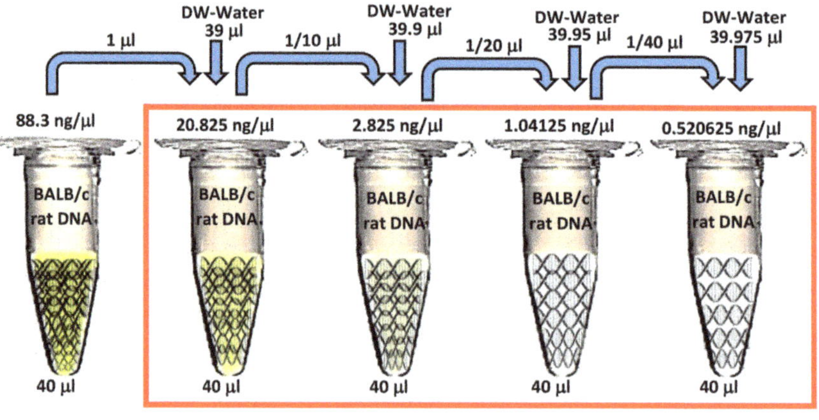

Figure 7.21. Serial dilution of BALB/c rat DNA.

Figure 7.22. Gold nanolayers with deposited dried BALB/c rat DNA at concentrations of (a) 88.3 ng/40 μl, (b) 44.15 ng/40 μl, and (c) 22.075 ng/40 μl.

hydration water network, the permittivity of wet DNA will change over time. Furthermore, the morphology of wet DNA is time-dependent as no flow exists over the gold thin film and the occurred dehydration constantly changes the molecular structures, and hence the measurand RI.

One advantage of fixture DNA over a gold nanolayer is its constant permittivity. Further, the DNA molecules take lattice structures after drying [142]. In our experiment, to make sure that the DNA is independent of environmental factors such as temperature and humidity during optical analysis, and also to ensure that the permittivity will not change due to lattice structure and the thickness is not a variable factor, we dry the DNA at a constant temperature of 60 °C for 24 h. The dried DNA substrate target in this experiment is 40 nm thick homogeneous gold film over glass.

The existence of surface waves in different wavelengths in the visible range when using a 30–70 nm thick homogeneous gold layer is a proven phenomenon in the literature [144–150]. Using the sputtering technique, we deposit a 40 nm thick gold layer over a glass film. Using a micropipette, we prepare the BALB/c rat DNA substances in microtubes corresponding to 88.3 ng/40 μl, 44.15 ng/40 μl, and 22.075 ng/40 μl concentrations dripped over the gold thin films, as shown in figure 7.22(a)–(c). We take the reference sample as the one without DNA thin film deposition over the gold film thin film for comparing the plasmonic effect when the air is replaced with the DNA layers. We then dry all films using dry air at a constant temperature of 60 °C in an incubator for a period of 24 h. The fading stains of the dried DNA appearing on the surface of gold thin films in figure 7.22 demonstrate the outcome of serial dilution.

7.5 Experimental broadband plasmonic spectroscopy

Precise broadband plasmonic spectrometry requires the use of optics, electronics, control systems, computer programming and mechanical engineering. We developed an opto-electro-mechanical stage to control simultaneously the sweeping angle of the optical source and spectrometer arms. Figure 7.23 exhibits the top view of the designed stage. Herein, we introduce fundamental features of this optical setup.

Similar to previous explanations about the Glan–Taylor prism, we are dealing with the *s*- or *p*- wave at every sweep. The gear box of this system automatically rotates the Glan–Taylor prism inside the optical source tube, which, at the beginning of sweep, has its scope at an angle between 10° and 70°. The light source holder is

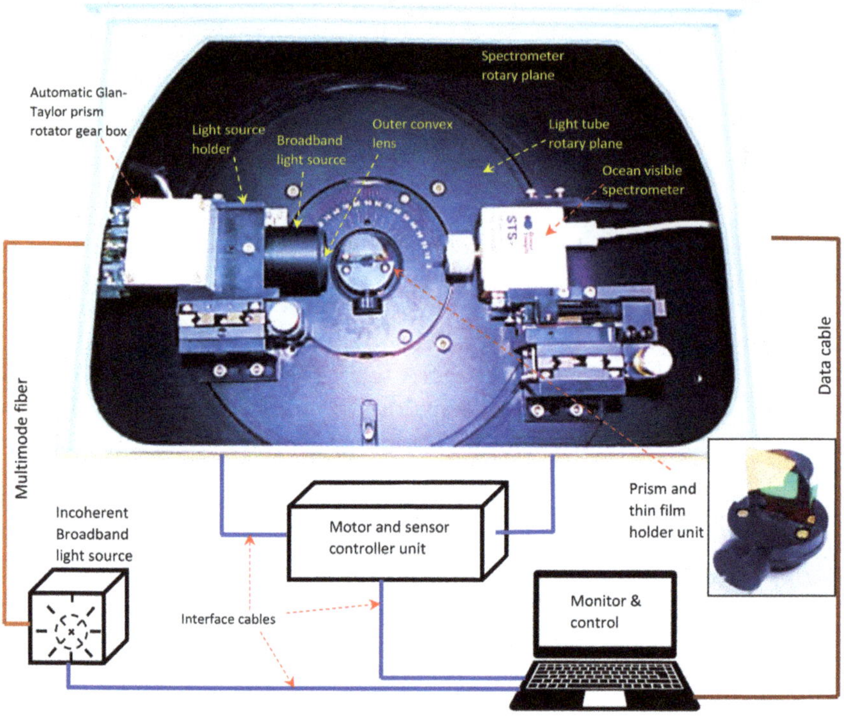

Figure 7.23. Schematic and top view of the designed system; experimental setup for the opto-electro-mechanical broadband plasmonic spectroscopy.

mounted over the internal rotary plane. Both the internal and external rotary planes can rotate in the reverse direction around a fixed grading plane. The outer convex lens at one end of the light source tube converges light before it impinges on the half-circle glass prism mounted over the fixed grading plane.

The designed disordered meta-biosensor in the form of gold nanolayer thin film with deposited dried DNA measurand of different concentrations must stick over the flat surface of the half-circular prism through the index matching oil. The incoherent light from light source is transferred to the light tube arm through a broadband multimode fiber. The miniaturized spectrometer (STS Ocean) is specifically engineered for spectrometry of visible broadband light. The data from spectrometer after every sweep at a step of 0.5° angle are sent to the monitoring and controlling software through the serial data cable link. The remaining cables in figure 7.23 are the controlling interface cables responsible for driving motors and controlling rotations by the software.

7.5.1 Results and analysis

Using the above, we are able to perform all spectral measurements in real time. In this experiment, the main aim is to capture the signature of the broadband plasmonic characteristics for the BALB/c rat's dried DNA sample. We analysed the big data as provided by the spectrometer to the software. Herein, we analyse and discuss the spectral results acquired from the opto-electro-mechanical setup, which principally provides the reflectance characteristics of the DNA samples deposited over the gold nanolayer surface. We considered the *p*-polarized incidence radiation and, as stated before, we varied the angle of incidence at a step of 0.5° starting from 10°; we kept the upper limit of incidence angle as 50°.

Figure 7.24 provides illustrative plots of broadband reflection spectrometry analysis results related to BALB/c rat dried DNA in the form of reflection intensity against wavelength. Here, we assume different fixed values of the angle of incidence while simultaneously including different DNA concentrations, namely 1/10, 1/20, and 1/40 ng μl^{-1} in our analysis. In particular, the panels in figure 7.24(a)–(j), respectively, correspond to the used angular values of 20.1°, 31.97°, 36.49°, 39.88°, 41.01°, 43.84°, 46.1°, 47.79°, and 48.92°. In all these figures, we also incorporate the case of no DNA deposition over the gold film as the reference sample—the situation that we refer to as 'air' (figure 7.24) because the ambience of gold film is free space.

In our experiment, the initial angle is 20.1° as the tangible presence of plasmonic outcome appears at this angular value. The perceptible fact in figure 7.24(a) is the intensity spectrum analogous to the most diluted DNA sample, which exhibits maximum reflection at 20.1°. The obtained results indicate that the less diluted DNA samples do not reveal plasmonic resonance to the extent that the case of the most

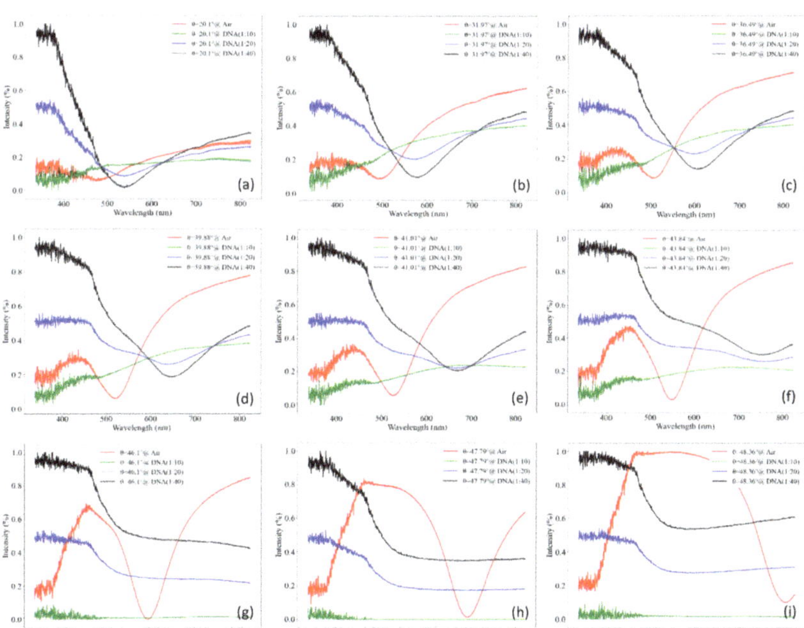

Figure 7.24. Intensity reflection spectra of the *p*-polarized light from the DNA thin film samples at different incidence angles, namely (a) 20.1°, (b) 31.97°, (c) 36.49°, (d) 39.88°, (e) 41.01°, (f) 43.84°, (g) 46.1°, (h) 47.79°, (i) 48.36°, and (h) 48.92°. The figure also incorporates the spectral response in the absence of measurand.

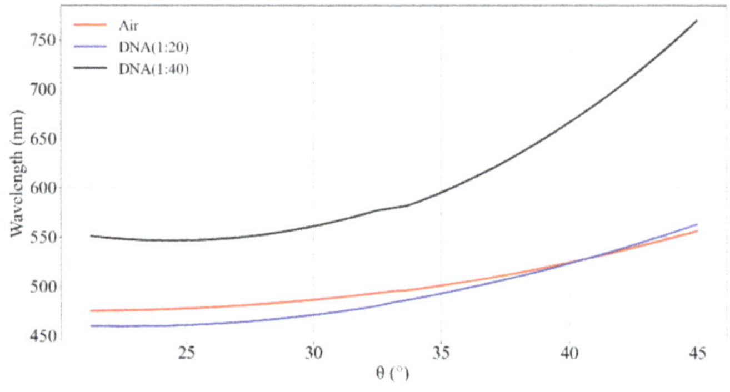

Figure 7.25. Plots of resonance wavelengths against incidence angle corresponding to the situations of the presence and absence (i.e. the case of 'air') of DNA samples.

diluted DNA sample (in our investigation) shows. Other reports reveal that, in the case of homogeneous gold layer of 40 nm thickness, the plasmonic resonance happens at higher angles [150].

The plasmonic signature of the BALB/c rat dried DNA is evident from figure 7.24 (a)–(e). At very small increase in incidence angle, the resonance wavelengths undergo sluggish redshifts either in the case of the presence of DNA or the air measurands in figure 7.24(a)–(j). In spite of the type of measurands, at larger incidence angles redshifts exist in every spectrum, but for air as the measurand, the reflection minima become more conspicuous. The other important fact in these results is that the DNA samples with 1:10 dilution ratio do not unveil plasmonic resonance at any of the incidence angles.

The more effective way to present the shifts in resonance wavelength is to obtain the sensitivity S of the device, which can be expressed as $S = \Delta\lambda/\Delta\theta$. We plot figure 7.25 based on the analysis of big data measured at every value of incidence angle. This figure shows the sensitivity comparison between air as a default measurand and diluted DNA with two different ratios of 1:20 and 1:40 considering the resonance wavelength against incidence angle. Typically, for relatively small incidence angles, the device yields $S = 7.2$ nm/degree and $S = 3.2$ nm/degree, respectively, in the cases of the presence and absence (i.e. air) of DNA sample. For large angles, however, the same situations provide the respective sensitivity values of $S = 22.4$ nm/degree and $S = 4.8$ nm/degree. The results demonstrate that the described disordered meta-biosensor-based plasmonic configuration can be used to detect highly diluted DNA samples.

7.6 Concluding remarks

This chapter reviewed in brief the biosensing application of ordered and disordered structures to be used as metamaterials. As an illustrative example, the stadium nanocomb grating structure has been considered. The nanocomb grating array configuration has been fabricated and experimentally demonstrated for glucose sensing, with the measurand being of different concentrations (of glucose aqueous solution). The obtained results demonstrate the usefulness of the structure in biosensing. For disordered metastructures, BALB/c rat dried DNA thin films have been used. In particular, gold thin films have been experimented with, using different DNA concentrations. Spectral characteristics have been evaluated under plasmonic conditions, which determine the approach to be viable in biosensing.

References

[1] Ou J -Y, Plum E, Zhang J and Zheludev N I 2013 An electro-mechanically reconfigurable plasmonic metamaterial operating in the near-infrared *Nat. Nanotech.* **8** 252–5

[2] Lee K, Gatensby R, McEvoy N, Hallam T and Duesberg G S 2013 High-performance sensors based on molybdenum disulfide thin films *Adv. Mater.* **25** 6699–702

[3] Al Shakhs M H, Ott P and Chau K J 2014 Band diagrams of layered plasmonic metamaterials *J. Appl. Phys.* **116** 173101

[4] Ghasemi M, Choudhury P K and Dehzangi A 2015 Nanoengineered thin films of copper for the optical monitoring of urine—a comparative study of the helical and columnar nanostructures *J. Electromagn. Waves Appl.* **29** 2321–9

[5] Ghasemi M, Baqir M A and Choudhury P K 2016 On the metasurface-based comb filters *IEEE Photon. Technol. Lett.* **28** 1100–3

[6] Ghasemi M and Choudhury P K 2016 Nanostructured concentric gold ring resonator-based metasurface filter device *Optik* **127** 9932–6

[7] Zabihpour T, Shahidi S A, Karimi-Maleh H and Ghorbani-Hasan Saraei A 2020 Voltametric food analytical sensor for determining vanillin based on amplified $NiFe_2O_4$ nanoparticle/ionic liquid sensor *J. Food Meas. Charact.* **14** 1039–45

[8] Sohrabi F, Saeidifard S, Ghasemi M, Asadishad T, Hamidi S M and Hosseini S M 2021 Role of plasmonics in detection of deadliest viruses: a review *Eur. Phys. J. Plus* **136** 675

[9] Valsecchi C and Brolo A G 2013 Periodic metallic nanostructures as plasmonic chemical sensors *Langmuir* **29** 5638–49

[10] Gonidec M 2017 Concept of non-periodic metasurfaces based on positional gradients applied to IR-flat lenses *Opt. Mater. Express* **7** 2346–51

[11] Fusco Z *et al* 2020 Non-periodic epsilon-near-zero metamaterials at visible wavelengths for efficient non-resonant optical sensing *Nano Lett.* **20** 3970–7

[12] Pourmand M and Choudhury P K 2020 Wideband THz filtering by graphene-over-dielectric periodic structures with and without MgF_2 defect layer *IEEE Access* **8** 137385–94

[13] Pourmand M, Choudhury P K and Mohamed M A 2021 Porous gold nanolayer coated halide metal perovskite-based broadband metamaterial absorber in the visible and near-IR regime *IEEE Access* **9** 8912–9

[14] Pourmand M, Choudhury P K and Mohamed M A 2021 Tunable absorber embedded with GST mediums and trilayer graphene strip microheaters *Sci. Rep.* **11** 3603

[15] Dang H *et al* 2021 Reproducible and sensitive plasmonic sensing platforms based on Au-nanoparticle-internalized nanodimpled substrates *Adv. Funct. Mater.* **31** 2105703

[16] Das A, Kumar K and Dhawan A 2021 Periodic arrays of plasmonic crossed-bowtie nanostructures interspaced with plasmonic nanocrosses for highly sensitive LSPR based chemical and biological sensing *RSC Adv.* **11** 8096–106

[17] Borowicz P, Latek M, Rzodkiewicz W, Łaszcz A, Czerwinski A and Ratajczak J 2012 Deep-ultraviolet Raman investigation of silicon oxide: thin film on silicon substrate versus bulk material *Adv. Nat. Sci.: Nanosci. Nanotechnol.* **3** 045003

[18] Artigas D and Torner L 2005 Dyakonov surface waves in photonic metamaterials *Phys. Rev. Lett.* **94** 013901

[19] Homola J 2008 Surface plasmon resonance sensors for detection of chemical and biological species *Chem. Rev.* **108** 462–93

[20] Dockrey J A *et al* 2013 Thin metamaterial Luneburg lens for surface waves *Phys. Rev.* B **87** 125137

[21] Takayama O, Bogdanov A A and Lavrinenko A V 2017 Photonic surface waves on metamaterial interfaces *J. Phys.* **29** 463001

[22] Cai R, Jin Y, Rabczuk T, Zhuang X and Djafari-Rouhani B 2021 Propagation and attenuation of Rayleigh and pseudo surface waves in viscoelastic metamaterials *J. Appl. Phys.* **129**

[23] Huang Y, Das P K and Bhethanabotla V R 2021 Surface acoustic waves in biosensing applications *Sens. Actuat.* **3** 100041

[24] Yeh P, Yariv A and Hong C S 1977 Electromagnetic propagation in periodic stratified media. I-general theory *J. Opt. Soc. Am.* **67** 423–38

[25] Polo J, Mackay T and Lakhtakia A 2013 *Electromagnetic Surface Waves: A Modern Perspective* (Amsterdam: Elsevier)

[26] Robertson W M and May M S 1999 Surface electromagnetic wave excitation on one-dimensional photonic band-gap arrays *Appl. Phys. Lett.* **74** 1800–2

[27] Robertson W M 1999 Experimental measurement of the effect of termination on surface electromagnetic waves in one-dimensional photonic bandgap arrays *J. Lightwave Technol.* **17** 2013–7

[28] Villa-Villa F, Gaspar-Armenta J A and Mendoza-Suárez A 2007 Surface modes in one dimensional photonic crystals that include left handed materials *J. Electromagn. Waves Appl.* **21** 485–99

[29] Borowicz P *et al* 2012 Deep-ultraviolet Raman investigation of silicon oxide: thin film on silicon substrate versus bulk material *Adv. Natural Sci.: Nanosci. Nanotechnol.* **3** 045003

[30] Joannopoulos J D, Villeneuve P R and Fan S 1997 Photonic crystals *Solid State Commun.* **102** 165–73

[31] Lu Z H, Lockwood D J and Baribeau J M 1995 Quantum confinement and light emission in SiO_2/Si superlattices *Nature* **378** 258–60

[32] Yablonovitch E 1993 Photonic band-gap structures *J. Opt. Soc. Am.* B **10** 283–95

[33] Laude V, Achaoui Y, Benchabane S and Khelif A 2009 Evanescent Bloch waves and the complex band structure of phononic crystals *Phys. Rev.* B **80** 092301

[34] Johnson S G and Joannopoulos J D 2003 Introduction to photonic crystals: Bloch's theorem, band diagrams, and gaps (but no defects) *Photonic Cryst. Tutorial* 1–16

[35] Benesperi I, Michaels H and Freitag M 2018 The researcher's guide to solid-state dye-sensitized solar cells *J. Mater. Chem.* C **6** 11903–42

[36] Gralak B, Cassier M, Demésy G and Guenneau S 2020 Electromagnetic waves in photonic crystals: laws of dispersion, causality and analytical properties *Compendium on Electromagnetic Analysis* (Singapore: World Scientific) pp 205–41

[37] Chen Q, Miyata N, Kokubo T and Nakamura T 2000 Bioactivity and mechanical properties of PDMS-modified CaO–SiO_2–TiO_2 hybrids prepared by sol–gel process *J. Biomed. Mater. Res.* **51** 605–11

[38] Patolsky F, Zheng G and Lieber C M 2006 Fabrication of silicon nanowire devices for ultrasensitive, label-free, real-time detection of biological and chemical species *Nat. Protoc* **1** 1711–24
[39] Kim Y K, Kim G T and Ha J S 2007 Simple patterning via adhesion between a buffered-oxide etchant-treated PDMS stamp and a SiO_2 substrate *Adv. Func. Mater.* **17** 2125–32
[40] Tamm I 1932 On the possible bound states of electrons on a crystal surface *Phys. Z. Sowjetunion* **1** 733–5
[41] De Stefano L *et al* 2004 Optical sensors for vapors, liquids, and biological molecules based on porous silicon technology *IEEE Trans. Nanotechnol.* **3** 49–54
[42] Chang H *et al* 2004 DNA-mediated fluctuations in ionic current through silicon oxide nanopore channels *Nano Lett.* **4** 1551–6
[43] Liu Y and Iqbal S M 2009 Silicon-based novel bio-sensing platforms at the micro and nano scale *ECS Trans.* **16** 25
[44] Ohno H *et al* 1990 Observation of 'Tamm states' in superlattices *Phys. Rev. Lett.* **64** 2555
[45] Resnick D J *et al* 2003 Imprint lithography: lab curiosity or the real NGL *Proc. SPIE* **5037** 12–23
[46] Hoff J D, Cheng L J, Meyhöfer E, Guo L J and Hunt A J 2004 Nanoscale protein patterning by imprint lithography *Nano Lett.* **4** 853–7
[47] Raj M K and Chakraborty S 2020 PDMS microfluidics: a mini review *J. Appl. Polymer Sci.* **137** 48958
[48] Victor A, Ribeiro J E and Araújo F F 2019 Study of PDMS characterization and its applications in biomedicine: a review *J. Mech. Eng. Biomechan* **4** 1–9
[49] Liu L and Sheardown H 2005 Glucose permeable poly (dimethyl siloxane) poly (N-isopropyl acrylamide) interpenetrating networks as ophthalmic biomaterials *Biomater.* **26** 233–44
[50] Johnston I D, McCluskey D K, Tan C K and Tracey M C 2014 Mechanical characterization of bulk Sylgard 184 for microfluidics and microengineering *J. Micromechan. Microeng.* **24** 035017
[51] Brounstein Z, Zhao J, Geller D, Gupta N and Labouriau A 2021 Long-term thermal aging of modified sylgard 184 formulations *Polymers* **13** 3125
[52] Sales F C, Ariati R M, Noronha V T and Ribeiro J E 2022 Mechanical characterization of PDMS with different mixing ratios *Proc. Struct. Integrity* **37** 383–8
[53] Sötebier C, Michel A and Fresnais J 2012 Polydimethylsiloxane (PDMS) coating onto magnetic nanoparticles induced by attractive electrostatic interaction *Appl. Sci.* **2** 485–95
[54] Gottesfeld S 1989 Ellipsometry: principles and recent applications in electrochemistry *Electroanal. Chem.* **15** 143–265
[55] Vedam K 1998 Spectroscopic ellipsometry: a historical overview *Thin Solid Films* **313–314** 1–9
[56] Theeten J B and Aspnes D E 1981 Ellipsometry in thin film analysis *Annu. Rev. Mater. Sci.* **11** 97–122
[57] Ghasemi M, Roostaei N, Sohrabi F, Hamidi S M and Choudhury P K 2020 Biosensing applications of all-dielectric SiO_2–PDMS meta-stadium grating nanocombs *Opt. Mater. Express* **10** 1018–33
[58] Belyaeva A, Galuza A, Kolenov I and Mizrakhy S 2021 Developments in terahertz ellipsometry: portable spectroscopic quasi-optical ellipsometer-reflectometer and its applications *J. Infrared, Millimeter, Terahertz Waves* **42** 130–53

[59] Wu P C *et al* 2017 Visible metasurfaces for on-chip polarimetry *ACS Photon* **5** 2568–73

[60] Shah Y D *et al* 2022 An all-dielectric metasurface polarimeter *ACS Photon* **9** 3245–52

[61] Losurdo M and Hingerl K 2013 *Ellipsometry at the Nanoscale* (Berlin: Springer)

[62] Aspnes D E 2014 Spectroscopic ellipsometry—past, present, and future *Thin Solid Films* **571** 334–4

[63] Liu Y C, Hsieh J H and Tung S K 2006 Extraction of optical constants of zinc oxide thin films by ellipsometry with various models *Thin Solid Films* **510** 32–8

[64] Archard J F and Taylor A M 1948 Improved Glan–Foucault prism *J. Sci. Instrum* **25** 407

[65] Gao P, Hu X, Guo G, Wang Y and Ai J 2018 Influence of the optical axis orientation of the Glan–Taylor prism and the incident angle on the extraordinary ray *J. Phys.: Conf. Series* **1053** 012074

[66] Bass M *et al* 2009 *Handbook of Optics II* (New York: McGraw-Hill)

[67] Fan J Y, Li H X and Wu F Q 2003 A study on transmitted intensity of disturbance for air-spaced Glan-type polarizing prisms *Opt. Commun.* **223** 11–6

[68] Bužavaitė-Vertelienė E *et al* 2022 Total internal reflection ellipsometry approach for Bloch surface waves biosensing applications *Biosensors* **12** 584

[69] Neal W E J 1977 Application of ellipsometry to surface films and film growth *Surf. Technol.* **6** 81–110

[70] Collins R W and Kim Y T 1990 Ellipsometry for thin-film and surface analysis *Anal. Chem.* **62** 887A–900A

[71] Fujiwara H 2007 *Spectroscopic Ellipsometry: Principles and Applications* (New York: Wiley)

[72] Chaves J 2008 *Introduction to Nonimaging Optics* (Boca Raton, FL: CRC Press)

[73] Ghasemi M *et al* 2017 Metamaterial absorber comprising chromium–gold nanorods-based columnar thin films *J. Nanophoton.* **11** 043505

[74] Malitson I H 1965 Interspecimen comparison of the refractive index of fused silica *J. Opt. Soc. Am.* **55** 1205–9

[75] Flynn T and Yoo S 2019 Change detection with the kernel cumulative sum algorithm *Proc. 2019 IEEE 58th Conf. on Decision and Control (CDC)* pp 6092–9

[76] Saeidifard S *et al* 2019 Two-dimensional plasmonic biosensing platform: cellular activity detection under laser stimulation *J. Appl. Phys.* **126** 104701

[77] Barer R and Tkaczyk S 1954 Refractive index of concentrated protein solutions *Nature* **173** 821–2

[78] bin Mat Yunus W M and bin Abdul Rahman A 1988 Refractive index of solutions at high concentrations *Appl. Opt.* **27** 3341–3

[79] Strop P and Brunger A T 2005 Refractive index-based determination of detergent concentration and its application to the study of membrane proteins *Protein Sci.* **14** 2207–11

[80] Jiménez Riobóo R J, Philipp M, Ramos M A and Krüger J K 2009 Concentration and temperature dependence of the refractive index of ethanol–water mixtures: influence of intermolecular interactions *Eur. Phys. J.* E **30** 19–26

[81] Savotchenko S E 2022 The surface waves propagating along the contact between the layer with the constant gradient of refractive index and photorefractive crystal *J. Opt.* **24** 045501

[82] Takayama O, Crasovan L, Artigas D and Torner L 2009 Observation of Dyakonov surface waves *Phys. Rev. Lett.* **102** 043903

[83] Yang Q, Qin L, Cao G, Zhang C and Li X 2018 Refractive index sensor based on graphene-coated photonic surface-wave resonance *Opt. Lett.* **43** 639–42
[84] Estes M K and Cohen J E A N 1989 Rotavirus gene structure and function *Microbiol. Rev.* **53** 410–49
[85] Brown J W, Daniels C J, Reeve J N and Konisky J 1989 Gene structure, organization, and expression in archaebacteria *Crit. Rev. Microbiol.* **16** 287–337
[86] Bohr V A 1991 Gene specific DNA repair *Carcinogenesis* **12** 1983–92
[87] Nadaud S, Bonnardeaux A, Lathrop M and Soubrier F 1994 Gene structure, polymorphism and mapping of the human endothelial nitric oxide synthase gene *Biochem. Biophys. Res. Commun.* **198** 1027–33
[88] Burset M and Guigo R 1996 Evaluation of gene structure prediction programs *Genomics* **34** 353–67
[89] Maddison W P 1997 Gene trees in species trees *Syst. Biol.* **46** 523–36
[90] Travers A and Muskhelishvili G 2015 DNA structure and function *FEBS J.* **282** 2279–95
[91] Bewick A J and Schmitz R J 2017 Gene body DNA methylation in plants *Curr. Opin. Plant Biol.* **36** 103–10
[92] Watson J D and Crick F H 1953 Molecular structure of nucleic acids: a structure for deoxyribose nucleic acid *Nature* **171** 1953
[93] Mukherjee S, Majumdar S and Bhattacharyya D Role of hydrogen bonds in protein–DNA recognition: effect of nonplanar amino groups *J. Phys. Chem.* B **109** 10484–92
[94] Soukup G A 2001 Nucleic acids: general properties *Encyclopedia of Life Science* (New York: Wiley)
[95] Khodaei A, Feizi-Derakhshi M R and Mozaffari-Tazehkand B 2020 A pattern recognition model to distinguish cancerous DNA sequences via signal processing methods *Soft. Comput.* **24** 16315–34
[96] Krishnamurthy R 2012 Role of pK_a of nucleobases in the origins of chemical evolution *Acc. Chem. Res.* **45** 2035–44
[97] Bottaro S, Di Palma F and Bussi G 2014 The role of nucleobase interactions in RNA structure and dynamics *Nucleic Acids Res.* **42** 13306–14
[98] Xu W, Chan K M and Kool E T 2017 Fluorescent nucleobases as tools for studying DNA and RNA *Nat. Chem.* **9** 1043–55
[99] Ranzau B L and Komor A C 2018 Genome, epigenome, and transcriptome editing via chemical modification of nucleobases in living cells *Biochemistry* **58** 330–5
[100] Cacchione S, De Santis P, Foti D, Palleschi A and Savino M 1989 Periodical polydeoxynucleotides and DNA curvature *Biochemistry* **28** 8706–13
[101] Brown T A 2023 *Genomes 5* (Boca Raton, FL: CRC Press)
[102] Kumari A 2018 Pyrimidine structure *Sweet Biochemistry—Remembering Structures, Cycles, and Pathways by Mnemonics* ed A Kumari (New York: Academic)
[103] Butler J M 2005 *Forensic DNA Typing: Biology, Technology, and Genetics of STR Markers* (Amsterdam: Elsevier)
[104] Turner B M 2008 *Chromatin and Gene Regulation: Molecular Mechanisms in Epigenetics* (New York: Wiley-Blackwell)
[105] Demeke T and Jenkins G R 2010 Influence of DNA extraction methods, PCR inhibitors and quantification methods on real-time PCR assay of biotechnology-derived traits *Anal. Bioanal. Chem.* **396** 1977–90

[106] Abdel-Latif A and Osman G 2017 Comparison of three genomic DNA extraction methods to obtain high DNA quality from maize *Plant Meth* **13** 1–9

[107] Dairawan M and Shetty P J 2020 The evolution of DNA extraction methods *Am. J. Biomed. Sci. Res.* **8** 39–45

[108] Ghasemi M, Choudhury P K and Dehzangi A 2015 Nanoengineered thin films of copper for the optical monitoring of urine—a comparative study of the helical and columnar nanostructures *J. Electromagn. Waves Appl.* **29** 2321–9

[109] Ross M B, Mirkin C A and Schatz G C 2016 Optical properties of one-, two-, and three-dimensional arrays of plasmonic nanostructures *J. Phys. Chem.* C **120** 816–30

[110] Ali A *et al* 2020 On the core–shell nanoparticle in fractional dimensional space *Materials* **13** 2400

[111] Baqir M A, Choudhury P K, Naqvi Q A and Mughal M J 2020 On the scattering and absorption by the SiO_2–VO_2 core–shell nanoparticles under different thermal conditions *IEEE Access* **8** 84850–7

[112] Maier S A 2007 *Plasmonics: Fundamentals and Applications* (New York: Springer)

[113] Zouhdi S, Sihvola A and Vinogradov A P 2009 *Metamaterials and Plasmonics: Fundamentals, Modelling, Applications* (New York: Springer)

[114] Sheta E M, Ibrahim A B M -A and Choudhury P K 2023 Metamaterial-based THz polarization-insensitive multi-controllable sensor comprising MgF_2–graphene periodic nanopillers over InSb thin film *IEEE Trans. Nanotechnol.* **22** 336–41

[115] Baqir M A and Choudhury P K 2022 Hyperbolic metamaterial-based optical biosensor for detecting cancer cells *IEEE Photon. Technol. Lett.* **35** 183–6

[116] Chen T, Li S and Sun H 2012 Metamaterials application in sensing *Sensors* **12** 2742–65

[117] Ebrahimi A, Withayachumnankul W, Al-Sarawi S and Abbott D 2013 High-sensitivity metamaterial-inspired sensor for microfluidic dielectric characterization *IEEE Sensors* J **14** 1345–51

[118] Withayachumnankul W, Tuantranont A, Fumeaux C and Abbott D 2013 Metamaterial-based microfluidic sensor for dielectric characterization *Sens. Actuat. A: Phys.* **189** 233–7

[119] Altintas O *et al* 2017 Fluid, strain and rotation sensing applications by using metamaterial based sensor *J. Electrochem. Soc.* **164** B567–73

[120] Xu W, Xie L and Ying Y 2017 Mechanisms and applications of terahertz metamaterial sensing: a review *Nanoscale* **9** 13864–78

[121] Zhang R *et al* 2019 Terahertz microfluidic metamaterial biosensor for sensitive detection of small-volume liquid samples *IEEE Trans. Terahertz Sci. Technol.* **9** 209–14

[122] Ghasemi M *et al* 2024 Effect of DNA serial dilution on the highly precise broadband plasmonic signature of a BALB/c rat's dried DNA deposited on gold thin film *Opt. Mater. Express* **14** 1420–9 2024

[123] Tang C *et al* 2021 Integrating terahertz metamaterial and water nanodroplets for ultra-sensitive detection of amyloid β aggregates in liquids *Sens. Actuat. B: Chem.* **329** 129113

[124] Qiu Z *et al* 2022 A metamaterial based microfluidic sensor for permittivity detection of liquid *J. Phys. D: Appl. Phys.* **55** 435001

[125] Kool E T, Morales J C and Guckian K M 2000 Mimicking the structure and function of DNA: insights into DNA stability and replication *Angew. Chem. Int. Ed. Engl.* **39** 990–1009

[126] Koo K M *et al* 2015 DNA–bare gold affinity interactions: mechanism and applications in biosensing *Anal. Methods* **7** 7042–54

[127] Bennett A J 1970 Influence of the electron charge distribution on surface-plasmon dispersion *Phys. Rev.* B **1** 203–7

[128] Kötz R, Kolb D M and Sass J K 1977 Electron density effects in surface plasmon excitation on silver and gold electrodes *Surf. Sci.* **69** 359–64
[129] David C and de Abajo G 2014 F. J. Surface plasmon dependence on the electron density profile at metal surfaces *ACS Nano* **8** 9558–66
[130] Raynie D E 2006 Modern extraction techniques *Anal. Chem.* **78** 3997–4004
[131] Hearn R P and Arblaster K E 2010 DNA extraction techniques for use in education *Biochem. Mol. Biol. Educ.* **38** 161–6
[132] Erickson R P 1996 Mouse models of human genetic disease: which mouse is more like a man? *Bioessays* **18** 993–8
[133] Taube S *et al* 2013 A mouse model for human norovirus *MBio* **4** e00450
[134] Fan Y *et al* 2020 Expression profile and bioinformatics analysis of COMMD10 in BALB/C mice and human *Cancer Gene Therapy* **27** 216–25
[135] Goossens W, Van Duppen V and Verwilghen R L 1991 K2-or K3-EDTA: the anti-coagulant of choice in routine haematology? *Clin. Lab. Haematol.* **13** 291–5
[136] Lam N Y, Rainer T H, Chiu R W and Lo Y D 2004 EDTA is a better anticoagulant than heparin or citrate for delayed blood processing for plasma DNA analysis *Clin. Chem.* **50** 256–7
[137] Banfi G, Salvagno G L and Lippi G 2007 The role of ethylenediamine tetraacetic acid (EDTA) as *in vitro* anticoagulant for diagnostic purposes *Clin. Chem. Lab. Med.* **45** 565–76
[138] Pizzorno J E and Murray M T 2020 *Textbook of Natural Medicine* (New York: Elsevier)
[139] Rio D C, Ares M Jr, Hannon G J and Nilsen T W 2010 Purification of RNA using TRIzol (TRI reagent) *Cold Spring Harbor Protocols* (New York: CSHL Press)
[140] Bo Y Y *et al* 2021 High-purity DNA extraction from animal tissue using picking in the TRIzol-based method *Biotechniques* **70** 186–90
[141] Clerget G, Bourguignon-Igel V and Rederstorff M 2015 Alcoholic precipitation of small non-coding RNAs *Methods Mol. Biol.* **1296** 11–6
[142] Devaraj A *et al* 2019 The extracellular DNA lattice of bacterial biofilms is structurally related to Holliday junction recombination intermediates *Proc. Natl Acad. Sci.* **116** 25068–77
[143] Innes R A and Sambles J R 1987 Optical characterisation of gold using surface plasmon-polaritons *J. Phys. F: Metal Phys.* **17** 277
[144] Ehler T T and Noe L J 1995 Surface plasmon studies of thin silver/gold bimetallic films *Langmuir* **11** 4177–9
[145] Perner M *et al* 1997 Optically induced damping of the surface plasmon resonance in gold colloids *Phys. Rev. Lett.* **78** 2192
[146] Kabashin A V and Nikitin P I 1998 Surface plasmon resonance interferometer for bio-and chemical-sensors *Opt. Commun.* **150** 5–8
[147] Lyon L A *et al* 1999 Surface plasmon resonance of colloidal Au-modified gold films *Sens. Actuat. B: Chem.* **54** 118–24
[148] Tao N J, Boussaad S, Huang W L, Arechabaleta R A and D'Agnese J 1999 High resolution surface plasmon resonance spectroscopy *Rev. Sci. Instrum.* **70** 4656–60
[149] Okamoto T, Yamaguchi I and Kobayashi T 2000 Local plasmon sensor with gold colloid monolayers deposited upon glass substrates *Opt. Lett.* **25** 372–4
[150] Gryczynski I, Malicka J, Gryczynski Z and Lakowicz J R 2004 Surface plasmon-coupled emission with gold films *J. Phys. Chem.* B **108** 12568–74

IOP Publishing

Ordered and Disordered Metamaterials
Design and applications
Pankaj K Choudhury and Tatjana Gric

Chapter 8

Functional disordered composites

Tatjana Gric

Herein, we are making a step forward in dealing with the novel theoretic and computational determination of the effective permittivity of composite disordered media. The presented methodology stands as a perfect tool allowing one to evaluate the permittivity tensor of the sample analytically. The presented theory is applicable for analysing biological media with no need for human intervention by performing an experimental analysis to measure the parameters of the sample. It should be noted that laboratory measurements of the permittivity are not required in this case either. The presented technique allows for the creation of phantom tissue models for further use in clinical applications. Doing so, we determine the tensor components for random multiphase composites. This is fertile ground for the detection and treatment of cancer.

8.1 Multiphase composites

Introduce the complex variable $z = x_1 + ix_2$ on the plane. Consider the unit square cell Q with N circular inclusions $D_k = \{z \in \mathbb{C}: |z - a_k| < r_k\}$, where a_k denotes the complex coordinate of the center and r_k its radius ($k = 1, 2, \ldots, N$). For simplicity, consider the case $r = r_k$. The concentration of inclusions has the form $f = N\pi r$. The host domain is denoted by D. The polygon curve ∂Q and the circles ∂D_k are oriented in the counterclockwise direction, hence $\partial D = \partial Q - \sum_{k=1}^{N} \partial D_k$.

The term *cell* is the established notion in the theory of composites as well as in biology, but with different meanings. In order to avoid confusion we use the notion unit cell Q for the homogenization terminology only in the present section. The biological term cell from the previous sections corresponds to the notion inclusion used in homogenization theory.

Let the permittivity of the host be normalized to unity and the permittivity of the kth inclusion be a complex number $\varepsilon_k = \varepsilon'_k + i\varepsilon''_k$, where $\varepsilon'_k = \mathrm{Re}\ \varepsilon_k$ and $\varepsilon''_k = \mathrm{Im}\ \varepsilon_k$. One can consider ε_k as the ratio of the permittivity of the kth inclusion to the permittivity of the matrix, where the dimension permittivities can be complex.

doi:10.1088/978-0-7503-5462-2ch8

One can assume that the constants ε_k take the values from a set J that contains fewer than N elements. Let $\#J = M$, i.e. the composite is $(M+1)-$ phases and $\varepsilon_k = \varepsilon^{(j)}$, if $j = 1, 2, \ldots, M$. In this case formulas similar to the formulas (4.2.26) and (4.2.27) from [1] can be derived for a multiphase composite by the method developed in [2]. These formulas can be considered as extensions of the methodological approach [3] to higher order contrast parameters terms.

Let $u = u' + iu''$ and $u_k = u_k' + iu''_k$ denote the potentials in D and $|z - a_k| < r_k$, respectively, where, for instance, $u' = \mathrm{Re}\, u$ and $u'' = \mathrm{Im}\, u$ in D. The complex functions u and u_k satisfy Laplace's equation in the corresponding domains and are continuously differentiable in their closures.

The perfect contact between the components is expressed by equations (8.1)

$$u(t) = u_k(t), \quad \frac{\partial u}{\partial \mathbf{n}}(t) = \varepsilon_k \frac{\partial u_k}{\partial \mathbf{n}}(t), \quad |t - a_k| = r_k \quad (k = 1, 2, \ldots, N), \tag{8.1}$$

where the normal derivative $\frac{\partial}{\partial \mathbf{n}}$ on the circles is used.

Following homogenization theory [4], we have to consider a composite in the plane torus topology. It is suggested that the biological tissue under consideration represents the considered structure. Hence, this structure can be considered as a doubly periodic representative unit cell Q of the considered composite; see Hashin's MMM principle [5] and its constructive application to random composites in chapter 3 of [1].

The function $u(t)$ satisfies the normalized jump conditions per unit cell Q

$$u(z+1) - u(z) = 1, \quad u(z+i) - u(z) = 0. \tag{8.2}$$

Condition (8.2) means that the external complex flux $\mathbf{q}_0 = (-1, 0)$ is applied. Here, the components of $\mathbf{q}_0$ are complex numbers. More precisely, for instance, the first relation (8.2) can be written in the real form as follows

$$\mathrm{Re}\, u(z+1) - \mathrm{Re}\, u(z) = 1, \ \mathrm{Im}\, u(z+1) - \mathrm{Im}\, u(z) = 0. \tag{8.3}$$

Two complex relations (8.1) can be written in the extended real form

$$\begin{aligned}
u'(t) = u_k'(t), \quad & \frac{\partial u'}{\partial \mathbf{n}}(t) = \varepsilon_k' \frac{\partial u_k'}{\partial \mathbf{n}}(t) - \varepsilon''_k \frac{\partial u''_k}{\partial \mathbf{n}}(t), \\
u''(t) = u''_k(t), \quad & \frac{\partial u''}{\partial \mathbf{n}}(t) = \varepsilon''_k \frac{\partial u_k'}{\partial \mathbf{n}}(t) + \varepsilon_k' \frac{\partial u''_k}{\partial \mathbf{n}}(t), \\
& |t - a_k| = r_k \quad (k = 1, 2, \ldots, N).
\end{aligned} \tag{8.4}$$

8.2 Complex potentials

We now reduce the problem (8.4) to an $\mathbb{R}$-linear problem. Introduce the vector complex potentials $\varphi(z)$ and $\varphi_k(z)$ analytic (meromorphic) in D and D_k, respectively, and the non-degenerate real matrix

$$\boldsymbol{\alpha}_k = \begin{pmatrix} \varepsilon'_k & -\varepsilon''_k \\ \varepsilon''_k & \varepsilon'_k \end{pmatrix}. \tag{8.5}$$

The harmonic and analytic functions are related by the vector equalities

$$\varphi(z) = \begin{pmatrix} u'(z) + iv'(z) \\ u''(z) + iv''(z) \end{pmatrix}, \quad z \in D \tag{8.6}$$

and

$$\varphi_k(z) = \frac{1}{2}(I + \boldsymbol{\alpha}_k)\begin{pmatrix} u'_k(z) + iv'_k(z) \\ u''_k(z) + iv''_k(z) \end{pmatrix}, \quad |z - a_k| < r_k. \tag{8.7}$$

where $v'(z)$, $v''(z)$ and $v'_k(z)$, $v''_k(z)$ denote the imaginary parts of the components of the analytic (meromorphic) vector functions $\varphi(z)$ and $\varphi_k(z)$, respectively. The complex flux is introduced as the derivatives $\psi(z) = \frac{d}{dz}\varphi(z)$ and $\psi_k(z) = \frac{d}{dz}\varphi_k(z)$. It can be expressed through the partial derivatives of u', u'' and u'_k, u''_k calculated by (8.6) and (8.7), respectively,

$$\psi(z) = \begin{pmatrix} \dfrac{\partial u'}{\partial x_1} - i\dfrac{\partial u'}{\partial x_2} \\ \dfrac{\partial u''}{\partial x_1} - i\dfrac{\partial u''}{\partial x_2} \end{pmatrix}, \quad z \in D, \tag{8.8}$$

and

$$\psi_k(z) = \frac{1}{2}(I + \boldsymbol{\alpha}_k)\begin{pmatrix} \dfrac{\partial u'_k}{\partial x_1} - i\dfrac{\partial u'_k}{\partial x_2} \\ \dfrac{\partial u''_k}{\partial x_1} - i\dfrac{\partial u''_k}{\partial x_2} \end{pmatrix}, \quad |z - a_k| < r_k. \tag{8.9}$$

Introduce the contrast parameter matrix

$$\boldsymbol{\varrho}_k = -(I - \boldsymbol{\alpha}_k)(I + \boldsymbol{\alpha}_k)^{-1} = \frac{1}{|1 + \varepsilon_k|^2}\begin{pmatrix} |\varepsilon_k|^2 - 1 & 2\varepsilon''_k \\ -2\varepsilon''_k & |\varepsilon_k|^2 - 1 \end{pmatrix}. \tag{8.10}$$

Then, condition (8.4) can be written in the form of the vector matrix $\mathbb{R}$-linear problem

$$\varphi(t) = \varphi_k(t) - \boldsymbol{\varrho}_k\overline{\varphi_k(t)}, \quad |t - a_k| = r_k \ (k = 1, 2, \dots, N). \tag{8.11}$$

The eigenvalues of the matrix $\boldsymbol{\varrho}_k$ are

$$\lambda_k = \frac{\varepsilon_k - 1}{\varepsilon_k + 1}, \quad \overline{\lambda_k} = \frac{\overline{\varepsilon_k} - 1}{\overline{\varepsilon_k} + 1}. \tag{8.12}$$

At least in the case $|\lambda_k| \leqslant 1$, one can derive asymptotic analytical formulas for the effective complex tensor.

Let $\|\varrho_k\|$ denote a matrix norm. Following [2], consider the lower order approximation in $\varrho = \max_{k=1,2,\dots,N}\|\varrho_k\|$, assuming that ϱ is a sufficiently small number. We are looking for $\varphi(z)$ and $\varphi_k(z)$ in the asymptotic form up to an additive constant vector

$$\varphi(z) = \begin{pmatrix} z \\ 0 \end{pmatrix} + \sum_{m=1}^{N} \varrho_m \varphi_m^{(1)}(z) + \cdots, \quad \varphi_k(z) = \begin{pmatrix} z \\ 0 \end{pmatrix} + \sum_{m=1}^{N} \varrho_m \varphi_{mk}^{(1)}(z) + \cdots, \tag{8.13}$$

where the vector functions $\varphi_m^{(1)}(z)$ are analytic in D and doubly periodic. Justification of the expansion (8.13) on contrast parameters and its relation to the generalized alternating method of Schwarz can be found in [1, 6].

By substituting these expansions into (8.11) and collecting terms with ϱ_m in (8.11), we obtain

$$\varphi_m^{(1)}(t) = \varphi_{mk}^{(1)}(t), \quad |t - a_k| = r_k, \quad k = 1, 2, \dots, N \ (k \neq m), \tag{8.14}$$

$$\varphi_m^{(1)}(t) = \varphi_{mm}^{(1)}(t) - \left(\frac{r_m^2}{\overline{z - a_m}} + a_m \right) \begin{pmatrix} 1 \\ 0 \end{pmatrix}, \quad |t - a_m| = r_m, \tag{8.15}$$

with m running over $1, 2, \dots, N$ in the above equations (8.14) and (8.15).

The equality (8.14) implies that the vector function $\varphi_m^{(1)}(z)$ is analytically continued into $|z - a_k| < r_k$ for $k \neq m$. The relation (8.15) implies that $\varphi_m^{(1)}(z)$ is analytically continued into $|z - a_m| < r_m$, except at the first coordinate, which has a pole at $z = a_m$ of first order. It follows from the theory of elliptic functions [7] and relations from [1] that

$$\varphi_m^{(1)}(z) = r_m^2 [\pi z - \zeta(z - a_m)] \begin{pmatrix} 1 \\ 0 \end{pmatrix}, \qquad z \in D, \tag{8.16}$$

$$\varphi_{mk}^{(1)}(z) = r_m^2 [\pi z - \zeta(z - a_m)] \begin{pmatrix} 1 \\ 0 \end{pmatrix}, \quad |z - a_k| < r_k \quad (k \neq m), \tag{8.17}$$

$$\varphi_{mm}^{(1)}(z) = r_m^2 \left[\pi z - \zeta(z - a_m) + \frac{1}{z - a_m} \right] \begin{pmatrix} 1 \\ 0 \end{pmatrix}, \quad |z - a_m| < r_m, \tag{8.18}$$

where $\zeta(z)$ stands for the Weierstrass elliptic function [7]. Therefore, (8.13) yields the asymptotic formula valid up to $O(\varrho^2)$

$$\varphi_k(z) \simeq \begin{pmatrix} z \\ 0 \end{pmatrix} + z \sum_{m=1}^{N} \mathbf{s}_m \pi r_m^2 - \sum_{m=1}^{N} \mathbf{s}_m r_m^2 \zeta(z - a_m) + \mathbf{s}_k \frac{r_k^2}{z - a_k}, \ |z - a_k| < r_k, \tag{8.19}$$

where the vector

$$\mathbf{s}_m = \varrho_m \begin{pmatrix} 1 \\ 0 \end{pmatrix} = \frac{1}{|1 + \varepsilon_m|^2} \begin{pmatrix} |\varepsilon_m|^2 - 1 \\ 2\varepsilon''_m \end{pmatrix}. \tag{8.20}$$

The complex flux $\psi_k(z) = \frac{d}{dz}\varphi_k(z)$ takes the form

$$\psi_k(z) \simeq \begin{pmatrix} 1 \\ 0 \end{pmatrix} + \sum_{m=1}^{N} \mathbf{s}_m \pi r_m^2 + \sum_{m=1}^{N} \mathbf{s}_m r_m^2 \wp(z - a_m) - \mathbf{s}_k \frac{r_k^2}{(z - a_k)^2}, \quad |z - a_k| < r_k, \tag{8.21}$$

where the Weierstrass function $\wp(z) = -\zeta'(z)$ is used [7].

8.3 Effective permittivity tensor

The effective permittivity tensor with complex components

$$\varepsilon_\perp = \begin{pmatrix} \varepsilon_{11} & \varepsilon_{12} \\ \varepsilon_{21} & \varepsilon_{22} \end{pmatrix} \tag{8.22}$$

can be calculated analogously to the real effective permittivity tensor [1] (chapter 3, section 2.2.2 in [2]). Introduce the averaged value of a function $F(z)$ over the unit cell Q

$$\langle F(z) \rangle = \int_Q F(x_1 + ix_2)\, dx_1 dx_2. \tag{8.23}$$

The complex effective permittivity tensor can be defined through the relation

$$\langle \varepsilon(z) \nabla u(z) \rangle = \varepsilon_e \langle \nabla u(z) \rangle, \tag{8.24}$$

where $\varepsilon(z) = 1$ in D and $\varepsilon(z) = \varepsilon_k$ in D_k.

Consider the vector equality with complex components

$$\langle \nabla u \rangle = \int_D \nabla u\, dx_1 dx_2 + \sum_{k=1}^{N} \int_{D_k} \nabla u_k dx_1 dx_2. \tag{8.25}$$

Transform the double integrals by the divergence theorem associated with the Ostrogradsky–Gauss theorem of real vector analysis

$$\int_D \nabla u\, dx_1 dx_2 = \int_{\partial Q} u\mathbf{n}\, ds + \sum_{k=1}^{N} \int_{\partial D_k} (u_k - u)\mathbf{n}\, ds, \tag{8.26}$$

where $\mathbf{n}$ denotes the outward normal unit vector to the boundary $\partial D = \partial Q - \sum_{k=1}^{N} \partial D_k$. It follows from the first relation (8.1) that the integrals over ∂D_k vanish. The integral over ∂Q is calculated by the jumps of u per unit cell Q given by (8.2) (see chapter 3 and remark 5 in section 2.2.2 of [1] for details). Ultimately, we have

$$\langle \nabla u(z) \rangle = \begin{pmatrix} 1 \\ 0 \end{pmatrix} \text{ and } \varepsilon_e \langle \nabla u(z) \rangle = \begin{pmatrix} \varepsilon_{11} \\ \varepsilon_{21} \end{pmatrix}. \tag{8.27}$$

We now proceed to calculate the left part of (8.24), which has to be equal to the second vector of (8.27)

$$\begin{pmatrix} \varepsilon_{11} \\ \varepsilon_{21} \end{pmatrix} = \int_D \nabla u \; dx_1 dx_2 + \sum_{k=1}^{N} \varepsilon_k \int_{D_k} \nabla u_k dx_1 dx_2. \tag{8.28}$$

Green's formula will be applied to the components of (8.28)

$$\int_S \left(\frac{\partial F}{\partial x_1} - \frac{\partial G}{\partial x_2} \right) dx_1 dx_2 = \int_{\partial S} G \; dx_1 + F \; dx_2. \tag{8.29}$$

The components of (8.28) can be written in the form

$$\varepsilon_{11} = \int_{\partial Q} u \; dx_2 + \sum_{k=1}^{N} (\varepsilon_k - 1) \int_{\partial D_k} u_k \; dx_2, \tag{8.30}$$

$$-\varepsilon_{21} = \int_{\partial Q} u \; dx_1 + \sum_{k=1}^{N} (\varepsilon_k - 1) \int_{\partial D_k} u_k \; dx_1. \tag{8.31}$$

Using the jump conditions per unit cell Q (8.2), we calculate the integrals over ∂Q

$$\int_{\partial Q} u \; dx_1 = -\int_{-\frac{1}{2}}^{\frac{1}{2}} \left[u\left(x_1 + \frac{i}{2}\right) - u\left(x_1 - \frac{i}{2}\right) \right] dx_1 = 0, \tag{8.32}$$

$$\int_{\partial Q} u \; dx_2 = \int_{-\frac{1}{2}}^{\frac{1}{2}} \left[u\left(\frac{1}{2} + ix_2\right) - u\left(-\frac{1}{2} + ix_2\right) \right] dx_1 = 1. \tag{8.33}$$

Again, applying Green's formula for D_k, we obtain

$$\begin{pmatrix} \varepsilon_{11} \\ \varepsilon_{21} \end{pmatrix} = \begin{pmatrix} 1 \\ 0 \end{pmatrix} + \sum_{k=1}^{N} (\varepsilon_k - 1) \int_{D_k} \nabla u_k \; dx_1 dx_2. \tag{8.34}$$

The mean value theorem for a function harmonic in a disk yields

$$\begin{pmatrix} \varepsilon_{11} \\ \varepsilon_{21} \end{pmatrix} = \begin{pmatrix} 1 \\ 0 \end{pmatrix} + \sum_{k=1}^{N} (\varepsilon_k - 1) \pi r_k^2 \; \nabla u_k(a_k). \tag{8.35}$$

Using the formula

$$2(I + \boldsymbol{\alpha}_k)^{-1} = \frac{2}{|1 + \varepsilon_k|^2} \left(I + \boldsymbol{\alpha}_k^T\right) = \frac{2}{|1 + \varepsilon_k|^2} \begin{pmatrix} 1 + \varepsilon_k' & \varepsilon''_k \\ -\varepsilon''_k & 1 + \varepsilon_k' \end{pmatrix}. \tag{8.36}$$

and equation (8.9), we obtain

$$\begin{pmatrix} \dfrac{\partial u_k'}{\partial x_1} - i\dfrac{\partial u_k'}{\partial x_2} \\ \dfrac{\partial u''_k}{\partial x_1} - i\dfrac{\partial u''_k}{\partial x_2} \end{pmatrix} = \frac{2}{|1 + \varepsilon_k|^2} (I + \boldsymbol{\alpha}_k^T) \psi_k(z), \;\; |z - a_k| < r_k. \tag{8.37}$$

Substitute (8.21) into the right part of (8.37) and find

$$\begin{aligned}
&\begin{pmatrix} \dfrac{\partial u_k'}{\partial x_1} - i\dfrac{\partial u_k'}{\partial x_2} \\ \dfrac{\partial u''_k}{\partial x_1} - i\dfrac{\partial u''_k}{\partial x_2} \end{pmatrix} = \frac{2}{|1+\varepsilon_k|^2}\begin{pmatrix} 1+\varepsilon_k' \\ -\varepsilon''_k \end{pmatrix} \\
&+ \frac{2}{|1+\varepsilon_k|^2}(I+\boldsymbol{\alpha}_k^{\top})\left[\sum_{m=1}^{N}\mathbf{s}_m\pi r_m^2 + \sum_{m=1}^{N}\mathbf{s}_m r_m^2\wp(a_k-a_m)\right] + O(\varrho^2) = \\
&\frac{2}{|1+\varepsilon_k|^2}\begin{pmatrix} 1+\varepsilon_k' \\ -\varepsilon''_k \end{pmatrix} + \sum_{m=1}^{N}\frac{2\pi r_m^2}{|1+\varepsilon_k|^2|1+\varepsilon_m|^2}\left[1+\frac{1}{\pi}\wp(a_k-a_m)\right]\boldsymbol{\beta}_{km} + O(\varrho^2),
\end{aligned} \tag{8.38}$$

where the vector

$$\boldsymbol{\beta}_{km} = |1+\varepsilon_m|^2(I+\boldsymbol{\alpha}_k^{\top})\mathbf{s}_m = \begin{pmatrix} (\varepsilon_k'+1)(|\varepsilon_m|^2-1) + 2\varepsilon''_k\varepsilon''_m \\ -\varepsilon''_k(|\varepsilon_m|^2-1) + 2\varepsilon''_m(\varepsilon_k'+1) \end{pmatrix}. \tag{8.39}$$

Here, it is assumed for brevity that $\wp(a_k - a_m)$ vanishes, when a_m coincides with a_k. It follows from (8.38) and (8.39) that

$$\begin{aligned}
\nabla u_k(a_k) &= \left.\begin{pmatrix} \dfrac{\partial u_k'}{\partial x_1} + i\dfrac{\partial u''_k}{\partial x_1} \\ \dfrac{\partial u_k'}{\partial x_2} + i\dfrac{\partial u''_k}{\partial x_2} \end{pmatrix}\right|_{z=a_k} = \begin{pmatrix} \dfrac{2}{1+\varepsilon_k} \\ 0 \end{pmatrix} + \sum_{m=1}^{N}\pi r_m^2\begin{pmatrix} \dfrac{2(\varepsilon_m-1)}{(1+\varepsilon_k)(1+\varepsilon_m)} \\ 0 \end{pmatrix} \\
&+ \sum_{m=1}^{N} r_m^2\frac{2(\varepsilon_m-1)}{(1+\varepsilon_k)(1+\varepsilon_m)}\begin{pmatrix} \mathrm{Re}\,\wp(a_k-a_m) \\ -\mathrm{Im}\,\wp(a_k-a_m) \end{pmatrix} + O(\varrho^2).
\end{aligned} \tag{8.40}$$

Substitute (8.40) into (8.35)

$$\begin{aligned}
\begin{pmatrix} \varepsilon_{11} \\ \varepsilon_{21} \end{pmatrix} &= \left(1 + 2\sum_{k=1}^{N}\lambda_k\pi r_k^2 + 2\sum_{k=1}^{N}\sum_{m=1}^{N}\lambda_k\lambda_m\pi r_k^2\pi r_m^2\right)\begin{pmatrix} 1 \\ 0 \end{pmatrix} \\
&+ 2\sum_{k=1}^{N}\sum_{m=1}^{N}\pi r_k^2 r_m^2\lambda_k\lambda_m\begin{pmatrix} \mathrm{Re}\,\wp(a_k-a_m) \\ -\mathrm{Im}\,\wp(a_k-a_m) \end{pmatrix} + O(\varrho^3),
\end{aligned} \tag{8.41}$$

where the eigenvalues (8.12) are used.

Let $f = N\pi r^2$ denote the concentration of inclusion per unit cell Q in the case of equal radii $r_k = r$. Then, (8.41) becomes

$$\begin{aligned}
\begin{pmatrix} \varepsilon_{11} \\ \varepsilon_{21} \end{pmatrix} &= \left[1 + 2f\frac{1}{N}\sum_{k=1}^{N}\lambda_k + 2\left(f\frac{1}{N}\sum_{k=1}^{N}\lambda_k\right)^2\right]\begin{pmatrix} 1 \\ 0 \end{pmatrix} \\
&+ 2f^2\frac{1}{\pi N^2}\sum_{k=1}^{N}\sum_{m=1}^{N}\lambda_k\lambda_m\begin{pmatrix} \mathrm{Re}\,\wp(a_k-a_m) \\ -\mathrm{Im}\,\wp(a_k-a_m) \end{pmatrix} + O(\varrho^3).
\end{aligned} \tag{8.42}$$

Similar to (8.42) we calculate the vector $\begin{pmatrix} \varepsilon_{12} \\ -\varepsilon_{22} \end{pmatrix}$. Its value can be obtained from (8.42) by rotation of the points a_k about 90°. It is equivalent to replacement of a_k by ia_k. Using the property of the Weierstrass function $\wp(iz) = -\wp(z)$, we obtain from (8.42)

$$\begin{pmatrix} \varepsilon_{22} \\ -\varepsilon_{12} \end{pmatrix} = \left[1 + 2f\frac{1}{N}\sum_{k=1}^{N}\lambda_k + 2\left(f\frac{1}{N}\sum_{k=1}^{N}\lambda_k\right)^2\right]\begin{pmatrix} 1 \\ 0 \end{pmatrix}$$
$$-2f^2\frac{1}{\pi N^2}\sum_{k=1}^{N}\sum_{m=1}^{N}\lambda_k\lambda_m\begin{pmatrix} \mathrm{Re}\ \wp(a_k - a_m) \\ -\mathrm{Im}\ \wp(a_k - a_m) \end{pmatrix} + O(\varrho^3). \tag{8.43}$$

Introduce the mean value $\langle\lambda\rangle = \frac{1}{N}\sum_{k=1}^{N}\lambda_k$. The vector equations (8.42) and (8.43) can be written in the matrix form

$$\varepsilon_\perp \equiv \begin{pmatrix} \varepsilon_{11} & \varepsilon_{12} \\ \varepsilon_{21} & \varepsilon_{22} \end{pmatrix} = [1 + 2f\langle\lambda\rangle + 2f^2\langle\lambda\rangle^2]I$$
$$+2f^2\frac{1}{\pi N^2}\sum_{k=1}^{N}\sum_{m=1}^{N}\lambda_k\lambda_m\begin{pmatrix} \mathrm{Re}\ \wp(a_k - a_m) & -\mathrm{Im}\ \wp(a_k - a_m) \\ -\mathrm{Im}\ \wp(a_k - a_m) & -\mathrm{Re}\ \wp(a_k - a_m) \end{pmatrix} + O(\varrho^3). \tag{8.44}$$

Introduce the Eisenstein function for the square array $E_2(z) = \pi + \wp(z)$ following [1]. It is assumed that $E_2(a_k - a_m) = \pi$, when a_m coincides with a_k. The tensor (8.44) can be written in the form

$$\varepsilon_\perp = (1 + 2f\langle\lambda\rangle)I$$
$$+2f^2\frac{1}{\pi N^2}\sum_{k=1}^{N}\sum_{m=1}^{N}\lambda_k\lambda_m\begin{pmatrix} \mathrm{Re}\ E_2(a_k - a_m) & -\mathrm{Im}\ E_2(a_k - a_m) \\ -\mathrm{Im}\ E_2(a_k - a_m) & -\mathrm{Re}\ E_2(a_k - a_m) \end{pmatrix} + O(\varrho^3). \tag{8.45}$$

Application of the Padé approximation to the first term in (8.44) yields

$$\varepsilon_\perp \equiv \begin{pmatrix} \varepsilon_{11} & \varepsilon_{12} \\ \varepsilon_{21} & \varepsilon_{22} \end{pmatrix} = \frac{1 + \langle\lambda\rangle f}{1 - \langle\lambda\rangle f}I$$
$$+2f^2\frac{1}{\pi N^2}\sum_{k=1}^{N}\sum_{m=1}^{N}\lambda_k\lambda_m\begin{pmatrix} \mathrm{Re}\ \wp(a_k - a_m) & -\mathrm{Im}\ \wp(a_k - a_m) \\ -\mathrm{Im}\ \wp(a_k - a_m) & -\mathrm{Re}\ \wp(a_k - a_m) \end{pmatrix} + O(\varrho^3). \tag{8.46}$$

The first invariant of the tensor $\varepsilon_\perp$ takes the form

$$\frac{1}{2}(\varepsilon_{11} + \varepsilon_{22}) = \frac{1 + \langle\lambda\rangle f}{1 - \langle\lambda\rangle f} + O(\varrho^3). \tag{8.47}$$

One can see that it does not depend on the location of inclusions.

The normalized longitudinal permittivity of the considered fibrous composite is calculated by the mean value, see formula (3.142) from [8]

$$\varepsilon_\| = \sum_{k=1}^{N}\varepsilon_k f_k + 1 - f, \tag{8.48}$$

where f_k stands for the concentration of the phase of permittivity ε_k. Equation (8.44) determines the effective permittivity tensor up to $O(\varrho^3)$. In the considered case of circular inclusions following [9] one can justify that the precision in contrast parameter coincides with the precision in concentration $O(f^3)$. The next terms in the expansions in ϱ and in f differ. The formula (8.44) with positive constants ε_k $(-1 < \lambda_k < 1)$, coincides with the formula derived in [2].

References

[1] Gluzman S, Mityushev V and Nawalaniec W 2017 *Computational Analysis of Structured Media* (New York: Academic)

[2] Mityushev V 2001 Transport properties of doubly periodic arrays of circular cylinders and optimal design problems *Appl. Math. Optim.* **44** 17–31

[3] Sihvola A H and Kong J A 1988 Effective permittivity of dielectric mixtures *IEEE Trans. Geosci. Remote Sens.* **26** 420–9

[4] Bakhvalov N S and Panasenko G 2012 *Homogenisation: Averaging Processes in Periodic Media: Mathematical Problems in the Mechanics of Composite Materials* **vol 36** (Berlin: Springer)

[5] Hashin Z 1983 Analysis of composite materials—a survey *J. Appl. Mech.* **50** 481–505

[6] Mityushev V and Rylko N 2022 Effective properties of two-dimensional dispersed composites. Part I. Schwarz's alternating method *Comput. Math. Appl.* **111** 50–60

[7] Akhiezer N 1990 *Elements of the Theory of Elliptic Functions* (Providence, RI: AMS)

[8] Wojnar R, Bytner S and Gałka A 1999 Effective properties of elastic composites subject to thermal fields *Thermal Stresses V* (Rochester: Lastran Corporation) 257–65

[9] Mityushev V and Rylko N 2013 Maxwell's approach to effective conductivity and its limitations *Q. J. Mech. Appl. Math.* **66** 241–51

Chapter 9

Enhanced chiral sensing with achiral photonic metasurfaces

Sotiris Droulias

Sensing molecular chirality at the nanoscale has been a long-standing challenge, due to the inherently weak nature of the associated chiroptical signals. In this chapter we demonstrate how, with the aid of achiral photonic metasurfaces, it is possible to detect weak chirality. We study both isotropic and anisotropic metasurfaces, and we demonstrate how such metasurfaces enable the enhanced detection, unambiguous enantiomer differentiation and complete measurement of the total chirality (magnitude and sign of both its real and imaginary part). We examine the additional degrees of freedom offered by anisotropic metasurfaces with respect to their isotropic counterparts, which enable chiroptical signals with unprecedented enhancement, and the possibility for absolute chirality measurements, i.e. measurements without the need for sample removal, using a crucial signal reversal (excitation with reversed polarization). Last, we introduce gain, and we explore the possible coupling paths between the gain, the metasurface and the chiral inclusions, to further enhance chiral sensing.

9.1 Introduction

Chirality, a geometric property in which an object is nonsuperimposable with its mirror image, is a fundamental property of life: the chemistry of life and the functionalities of its building blocks are largely stereospecific, organisms ranging from protists to plants and animals possess morphological asymmetries with respect to their left–right axis, and biomolecular structures fundamental to life, such as amino acids, sugars, RNA and DNA, are both chiral and single-handed. Chiral biomolecules and chemical compounds exist in right- and left-handed forms, known as *enantiomers*. Their functionality is often determined by their handedness and, therefore, the development of sensitive chiral sensing techniques has been vital for a wide range of scientific disciplines, and has enabled the study of fundamental

doi:10.1088/978-0-7503-5462-2ch9

symmetries of the universe [1], determination of biomolecular structures [2–4] and even the development of safe and effective drugs [5, 6], to name a few of the most important applications.

9.1.1 Optical activity: definitions and effects

Optical activity, a characteristic light–matter interaction unique to chiral materials, manifests via the effects of *optical rotation* (OR) and *circular dichroism* (CD). OR (also known as circular birefringence) is the rotation of the orientation of the plane of polarization about the optical axis of linearly polarized light as it travels through a chiral medium. CD is dichroism involving circularly polarized light, i.e. the differential absorption between left and right circularly polarized (LCP/RCP) light. OR and CD originate from the different refractive index that LCP and RCP waves experience in a chiral medium [7, 8], and their effects on the polarization of a linearly polarized wave are shown in figure 9.1.

By 1815 Fresnel had discovered that linearly polarized light can be regarded as a superposition of LCP/RCP light. This finding led to his proposal of the first phenomenological theory about optical activity, which has the general form

$$\phi = \frac{\pi l}{\lambda}(n_- - n_+), \tag{9.1}$$

where n_- and n_+ are the indices of refraction of a chiral medium for LCP and RCP, respectively, λ is the vacuum wavelength of light and l is the optical path length of the medium.

With the advent of electromagnetism, a macroscopic description of natural optically active materials became possible and is now contained in the material parameters of the constitutive relations. These, along with Maxwell's equations, are written according to Condon [7, 8] as (here and throughout this work, we follow the $e^{+i\omega t}$ convention)

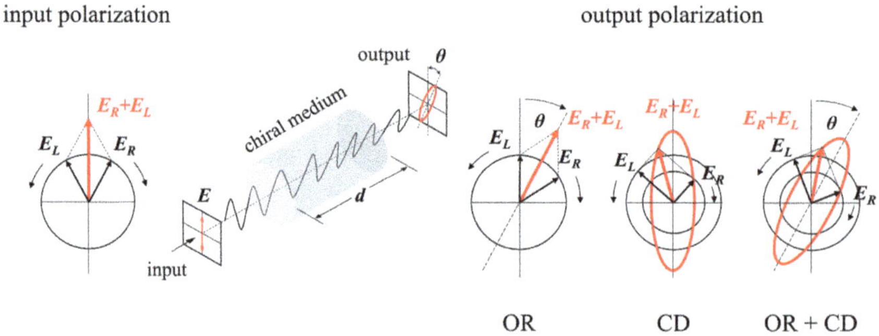

Figure 9.1. Effects of optical activity on a linearly polarized wave passing through a chiral medium. Optical rotation (OR) rotates the polarization plane, circular dichroism (CD) makes it elliptical and, under both effects, the incident polarization becomes a rotated ellipse. The total wave is expressed as a superposition of RCP (E_R field) and LCP (E_L field) light.

$$\nabla \times \mathbf{E} = -i\omega\mathbf{B}, \tag{9.2}$$

$$\nabla \times \mathbf{H} = +i\omega\mathbf{D}, \tag{9.3}$$

$$\mathbf{D} = \epsilon_0\epsilon_r\mathbf{E} - i\,(\kappa/c)\mathbf{H}, \tag{9.4}$$

$$\mathbf{B} = \mu_0\mu_r\mathbf{H} + i\,(\kappa/c)\mathbf{E}, \tag{9.5}$$

where ϵ_0, μ_0 are the vacuum permittivity and permeability, ϵ_r, μ_r are the relative material permittivity and permeability, respectively, and c is the vacuum speed of light; κ is the chirality (also known as 'Pasteur') parameter, which expresses the chiral molecular response. The eigenwaves of such a medium are RCP/LCP (or $\pm$) waves, which propagate with refractive indices $n_\pm = n_c \pm \kappa$, respectively, where $(n_+ + n_-)/2 = n_c \equiv \sqrt{\epsilon_r\mu_r}$ is the average (background) refractive index.

The chirality parameter, κ, is a dimensionless complex quantity, which expresses the chiral molecular response and is, in general, dispersive; its sign, sgn (κ), is associated with the handedness of the enantiomer and its magnitude, $|\kappa|$, is a function of the molecular properties and the concentration of the chiral medium. All things considered, equation (9.1) is written as $\phi = (2\pi l/\lambda) \cdot \kappa$, i.e. the real part of κ, Re(κ), is associated with effects of OR, while the imaginary part of κ, Im(κ), is associated with CD.

9.1.2 Chiral sensing techniques

Chiral sensing refers to a sensing scheme (apparatus and measurement technique) that is able to detect and distinguish enantiomers. Although, ideally, one would like to retrieve the entire unknown chirality parameter κ of a chiral medium, in some cases it suffices to just identify its handedness. To be able to distinguish enantiomers, a chiral sensing scheme must be sensitive to sgn (κ), and this is evident in the form of the constitutive relations, which expresses the inherent dependence of κ on the handedness of the chiral medium; under the transformation $\mathbf{r} \to -\mathbf{r}$, where $\mathbf{r}$ is the position vector, $\mathbf{E}$ and $\mathbf{D}$ change sign (polar vectors), while $\mathbf{B}$ and $\mathbf{H}$ remain invariant (axial vectors) and, hence, κ must change sign [see equations (9.4) and (9.5)]. Additionally, because $|\kappa|$ is a function of the molecular properties (i.e. polarizability) and the concentration of the chiral medium [9], a chiral sensing scheme must be also sensitive to $|\kappa|$.

Traditionally, κ is measured using the polarimetric techniques of optical rotatory dispersion (ORD) and CD [9]. These techniques allow for the direct detection of the real part and the imaginary part of ϕ, and, thus, of both Re(κ) and Im(κ) of a natural optically active medium [10]. They detect changes in the polarization state of waves propagating through the chiral medium, which accumulate with the propagation distance and, hence, the larger the chiral sample, the more pronounced the changes and the corresponding measured signals; these signals are commonly referred to as *chiroptical*, and include the rotation θ and ellipticity η of the output wave polarization (measured in degrees). For example, typical chiral solutions of mm extent provide chiroptical signals of a few degrees. However, chiral samples of low concentration and, in particular, small extent (e.g. thin films of nanometer size or protein monolayers) may yield chiroptical signals of a few μdeg, which are comparable to or even below the detection limits of modern instrumentation [11], and are, thus, hardly detectable.

In the past few years, the idea of using resonant near fields, instead of propagating waves, has led to the development of nanophotonic platforms to enhance weak chiralities, which would be otherwise undetectable with traditional ORD and CD [12–25]. Nanophotonic platforms supporting modes with strong and sustainable near fields may offer long interaction times between the chiral medium and the probing wave, thus enhancing their interaction and leading to detectable signals. Due to the localized nature of near fields, the detectable volume can be significantly reduced, practically in the order of the spatial extent of the near fields, enabling detection even in the ~nm scale. The underlying mechanism for chirality enhancement is based on the fact that chiral matter–wave interactions are enabled by a nonvanishing pseudoscalar product between **E** and **H**. Therefore, nanophotonic platforms are typically designed to support two orthogonal modes where, at the same frequency, the electric field of one mode is parallel to the magnetic field of the other [18–20, 22–25]; for simplicity, we will refer to these modes as *electric* and *magnetic*, respectively. When the nanophotonic platform is illuminated with light of circular polarization, both modes are excited, i.e. $\mathbf{E} \cdot \mathbf{H} \neq 0$, and the nanophotonic platform can couple to the chiral medium. Due to the strong nature of near fields, the optical chirality density [26] can be higher than that of the same wave in vacuum, leading to *superchiral* near fields, which enable the enhancement of far-field chiroptical signals, as demonstrated in several works involving different nanophotonic platforms, such as propagating surface plasmons [12], plasmonic particles [14–17], chiral metasurfaces [18–21] and, recently, achiral metasurfaces [22–24].

In general, nanophotonic-based approaches have proven to be a powerful means for granting access to weak chiroptical signals, not previously attainable with traditional polarimetric techniques. The general principle of operation behind almost all contemporary nanophotonic-based chiral sensing approaches relies on CD measurements of an optically active chiral substance, which is embedded in the nanophotonic platform. To achieve this, RCP and LCP waves are utilized to excite the nanophotonic platform, generate superchiral fields and enable CD measurements in transmission. While several works have attributed the resulting CD signal to be proportional only to $\mathrm{Im}(\kappa)$ [22, 27, 28], in reality, as supported by past and recent experimental results [15, 24, 29, 30], the observed CD signals depend on both $\mathrm{Re}(\kappa)$ and $\mathrm{Im}(\kappa)$. This is due to the fact that the employed nanophotonic platforms typically have intrinsic chiroptical responses that contribute to the total signal, often precluding direct quantitative measurement of chirality [16, 27, 31–34]. Thus, with most contemporary nanophotonic approaches, sensing of the magnitude and sign of both the real and imaginary part of the chirality parameter of a natural optically active substance (*complete* measurement) has, so far, not been possible.

From this discussion it becomes apparent that a chiral sensing scheme should be able to discriminate between the contributions of $\mathrm{Re}(\kappa)$ and $\mathrm{Im}(\kappa)$, especially far from the chiral molecular resonances, where Im(κ) is weak and Re(κ) is dominant. Additionally, because the magnitude of the chirality parameter, $|\kappa|$, is a function of the molecular properties and its concentration, while its sign, sgn(κ), depends on the handedness of the chiral medium, the sensing scheme should be sensitive to both $|\kappa|$ and sgn(κ). In this work, we propose schemes for the unambiguous determination of an unknown chirality, and we explain how it is possible to detect and distinguish Re (κ) and Im(κ) using linearly polarized waves to measure effects of rotation θ and ellipticity η, instead of using circularly polarized waves to measure effects of CD. In what follows, in section 9.2, we first establish the analytical framework that we use to provide insights into the enhancement mechanism. In section 9.3 we apply our framework to achiral isotropic metasurfaces and in section 9.4 to achiral anisotropic metasurfaces. Last, in section 9.5 we introduce gain to further enhance the chiral detection.

9.2 Electromagnetic framework for studying chiral sensing with achiral metasurfaces

In this section we present the analytical framework for studying chiral sensing with achiral, electrically thin, photonic metasurfaces. The chiral medium, characterized by κ, is embedded in the metasurface, which supports one electric and one magnetic mode; we refer to the chiral medium as *chiral inclusion*. The chiral inclusion induces weak coupling between the two otherwise orthogonal modes that, upon detection, provides information about the properties of the unknown κ. Because the metasurface is achiral, the magneto-electric coupling comes entirely from the coupling of the two modes via the chiral inclusion. The sensing scheme is based on using the achiral metasurface judiciously to (a) enhance the chiroptical signals with respect to the respective signals that can be retrieved from the chiral inclusion alone, (b) differentiate between left and right enantiomers and (c) unambiguously determine the chirality parameter of the unknown chirality, i.e. retrieve both Re(κ) and Im(κ).

Here, we employ a less familiar but more convenient formalism, which is based on replacing the actual metasurface with a polarizable sheet that has the same far-field response with the metasurface. As a result, an observer in the far-field cannot tell whether the scattered field is coming from the metasurface or the equivalent sheet and, hence, we can model the metasurface by means of the equivalent sheet. We start by establishing a connection between the bulk effective material parameters of the combined system of metasurface with chiral inclusions, and the corresponding material parameters of the sheet, i.e. surface complex conductivities. The surface conductivities are associated with surface currents of electric and magnetic type that correspond to the electric and magnetic mode of the metasurface. To understand how the metasurface's resonant modes are associated with the chirality-induced magneto-electric conductivity of the sheet, we introduce a coupled oscillator model to describe the chirality-induced coupling of the sheet's electric and magnetic currents that represent the metasurface's resonant modes. Last, we solve Maxwell's equations for the polarizable sheet, to associate the scattering amplitudes of the interrogating wave with the surface conductivities of the sheet. The analysis establishes a connection between microscopic quantities, such as the driving forces and induced currents in the coupled oscillator model, and macroscopically observed quantities, such as the amplitudes of the waves scattered by the metasurface and the chirality parameter κ of the inclusions.

9.2.1 From bulk material parameters to surface conductivities

Let us start with an achiral isotropic metasurface, homogeneously embedded with a chiral inclusion, as shown schematically in figure 9.2. The metasurface supports one electric and one magnetic mode, associated with the surface currents j_e and j_m, respectively, of the equivalent sheet. The presence of the chiral inclusion induces weak coupling between the two modes of the metasurface, which in the sheet formalism is expressed by the magneto-electric conductivity σ_{em}. In order to understand the coupling mechanism, we replace the homogenizable metasurface by an electrically thin slab of thickness D_{slab}, with bulk effective susceptibilities χ_{ij}^{eff}, where $\{i, j\} = \{e, m\}$. Inside the homogeneous slab, the constitutive relations equations (9.4) and (9.5) are expressed as

$$\mathbf{D} = \epsilon_0 \epsilon_{\text{eff}} \mathbf{E} - i(\kappa_{\text{eff}}/c)\mathbf{H}, \tag{9.6}$$

$$\mathbf{B} = \mu_0 \mu_{\text{eff}} \mathbf{H} + i(\kappa_{\text{eff}}/c)\mathbf{E}, \tag{9.7}$$

where ϵ_{eff}, μ_{eff} and κ_{eff} are the effective permittivity, permeability and chirality of the slab. At frequencies far from the resonant frequency of the two modes, the effective chirality reduces to the chirality of the chiral inclusion alone. Therefore, to study the chirality enhancement due to metasurface's resonant modes, it is convenient to separate their contribution to the total effective material parameters from possible background contributions, e.g. originating from other modes located at other spectral ranges. Along these lines, we write the bulk effective parameters as a sum of background (non-resonant) and resonant contributions, i.e. $\epsilon_{\text{eff}} = \epsilon_0(1 + \chi_{ee}^{\text{bgnd}} + \chi_{ee}^{\text{res}})$, $\mu_{\text{eff}} = \mu_0(1+$

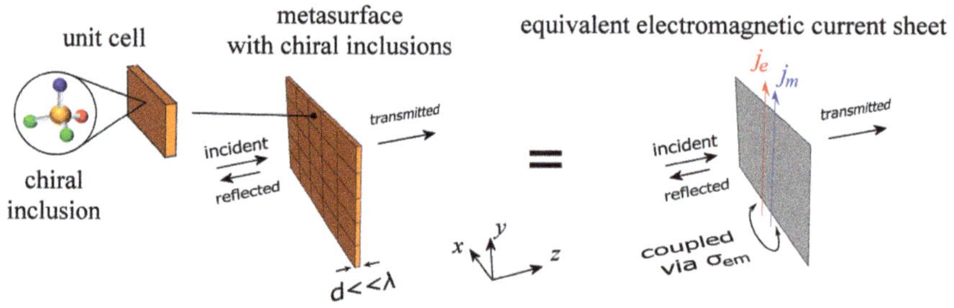

Figure 9.2. Schematic illustration of achiral metasurface for enhanced sensing of chiral molecules, embedded in the metasurface (chiral inclusions). The metasurface supports an electric and a magnetic mode, which are coupled via the chiral inclusions. Due to its small thickness $d \ll \lambda$, the composite system (metasurface and chiral inclusions) is equivalently described by an infinite electromagnetic current sheet that supports electric and magnetic surface currents, j_e and j_m, respectively. The surface currents are coupled via a magneto-electric conductivity, σ_{em}, which is induced by the chiral inclusions. Reprinted (figure) with permission from [30], Copyright (2020) by the American Physical Society.

$\chi_{mm}^{\text{bgnd}} + \chi_{mm}^{\text{res}}$) and $\kappa_{\text{eff}} = \kappa + \chi_{em}^{\text{res}} = \kappa + \chi_{me}^{\text{res}}$, where κ is the bulk chirality parameter of the chiral inclusions. The resonant susceptibilities χ_{ee}^{res}, χ_{mm}^{res}, χ_{em}^{res}, χ_{me}^{res} are the bulk susceptibilities (dimensionless) due to the resonant modes, while the non-resonant susceptibilities χ_{ee}^{bgnd}, χ_{mm}^{bgnd} account for background contributions from other modes. Further, $\chi_{em}^{\text{res}} = \chi_{me}^{\text{res}}$ is the contribution to the chirality parameter originating from the coupling of the resonant modes with the chiral inclusions, which, in the absence of coupling vanishes, and κ_{eff} reduces to κ. We note here that the term 'resonant' refers to the metasurface and not to the molecular resonance of the chiral inclusion. Accordingly, the constitutive equations are written as

$$\begin{pmatrix} \mathbf{D} \\ \mathbf{B} \end{pmatrix} = \overbrace{\begin{pmatrix} \epsilon_0\left(1 + \chi_{ee}^{\text{bgnd}}\right) & -i\dfrac{\kappa}{c} \\ +i\dfrac{\kappa}{c} & \mu_0\left(1 + \chi_{mm}^{\text{bgnd}}\right) \end{pmatrix}}^{\text{background}} \begin{pmatrix} \mathbf{E} \\ \mathbf{H} \end{pmatrix} + \overbrace{\begin{pmatrix} \epsilon_0 \chi_{ee}^{\text{res}} & -i\dfrac{\chi_{em}^{\text{res}}}{c} \\ +i\dfrac{\chi_{me}^{\text{res}}}{c} & \mu_0 \chi_{mm}^{\text{res}} \end{pmatrix}}^{\text{resonant}} \begin{pmatrix} \mathbf{E} \\ \mathbf{H} \end{pmatrix}. \tag{9.8}$$

All χ_{ij}^{res}, with the subscript $\{i, j\} = \{e, m\}$ denoting the electric (e) or magnetic (m) character of susceptibilities, are bulk quantities and are related to the surface susceptibilities χ_{ij} of our thin polarizable sheet model as $\chi_{ij}^{\text{res}} = \chi_{ij}/D_{\text{slab}}$, i.e. χ_{ij} are measured in units of [m]. The resonant part of equation (9.8) is then expressed in terms of the surface susceptibilities as

$$\begin{pmatrix} \langle \tilde{p}_e \rangle \\ \langle \tilde{p}_m \rangle \end{pmatrix} = \begin{pmatrix} \epsilon_0 \chi_{ee} & -i\dfrac{\chi_{em}}{c} \\ +i\dfrac{\chi_{me}}{c} & \mu_0 \chi_{mm} \end{pmatrix} \begin{pmatrix} \tilde{E}_{\text{loc}} \\ \tilde{H}_{\text{loc}} \end{pmatrix} \equiv \begin{pmatrix} \langle \tilde{p}_{ee} \rangle & \langle \tilde{p}_{em} \rangle \\ \langle \tilde{p}_{me} \rangle & \langle \tilde{p}_{mm} \rangle \end{pmatrix} \begin{pmatrix} \tilde{E}_{\text{loc}} \\ \tilde{H}_{\text{loc}} \end{pmatrix}, \tag{9.9}$$

where the brackets denote spatial average and the tildes account for the same parameters in the frequency domain; p is the dipole moment, with the subscript denoting its electric (e) or magnetic (m) character. Because we consider isotropic metasurfaces, without loss of generality, in equation (9.9) we have expressed the constitutive equations in terms of scalar quantities, among which $\tilde{E}_{\text{loc}}$, $\tilde{H}_{\text{loc}}$ are the local (surface) fields (at the location of the polarizable sheet). Using

$$\langle \tilde{j}_{ee} \rangle = i\omega \langle \tilde{p}_{ee} \rangle = i\omega \epsilon_0 \chi_{ee} \tilde{E}_{\text{loc}}, \tag{9.10}$$

$$\langle \tilde{j}_{mm} \rangle = i\omega \langle \tilde{p}_{mm} \rangle = i\omega \mu \chi_{mm} \tilde{H}_{\text{loc}}, \tag{9.11}$$

$$\langle \tilde{j}_{em} \rangle = i\omega \langle \tilde{p}_{em} \rangle = i\omega (-\chi_{em}/c) \tilde{H}_{\text{loc}}, \tag{9.12}$$

$$\langle \tilde{j}_{me} \rangle = i\omega \langle \tilde{p}_{me} \rangle = i\omega (+\chi_{me}/c) \tilde{E}_{\text{loc}}, \tag{9.13}$$

we can express equation (9.9) in terms of the surface conductivities as

$$\begin{pmatrix} \langle \tilde{j}_e \rangle \\ \langle \tilde{j}_m \rangle \end{pmatrix} = \begin{pmatrix} \sigma_{ee} & \sigma_{em} \\ \sigma_{me} & \sigma_{mm} \end{pmatrix} \begin{pmatrix} \tilde{E}_{\text{loc}} \\ \tilde{H}_{\text{loc}} \end{pmatrix}, \tag{9.14}$$

where the averaged surface electric and magnetic currents, $\langle \tilde{j}_e \rangle$ and $\langle \tilde{j}_m \rangle$, are measured in A m^{-1} and V m^{-1}, respectively, the conductivities $\sigma_{ee} = i\omega\epsilon_0\chi_{ee}$, $\sigma_{mm} = i\omega\mu_0\chi_{mm}$ are measured in units of S and Ω, respectively, and the magneto-electric conductivities $\sigma_{em} = i\omega(-i\chi_{em}/c)$ and $\sigma_{me} = i\omega(+i\chi_{me}/c)$ are dimensionless.

9.2.2 Coupled oscillator model

The magneto-electric conductivities σ_{em}, σ_{me} of the sheet express the coupling of the metasurface's resonant modes induced by the chiral inclusions and, consequently, should depend on both the strength of the resonant modes and the strength of the chirality. Therefore, there must be a connection between σ_{em}, σ_{me} and κ, σ_{ee}, σ_{mm}. In order to understand the coupling mechanism, which is shown schematically in figure 9.3, we can write a coupled oscillator model for j_e and j_m to retrieve the connection between all involved parameters.

Let us start with the metasurface without chiral inclusions. In this case, we may write a set of two uncoupled equations, one for each current density

$$\frac{d^2 j_e(t)}{dt^2} + \gamma_e \frac{dj_e(t)}{dt} + \omega_e^2 j_e(t) = f_e(t), \tag{9.15}$$

$$\frac{d^2 j_m(t)}{dt^2} + \gamma_m \frac{dj_m(t)}{dt} + \omega_m^2 j_m(t) = f_m(t), \tag{9.16}$$

where ω_e, ω_m are the resonant frequencies, γ_e, γ_m the damping rates and f_e, f_m the driving forces of j_e and j_m, respectively. For harmonic currents of the form $j_e(t) = \tilde{j}_e(\omega)e^{+i\omega t}$, $j_m(t) = \tilde{j}_m(\omega)e^{+i\omega t}$, the solution of equations (9.15) and (9.16) in the frequency domain yields

$$\tilde{j}_e(\omega) = \frac{1}{D_e(\omega)}\tilde{f}_e(\omega), \tag{9.17}$$

$$\tilde{j}_m(\omega) = \frac{1}{D_m(\omega)}\tilde{f}_m(\omega), \tag{9.18}$$

where $D_e(\omega) = \omega_e^2 - \omega^2 + i\gamma_e\omega$ and $D_m(\omega) = \omega_m^2 - \omega^2 + i\gamma_m\omega$.

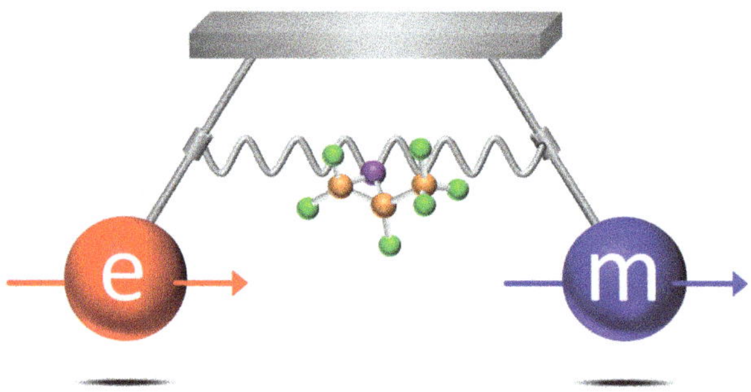

Figure 9.3. Coupled oscillator model for the chirality-induced coupling of the electric and magnetic surface currents, j_e and j_m, respectively. The surface currents of the sheet are associated with the collinear electric **e** and magnetic **m** moments, which are induced by the electric and magnetic modes of the metasurface, respectively.

We can now use equations (9.17) and (9.18) with equation (9.14) to find the surface conductivities σ_{ee}, σ_{mm} in the absence of magneto-electric coupling, by relating the microscopic quantities $\tilde{j}_e$, $\tilde{j}_m$, $\tilde{f}_e$, $\tilde{f}_m$ to the macroscopic quantities $\langle\tilde{j}_e\rangle$, $\langle\tilde{j}_m\rangle$, $\tilde{E}_{\text{loc}}$, $\tilde{H}_{\text{loc}}$. Taking into account that j_e, j_m is the contribution of each constituent meta-atom to the total current density of the metasurface, for the average surface current density we may write $\langle\tilde{j}_e\rangle = n\tilde{j}_e$ and $\langle\tilde{j}_m\rangle = n\tilde{j}_m$, respectively, where n accounts for the atoms per unit of surface area. The microscopic driving forces f_e, f_m can be associated with the local fields as $\tilde{f}_e = i\omega C_e\tilde{E}_{\text{loc}}$ and $\tilde{f}_m = i\omega C_m\tilde{H}_{\text{loc}}$, taking into account that they are proportional to the derivatives of the (macroscopic) fields E_{loc} and H_{loc}, respectively, rather than the fields themselves, as they drive currents rather than polarizations. The unknown proportionality constants C_e, C_m can be determined using equation (9.14). For example, for C_e we may write $\langle\tilde{j}_e\rangle = \sigma_{ee}\tilde{E}_{\text{loc}}$ or $n\tilde{j}_e = \sigma_{ee}\tilde{E}_{\text{loc}}$ or $nD_e^{-1}\tilde{f}_e = i\omega\epsilon_0\chi_{ee}\tilde{E}_{\text{loc}}$, where we have used the result of equation (9.17) and the relation $\sigma_{ee} = i\omega\epsilon_0\chi_{ee}$ which connects the surface electric conductivity with the respective surface susceptibility. By simple inspection of $\tilde{f}_e = n^{-1}D_e i\omega\epsilon_0\chi_{ee}\tilde{E}_{\text{loc}}$ and $\tilde{f}_e = i\omega C_e\tilde{E}_{\text{loc}}$, we obtain in the static limit ($\omega \to 0$): $C_e = n^{-1}\omega_e^2\epsilon_0\chi_{ee}^{\text{static}}$, as has also been retrieved previously in [35]. Similarly, for C_m, we find $C_m = n^{-1}\omega_m^2\mu_0\chi_{mm}^{\text{static}}$ and we can now write

$$\sigma_{ee} = \frac{\langle\tilde{j}_e\rangle}{\tilde{E}_{\text{loc}}} = \frac{n\tilde{j}_e}{(i\omega C_e)^{-1}\tilde{f}_e} = \frac{i\omega\alpha_e}{D_e(\omega)}, \tag{9.19}$$

$$\sigma_{mm} = \frac{\langle\tilde{j}_m\rangle}{\tilde{H}_{\text{loc}}} = \frac{n\tilde{j}_m}{(i\omega C_m)^{-1}\tilde{f}_m} = \frac{i\omega\alpha_m}{D_m(\omega)}, \tag{9.20}$$

where $\alpha_e \equiv nC_e = \omega_e^2\epsilon_0\chi_{ee}^{\text{static}}$ and $\alpha_m \equiv nC_m = \omega_m^2\mu_0\chi_{mm}^{\text{static}}$.

Having established the surface conductivities in the absence of chirality, we may now introduce a weak coupling coefficient κ_c, to account for the magneto-electric coupling between j_e and j_m due to κ. The system of equations (9.15) and (9.16) is slightly modified as

$$\frac{d^2j_e(t)}{dt^2} + \gamma_e\frac{dj_e(t)}{dt} + \omega_e^2 j_e(t) - \frac{\kappa_c}{nC_m}j_m(t) = f_e(t), \tag{9.21}$$

$$\frac{d^2j_m(t)}{dt^2} + \gamma_m\frac{dj_m(t)}{dt} + \omega_m^2 j_m(t) + \frac{\kappa_c}{nC_e}j_e(t) = f_m(t), \tag{9.22}$$

where the normalization in κ_c is chosen to preserve the correct units among the involved quantities, that is: j_e[A × m], j_m[V × m], n[1/m^2], C_e[S × Hz × m^2], C_m[Ω × Hz × m^2] and κ_c[Hz^2]. (Note that κ_c connects electric and magnetic quantities, rather than only electric quantities.) The $\mp$ sign in front of κ_c reflects its connection with κ (j_e and j_m are coupled via κ_c similarly to how **E** and **D** are coupled to **B** and **H**, respectively, via κ). The solution of equations (9.21) and (9.22) in the frequency domain yields

$$\tilde{j}_e(\omega) = \overbrace{\frac{1}{D_e(\omega)}\tilde{f}_e(\omega)}^{\tilde{j}_{ee}} + \overbrace{\frac{\kappa_c \alpha_m^{-1}}{D_e D_m}\tilde{f}_m(\omega)}^{\tilde{j}_{em}}, \tag{9.23}$$

$$\tilde{j}_m(\omega) = \underbrace{\frac{1}{D_m(\omega)}\tilde{f}_m(\omega)}_{\tilde{j}_{mm}} + \underbrace{\frac{\kappa_c \alpha_e^{-1}}{D_e D_m}\tilde{f}_e(\omega)}_{\tilde{j}_{me}}, \tag{9.24}$$

where we have used the approximation $D_e D_m + \kappa_c^2 \alpha_e^{-1} \alpha_m^{-1} \sim D_e D_m$, taking advantage of the extremely weak magneto-electric coupling (for stronger coupling, as in chiral metamaterials for example, this approximation does not hold). Under this approximation, $\tilde{j}_{ee}, \tilde{j}_{mm}$ become identical to $\tilde{j}_e, \tilde{j}_m$ of the uncoupled case (equations (9.17) and (9.18)) and σ_{ee}, σ_{mm} are therefore given by equations (9.19) and (9.20). For the remaining σ_{em}, σ_{me} we may write

$$\sigma_{em} = \frac{\langle \tilde{j}_{em} \rangle}{\tilde{H}_{\text{loc}}} = \frac{n\tilde{j}_{em}}{(i\omega C_m)^{-1}\tilde{f}_m} = +\frac{i\omega\kappa_c}{D_e(\omega)D_m(\omega)}, \tag{9.25}$$

$$\sigma_{me} = \frac{\langle \tilde{j}_{me} \rangle}{\tilde{E}_{\text{loc}}} = \frac{n\tilde{j}_{me}}{(i\omega C_e)^{-1}\tilde{f}_e} = -\frac{i\omega\kappa_c}{D_e(\omega)D_m(\omega)}. \tag{9.26}$$

With the set of equations (9.19), (9.20), (9.25) and (9.26) we have now established a direct connection between the surface conductivities of the sheet and the properties of the metasurface's resonant modes, which enter the coupled oscillator model via $D_{e,m}(\omega)$ and $\alpha_{e,m}$. Therefore, by engineering the metasurface's modes, not only do we tailor σ_{ee}, σ_{mm}, but also σ_{em}, σ_{me}.

9.2.3 Macroscopic description of chirality enhancement

To understand how σ_{em}, σ_{me} are explicitly associated with σ_{ee}, σ_{mm}, we can combine equations (9.25) and (9.26) with equations (9.19) and (9.20) to eliminate the terms D_e,D_m, obtaining the result

$$\sigma_c \equiv \sigma_{em} = -\sigma_{me} = \frac{i\omega\kappa_c}{(i\omega\alpha_e)(i\omega\alpha_m)}\sigma_{ee}\sigma_{mm}. \tag{9.27}$$

Clearly, in the weak coupling regime, the magneto-electric conductivity σ_c is proportional to the product of σ_{ee}, σ_{mm}. This result explains the origin of the magneto-electric enhancement and provides guidelines on how to tailor the chiral–matter interaction. In equation (9.27) all quantities are macroscopic, except for the coupling constant, κ_c, in the coupled oscillator model. To also express κ_c in terms of macroscopic quantities, we need to find its connection with the macroscopic fields and the chirality κ.

Let us denote with ($\mathbf{E}_e$, $\mathbf{H}_e$) and ($\mathbf{E}_m$, $\mathbf{H}_m$) the fields of the electric (e) and the magnetic mode (m) that are associated with the surface currents j_e and j_m, respectively. While without chirality ($\kappa = 0$) the two modes are orthogonal and j_e, j_m are uncoupled, the onset of chirality ($\kappa \neq 0$) induces perturbative polarization densities, which couple the two modes and, therefore, j_e with j_m. The polarization densities are of electric type $\delta\mathbf{P} = -i(\kappa/c)\mathbf{H}$ and of magnetic type $\delta\mathbf{M} = +i(\kappa/c)\mathbf{E}$ (see equations (9.4) and (9.5)) and the power transferred from mode e to mode m is caused by the polarization currents $i\omega\delta\mathbf{P}_{me} = +(\kappa\omega/c)\mathbf{H}_e$ and $i\omega\delta\mathbf{M}_{me} = -(\kappa\omega/c)\mathbf{E}_e$. The total power transferred from mode e to mode m is calculated by integrating the quantity $i\omega\delta\mathbf{P}_{me}\cdot\mathbf{E}_m^* + i\omega\delta\mathbf{M}_{me}\cdot\mathbf{H}_m^*$ across the whole volume $V_{\text{inclusion}}$ where the chiral substance extends [36]. Similarly, the total power transferred from mode m to mode e corresponds to exchanging $e \leftrightarrow m$ in the subscripts. Taking into account that the surface current densities are proportional to the local fields, the coupling coefficient κ_c will be

$$\kappa_c \propto \int_{V_{\text{inclusion}}} \kappa\frac{\omega}{c}\left(\mathbf{E}_m^*\cdot\mathbf{H}_e - \mathbf{E}_e\cdot\mathbf{H}_m^*\right)dV. \tag{9.28}$$

Hence, the coupling coefficient is proportional to the chirality parameter κ and an overlap integral of the fields [37]. In view of this result, the magneto-electric coupling of equation (9.27) takes the simple form

$$\sigma_c \equiv \sigma_{em} = -\sigma_{me} = \kappa C_0\sigma_{ee}\sigma_{mm}, \tag{9.29}$$

where C_0 is a constant, containing the overlap integral of the fields and the prefactors in equations (9.27) and (9.28). This result is of central importance; it explains how the effective chirality of the combined chiral inclusion–metasurface, σ_c, is expressed in terms of the individual contributions from the chiral inclusion and the metasurface: σ_c is proportional to the chirality parameter κ of the chiral inclusion, the spatial overlap of the electric and magnetic mode, and the product of their corresponding individual conductivities, σ_{ee}, σ_{mm}. Note that C_0, σ_{ee} and σ_{mm} are properties of the metasurface and are independent of κ, which is a property of the chiral inclusion alone. Hence, they can be tailored independently from the chiral inclusion to increase the product $C_0\sigma_{ee}\sigma_{mm}$, and enhance the weak κ.

To demonstrate how the effective chirality of the composite metasurface–chiral inclusion can be tailored by the properties of the metasurface's resonant modes, let us model the resonant conductivities σ_{ee}, σ_{mm} with a Lorentzian function

$$\sigma_{ee/mm}(\omega) = \frac{i a_{e/m}\omega}{\omega_{e/m}^2 - \omega^2 + i\gamma_{e/m}\omega}, \tag{9.30}$$

where the subscript e/m denotes the respective parameters for the electric/magnetic conductivity (the modification of the Lorentzian is because the current is the time derivative of the dielectric polarization).

For simplicity we choose $\alpha_e/\omega_0 = \alpha_m/\omega_0 = 1$, $\gamma_e/\omega_0 = \gamma_m/\omega_0 = 0.1$, and we detune the resonant frequencies around the central frequency ω_0 by $\delta\omega$, as $\omega_e = \omega_0 - \delta\omega$ and $\omega_m = \omega_0 + \delta\omega$. In figure 9.4(a) we plot σ_{ee} and σ_{mm} for $\delta\omega = 0$, and in figure 9.4(b) we plot $\text{Re}(\sigma_{ee})$ and $\text{Re}(\sigma_{mm})$, as a function of the normalized frequency detuning $\delta\omega/\omega_0$. In figure 9.4(c) we plot the magnitude of the product

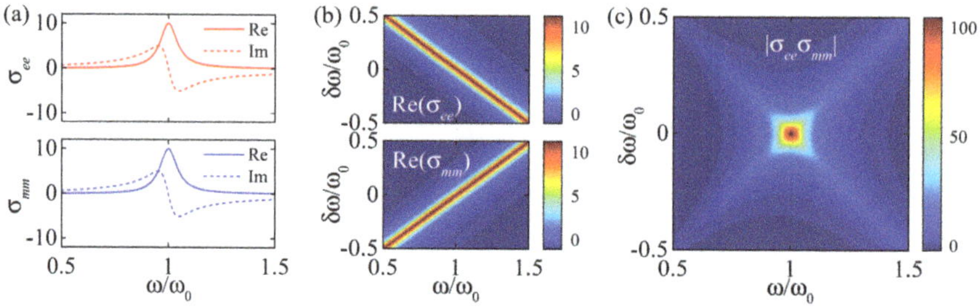

Figure 9.4. Enhancement of magneto-electric coupling as a function of the frequency detuning, $\delta\omega$, between the electric (σ_{ee}) and magnetic (σ_{mm}) conductivity. (a) σ_{ee} and σ_{mm} for zero frequency detuning. (b) $\mathrm{Re}(\sigma_{ee})$ and $\mathrm{Re}(\sigma_{mm})$ as a function of $\delta\omega$. (c) Magnitude of product $\sigma_{ee}\sigma_{mm}$ as a function of $\delta\omega$. For zero detuning ($\delta\omega = 0$), the product $|\sigma_{ee}\sigma_{mm}|$ is maximized, consequently maximizing the magneto-electric conductivity σ_c.

$\sigma_{ee}\sigma_{mm}$. We see that when the two modes have the same resonant frequency ($\delta\omega = 0$), the product of the individual conductivities is maximized, acquiring the value $\alpha_e\alpha_m/\gamma_e\gamma_m$. Therefore, we expect that, for zero detuning between the resonant modes, the magneto-electric coupling expressed by equation (9.29) is also maximized for a certain κ. Because the enhancement occurs in the near field of the metasurface's resonant modes, it would be practical to have some far-field proxy for the detection of the chiral–matter interaction enhancement. This is what we will examine next.

9.2.4 Far-field probing of near-field chiral enhancement

The magneto-electric conductivity σ_c is an effective material property associated with the enhancement of the chiral–matter interaction due to the chirality-induced coupling of the metasurface's modes. While this enhancement occurs in the vicinity of the metasurface, where the chiral inclusion resides and the modes oscillate, it would be practical to have a sensing scheme that can detect the presence of the chiral inclusion through measurements in the far field, which are easily accessible. At this point we may recall that the polarization of a linearly polarized wave passing through a chiral medium is known to undergo rotation and/or to become elliptical (see figure 9.1). Therefore, such a sensing scheme can be based on the detection of changes in the polarization state of a linearly polarized wave that is scattered by the metasurface.

We consider a metasurface extending on the xy-plane and a wave propagating along the $+z$ direction, as shown in figure 9.5(a). The wave is incident on the metasurface from the semi-infinite space (1), characterized by the permittivity ϵ_1 and permeability μ_1. The incident wave is partly reflected and partly transmitted to the semi-infinite space (2), characterized by the permittivity ϵ_2 and permeability μ_2. The boundary conditions for the equivalent electromagnetic current sheet are written as $\mathbf{n} \times (\mathbf{E}_2 - \mathbf{E}_1) = -\langle \mathbf{j}_m \rangle$, $\mathbf{n} \times (\mathbf{H}_2 - \mathbf{H}_1) = +\langle \mathbf{j}_e \rangle$, where $\mathbf{n}$ is the surface normal of the current sheet pointing from region 1 to region 2. Application of the boundary conditions at $z = 0$, where the sheet is located, yields the system of equations

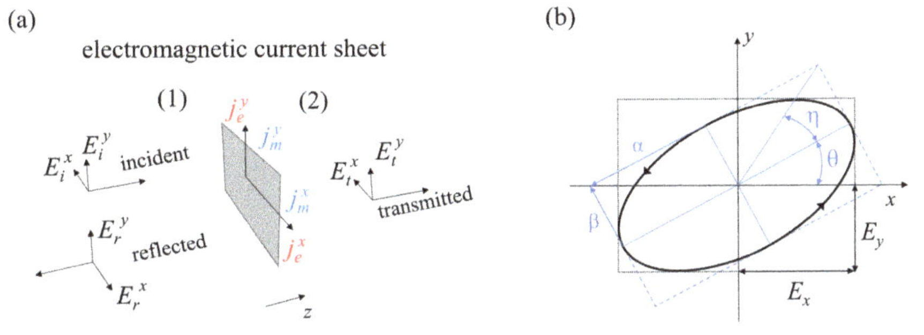

Figure 9.5. Plane waves scattered by an electromagnetic current sheet. (a) The incident field (components E_i^x, E_i^y) propagates along the $+z$ direction and is partly reflected (components E_r^x, E_r^y) by the sheet and partly transmitted (components E_t^x, E_t^y). The sheet is located at $z = 0$ and separates the surrounding homogeneous space into regions (1) and (2) with generally different properties. (b) Polarization analysis of scattered wave. In the most general case, the electric field vector of the wave draws out an ellipse, characterized by the rotation angle $\theta \in [-\frac{\pi}{2}, \frac{\pi}{2}]$, which is the angle between the ellipse major axis and the x-axis, and the ellipticity angle $\eta \in [-\frac{\pi}{4}, \frac{\pi}{4}]$, which is given in terms of the minor and major axes of the ellipse as $\tan \eta = \beta/\alpha$.

$$\begin{pmatrix} \frac{1}{\zeta_1} & -\frac{1}{\zeta_1} & -\frac{1}{\zeta_2} & \frac{1}{\zeta_2} & 0 & 0 & 0 & 0 \\ 0 & 0 & 0 & 0 & \frac{1}{\zeta_1} & -\frac{1}{\zeta_1} & -\frac{1}{\zeta_2} & \frac{1}{\zeta_2} \\ 0 & 0 & 0 & 0 & -1 & -1 & 1 & 1 \\ 1 & 1 & -1 & -1 & 0 & 0 & 0 & 0 \end{pmatrix} \begin{pmatrix} E_i^x \\ E_r^x \\ E_t^x \\ 0 \\ E_i^y \\ E_r^y \\ E_t^y \\ 0 \end{pmatrix} = \begin{pmatrix} \langle \tilde{j}_e^x \rangle \\ \langle \tilde{j}_e^y \rangle \\ \langle \tilde{j}_m^x \rangle \\ \langle \tilde{j}_m^y \rangle \end{pmatrix}, \tag{9.31}$$

where $\zeta_{1,2} = \sqrt{\frac{\mu_0 \mu_{1,2}}{\epsilon_0 \epsilon_{1,2}}}$ is the wave impedance of each of the two half-spaces. The surface currents are associated with the local fields at $z = 0$ via the conductivity tensor $\overleftrightarrow{\sigma}$ as

$$\begin{pmatrix} \langle \tilde{j}_e^x \rangle \\ \langle \tilde{j}_e^y \rangle \\ \langle \tilde{j}_m^x \rangle \\ \langle \tilde{j}_m^y \rangle \end{pmatrix} = \overleftrightarrow{\sigma} \begin{pmatrix} \tilde{E}_{loc}^x \\ \tilde{E}_{loc}^y \\ \tilde{H}_{loc}^x \\ \tilde{H}_{loc}^y \end{pmatrix}, \tag{9.32}$$

which extends equation (9.14) to generally anisotropic media. In its most general form, the conductivity tensor is a full matrix

$$\overleftrightarrow{\sigma} = \begin{pmatrix} \sigma_{ee}^{xx} & \sigma_{ee}^{xy} & \sigma_{em}^{xx} & \sigma_{em}^{xy} \\ \sigma_{ee}^{yx} & \sigma_{ee}^{yy} & \sigma_{em}^{yx} & \sigma_{em}^{yy} \\ \sigma_{me}^{xx} & \sigma_{me}^{xy} & \sigma_{mm}^{xx} & \sigma_{mm}^{xy} \\ \sigma_{me}^{yx} & \sigma_{me}^{yy} & \sigma_{mm}^{yx} & \sigma_{mm}^{yy} \end{pmatrix}, \tag{9.33}$$

containing elements $\sigma_{ij}^{\alpha\beta}$, with the subscript $\{i, j\} = \{e, m\}$ denoting the electric (e) or magnetic (m) character of the associated conductivities and fields, and the superscript $\{\alpha, \beta\} = \{x, y\}$ accounting for their associated components along the x, y directions.

The local fields involved in equation (9.32) are defined as an average across the current sheet

$$\begin{pmatrix} \tilde{E}_{loc}^{x} \\ \tilde{E}_{loc}^{y} \\ \tilde{H}_{loc}^{x} \\ \tilde{H}_{loc}^{y} \end{pmatrix} = \frac{1}{2}\begin{pmatrix} \tilde{E}_{loc,\,1}^{x} + \tilde{E}_{loc,\,2}^{x} \\ \tilde{E}_{loc,\,1}^{y} + \tilde{E}_{loc,\,2}^{y} \\ \tilde{H}_{loc,\,1}^{x} + \tilde{H}_{loc,\,2}^{x} \\ \tilde{H}_{loc,\,1}^{y} + \tilde{H}_{loc,\,2}^{y} \end{pmatrix} = \frac{1}{2}\begin{pmatrix} E_i^x + E_r^x + E_t^x \\ E_i^y + E_r^y + E_t^y \\ -\frac{1}{\zeta_1}E_i^y + \frac{1}{\zeta_1}E_r^y - \frac{1}{\zeta_2}E_t^y \\ +\frac{1}{\zeta_1}E_i^x - \frac{1}{\zeta_1}E_r^x + \frac{1}{\zeta_2}E_t^x \end{pmatrix}, \tag{9.34}$$

as both the electric and magnetic fields have discontinuities at the current sheet, due to the presence of both electric and magnetic currents [38, 39].

Combining equations (9.32)–(9.34) into equation (9.31) and solving for $E_t^x = t_{xx}E_i^x + t_{xy}E_i^y$, $E_t^y = t_{yy}E_i^y + t_{yx}E_i^x$, $E_r^x = r_{xx}E_i^x + r_{xy}E_i^y$ and $E_r^y = r_{yy}E_i^y + r_{yx}E_i^x$, we can now derive the analytical expressions for the scattering amplitudes t_{xx}, t_{yy}, t_{xy}, t_{yx}, r_{xx}, r_{yy}, r_{xy}, r_{yx} in terms of the surface conductivities, where the subscripts in $t_{\text{out,inc}}$, $r_{\text{out,inc}}$ denote the output (out) and incident (inc) E-field polarization, respectively. Up to this point we have assumed that the surface conductivities are known, in order to establish a general electromagnetic framework beyond any specific configuration. In the following sections we will employ our analytical framework to detect chiral molecules using examples of realistic metasurfaces.

Last, having calculated the components E_x, E_y of the scattered wave, its polarization state can be analysed in terms of its rotation θ and ellipticity η, as shown in figure 9.5(b). These angles can be calculated using the Stokes parameters, which are directly connected with practical polarization measurements, and are defined as

$$S_0 = E_xE_x^* + E_yE_y^*, \tag{9.35}$$

$$S_1 = E_xE_x^* - E_yE_y^*, \tag{9.36}$$

$$S_2 = E_xE_y^* + E_yE_x^*, \tag{9.37}$$

$$S_3 = i(E_xE_y^* - E_yE_x^*). \tag{9.38}$$

With the Stokes parameters known, we can calculate θ, η to obtain information about the polarization state of the scattered wave, as

$$\theta = \frac{1}{2}\arctan\frac{S_2}{S_1}, \tag{9.39}$$

$$\eta = \frac{1}{2}\arctan\frac{S_3}{\sqrt{S_1^2 + S_2^2}}. \tag{9.40}$$

9.3 Chiral sensing with isotropic metasurfaces

In this section we study the detection of an unknown chiral inclusion and the retrieval of its chirality parameter κ, using achiral isotropic metasurfaces. To detect the chiral inclusion we probe the metasurface, in which the chiral inclusion is embedded, with linearly polarized waves, and we measure changes in the transmitted polarization, which we characterize in terms of the rotation θ and ellipticity η of the polarization vector. Because the metasurface is achiral, any changes in the transmitted polarization are solely due to the embedded chirality. To understand the mechanism of enhanced sensing, we first need to establish a connection between the scattering amplitudes and the surface conductivities that express the chirality-induced coupling of the metasurface's electric and magnetic mode.

9.3.1 Scattering amplitudes

To calculate the scattering amplitudes we solve the system of equations (9.31)–(9.34) using the conductivity tensor that describes the response of achiral isotropic metasurfaces. To find the elements of the conductivity tensor, first, we take into account that for achiral metasurfaces, and in the absence of any chiral inclusions, the two off-diagonal 2×2 blocks in equation (9.33) are zero (all $\sigma_{ij}^{\alpha\beta}$ with $i \neq j$). Additionally, each of the two diagonal 2×2 blocks can be expressed in diagonal form, leaving only four quantities necessary to describe the metasurface, that is σ_{ee}^{xx}, σ_{ee}^{yy}, σ_{mm}^{xx} and σ_{mm}^{yy}. The presence of chirality ($\kappa \neq 0$) introduces magneto-electric coupling, rendering nonzero the components $\sigma_{ij}^{\alpha\beta}$ with $i \neq j$. We may write (a) $\sigma_{em}^{\alpha\beta} = -\sigma_{me}^{\alpha\beta} \equiv \sigma_{c}^{\alpha\beta}$ for $\alpha = \beta$ and (b) $\sigma_{em}^{\alpha\beta} = \sigma_{me}^{\alpha\beta} = 0$ for $\alpha \neq \beta$. Additionally, for isotropic metasurfaces there is no distinction between the x and y directions and, therefore, $\sigma_{ee}^{xx} = \sigma_{ee}^{yy} \equiv \sigma_{ee}$, $\sigma_{mm}^{xx} = \sigma_{mm}^{yy} \equiv \sigma_{mm}$ and $\sigma_{c}^{xx} = \sigma_{c}^{yy} \equiv \sigma_c$. All things considered, the conductivity tensor takes the simple form

$$\overleftrightarrow{\sigma} = \begin{pmatrix} \sigma_{ee} & 0 & \sigma_c & 0 \\ 0 & \sigma_{ee} & 0 & \sigma_c \\ -\sigma_c & 0 & \sigma_{mm} & 0 \\ 0 & -\sigma_c & 0 & \sigma_{mm} \end{pmatrix}. \tag{9.41}$$

Using equation (9.41) to solve the system of equations (9.31)–(9.34) yields the transmission and reflection amplitudes expressed in terms of the surface conductivities, as

$$t \equiv t_{xx} = t_{yy} = \frac{\rho - \hat{\sigma}_{ee}\hat{\sigma}_{mm}}{\dfrac{\rho+1}{2} + \hat{\sigma}_{ee} + \hat{\sigma}_{mm} + \dfrac{\rho+1}{2\rho}\hat{\sigma}_{ee}\hat{\sigma}_{mm}}, \tag{9.42}$$

$$t_c \equiv t_{xy} = -t_{yx} = \frac{\rho\sigma_c}{\frac{\rho+1}{2} + \hat{\sigma}_{ee} + \hat{\sigma}_{mm} + \frac{\rho+1}{2\rho}\hat{\sigma}_{ee}\hat{\sigma}_{mm}}, \tag{9.43}$$

$$r \equiv r_{xx} = r_{yy} = \frac{\hat{\sigma}_{mm} - \hat{\sigma}_{ee} + \frac{\rho-1}{2} + \frac{\rho-1}{2\rho}\hat{\sigma}_{ee}\hat{\sigma}_{mm}}{\frac{\rho+1}{2} + \hat{\sigma}_{ee} + \hat{\sigma}_{mm} + \frac{\rho+1}{2\rho}\hat{\sigma}_{ee}\hat{\sigma}_{mm}}, \tag{9.44}$$

$$r_c = r_{xy} = r_{yx} = 0, \tag{9.45}$$

where $\rho \equiv \zeta_2/\zeta_1 = \sqrt{\mu_2/\epsilon_2}/\sqrt{\mu_1/\epsilon_1}$ is the ratio of the wave impedance between the two half-spaces ($\rho = 1$ for a metasurface in uniform environment), and we have normalized the conductivities, which are now dimensionless, as $\hat{\sigma}_{ee} = \zeta_2\sigma_{ee}/2$, $\hat{\sigma}_{mm} = \sigma_{mm}/2\zeta_1$. To derive equations (9.42)–(9.45), we made the approximation $\sigma_c \ll \hat{\sigma}_{ee}, \hat{\sigma}_{mm}$, eliminating any σ_c^2 term; this approximation is valid, as long as the magneto-electric coupling is perturbative.

9.3.2 Enhancement of chiral–matter interaction

To study the enhancement of the magneto-electric coupling in isotropic metasurfaces, we solve equations (9.42)–(9.45) in terms of the conductivities

$$\hat{\sigma}_{ee} = \frac{\rho - \rho r - t}{1 + r + t}, \tag{9.46}$$

$$\hat{\sigma}_{mm} = \frac{\rho(1 + r - t)}{\rho - \rho r + t}, \tag{9.47}$$

$$\sigma_c = \frac{2t_c(1 + \rho + r - \rho r)}{(\rho - \rho r + t)(1 + r + t)}. \tag{9.48}$$

Let us consider the realistic achiral metasurface shown schematically in figure 9.6 (a). The metasurface consists of silicon nanodisks [23] with a radius of 150nm, which are periodically arranged on the xy-plane with periodicity $\alpha = 500$ nm. The nanodisks have cylindrical holes of 10 nm radius, in which the chiral inclusion is contained (refractive index 1.33 – 0.01i). For simplicity, the metasurface is suspended in air ($\rho = 1$), and the refractive index of the nanodisks is constant and equal to 3.5 – 0.01i, essentially accounting for the weak dispersion of silicon in the chosen spectral range. The disk thickness d is used for tuning the spectral separation between the two modes, which we label as ED (electric dipole) and MD (magnetic dipole). Their field distribution along the center of the nanodisk and their frequency detuning and individual Q factors are shown in figure 9.6(b). To retrieve the effective surface conductivities of the metasurface, first, we set $\kappa = 0$ for the chiral inclusion located in the nanohole, and we illuminate the metasurface with linearly polarized light (see figure 9.6(a)). From the numerically calculated reflection and transmission

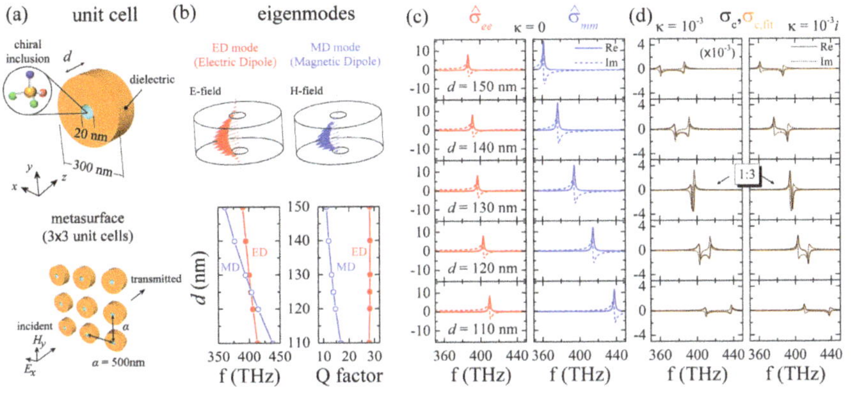

Figure 9.6. Enhanced chiral sensing using an achiral isotropic metasurface. (a) Schematic illustration of a single unit cell and the metasurface (nine unit cells shown here), which consists of dielectric nanodisks with cylindrical holes, in which a chiral inclusion is contained. (b) Eigenmodes, one of electric (ED—electric dipole) and one of magnetic (MD—magnetic dipole) type that are utilized for sensing. Top panel: field distribution of the modes along the axis of the disk, where the chiral inclusion is located. Bottom panel: spectral detuning (left panel) and Q factors (right panel) of the modes, as a function of the disk thickness, d. (c) Retrieved conductivities $\hat{\sigma}_{ee}$, $\hat{\sigma}_{mm}$ as a function of the disk thickness, d, for $\kappa = 0$. (d) Retrieved conductivity σ_c and its analytical fit $\sigma_{c,\text{fit}}$ using equation (9.29), as a function of the disk thickness, d, for $\kappa = 10^{-3}$ (left panel) and $\kappa = 10^{-3}i$ (right panel). The results of σ_c for $d = 130$ nm have been divided by a factor of 3 for easier comparison. Reprinted (figure) with permission from [30], Copyright (2020) by the American Physical Society.

amplitudes we retrieve the surface conductivities $\hat{\sigma}_{ee}$, $\hat{\sigma}_{mm}$ using equations (9.46) and (9.47), which we plot in figure 9.6(c), as a function of the disk thickness, ranging from 110 nm to 150 nm. Next, we set $\kappa \neq 0$ and repeat the calculations to retrieve σ_c with the aid of equation (9.48). To simulate a non-resonant chirality (detection far from the chiral molecular resonance), we use a purely real $\kappa = 10^{-3}$, which yields the retrieved magneto-electric conductivity shown in figure 9.6(d), left panel. For a resonant chirality (detection at the chiral molecular resonance) we use a purely imaginary $\kappa = 10^{-3}i$, which yields the retrieved magneto-electric conductivity shown in figure 9.6(d), right panel. We observe that σ_c is maximized when the detuning between $\hat{\sigma}_{ee}$ and $\hat{\sigma}_{mm}$ is minimized (for $d = 130$ nm), in accord with our previous observations in figure 9.4. Additionally, we observe that the change from purely real to purely imaginary κ results in $\sigma_c(\omega; \kappa = 10^{-3}i) = i\sigma_c(\omega; \kappa = 10^{-3})$, in accord with equation (9.29).

In figure 9.6(d), we also plot the analytical fit of σ_c, i.e. $\sigma_{c,\text{fit}}$ (orange lines), using equation (9.29), which is modified to account for background contributions from nearby modes as $\sigma_{c,\text{fit}} = \kappa C_{0,\text{fit}}(\hat{\sigma}_{ee} - i\beta_e\omega)(\hat{\sigma}_{mm} - i\beta_m\omega)$. In this expression $\hat{\sigma}_{ee}$, $\hat{\sigma}_{mm}$ are the numerically retrieved conductivities shown in figure 9.6(c), $\kappa = 10^{-3}$, $10^{-3}i$ is the chirality parameter used in the simulations, $C_{0,\text{fit}} = -0.268$ a constant that we use for the fitting, and $\beta_e = (2\pi)^{-1} \times 1.4$ fs, $\beta_m = 0$. The observed agreement between the numerically retrieved σ_c and its fit further verifies the validity of equation (9.29).

9.3.3 Enhanced detection of unknown chirality and enantiomer differentiation

To assess the enhancement of the chiroptical signals with respect to the respective signals of the chiral inclusion without the metasurface, we analyse the polarization state of the transmitted wave in the simulated examples of figure 9.6, using the Stokes parameters. The chiroptical signals θ, η for purely real and purely imaginary κ are presented in figure 9.7(a, b), respectively. We observe that, as we scan the disk thickness d, θ and η become weakest for $d = 130$ nm, where the magneto-electric coupling is maximized (see figure 9.6). At first sight, this seems to contradict our expectations; our previous example of figure 9.4 showed that, with reduced detuning between the electric and magnetic mode, σ_c increases and, therefore, θ, η are expected to be enhanced.

To understand this counter-intuitive behavior, we use our theoretical framework to calculate the chiroptical signals θ and η analytically. Inserting equations (9.42) and (9.43) in the Stokes parameters, we find that the chiroptical signals are expressed in terms of t and t_c as

$$\theta = \mathrm{Re}\left(-\frac{t_c}{t}\right) = \mathrm{Re}\left(\frac{\rho\sigma_c}{\hat{\sigma}_{ee}\hat{\sigma}_{mm} - \rho}\right), \tag{9.49}$$

$$\eta = \mathrm{Im}\left(-\frac{t_c}{t}\right) = \mathrm{Im}\left(\frac{\rho\sigma_c}{\hat{\sigma}_{ee}\hat{\sigma}_{mm} - \rho}\right), \tag{9.50}$$

where we have used $t_c \ll t$ to approximate $\tan^{-1}(\Phi) \sim \Phi$. We note here that, even for signals as large as $\sim 5°$, the error in this approximation is only 0.25% (in practice,

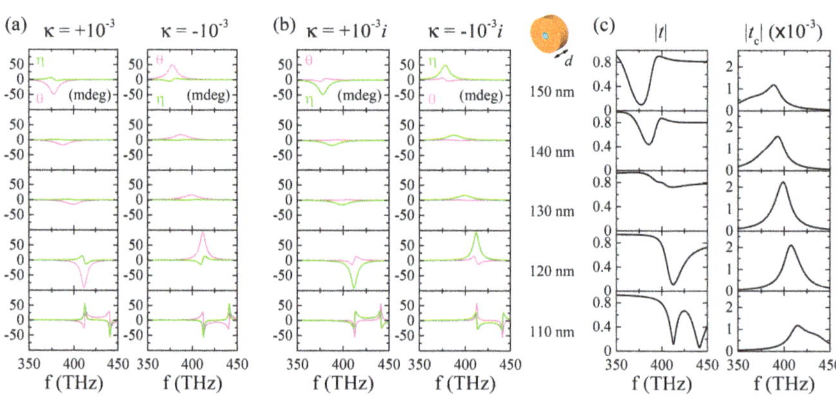

Figure 9.7. Enhanced detection and enantiomer differentiation of chiral inclusion, using the achiral isotropic metasurface of figure 9.6. Chiroptical signals θ (magenta lines) and η (green lines) for (a) $\kappa = \pm 10^{-3}$ and (b) $\kappa = \pm 10^{-3}i$. (c) Magnitude of co- and cross- transmission amplitudes t and t_c, respectively, for $\kappa = \pm 10^{-3}$ or $\kappa = \pm 10^{-3}i$. The numerical results satisfy $\theta(i\kappa) = -\eta(\kappa)$, $\eta(i\kappa) = +\theta(\kappa)$, in agreement with the theoretical predictions. Reprinted (figure) with permission from [30], Copyright (2020) by the American Physical Society.

the θ, η signals are of the order of some mdeg [25]). Note that, because the metasurface is achiral, if the inclusion is achiral as well ($\kappa = 0$) we find $\theta = \eta = 0$, i.e. the transmitted wave maintains its linear polarization. Therefore, nonzero chiroptical signals θ, η lead to the conclusion that $\kappa \neq 0$, and the more pronounced the chirality enhancement, the stronger the chiroptical signals.

The result of equations (9.49) and (9.50) predicts that the chiroptical signals θ, η are proportional to the ratio t_c/t. This means that, for a certain polarization conversion strength t_c, the far-field chiroptical signals are expected to change inversely with t. To interpret the simulated response according to this observation, in figure 9.7(c) we plot the magnitude of the co- and cross-transmission amplitudes t and t_c, respectively, for $\kappa = \pm 10^{-3}$ and $\kappa = \pm 10^{-3}i$, as a function of the disk thickness d. Note that $|t|$ remains practically unchanged when we switch κ on. Further, $|t_c|$ is the same for $\kappa = \pm 10^{-3}$, $\pm 10^{-3}i$; this can be understood from the form of equation (9.43), when combined with equation (9.29). In the simulations we observe that as the resonant frequencies of the modes become less detuned and the magneto-electric coupling (and therefore $|t_c|$) increases, the transmission amplitude $|t|$ increases exceedingly (see figure 9.7(c)). As a result, θ and η become weaker, as also verified numerically by calculating the real and imaginary parts of the ratio t_c/t. Hence, although the near field interaction is maximized, it is the far field signature that is reduced. Regardless, even in this case, the chiroptical signals are significantly stronger than those from the chiral inclusions alone, for both real and imaginary κ. For example, in the absence of the nanodisks, numerical calculations of pure chiral cylinders of radius 10 nm and thickness 130 nm, periodically arranged in air, yield $\theta = 0.06$ mdeg, $\eta = 0$ mdeg for $\kappa = 10^{-3}$ and $\theta = 0$ mdeg, $\eta = 0.06$ mdeg for $\kappa = 10^{-3}i$, at 400 THz. Evidently, the metasurface provides enhancement by a factor of at least 2 orders of magnitude.

Additionally, for all disk thicknesses examined here, we observe that $\theta(-\kappa) = -\theta(\kappa)$ and $\eta(-\kappa) = -\eta(\kappa)$. This property is predicted by the functional form of equation (9.43), combined with equation (9.29), and is essential for differentiating between left and right enantiomers.

Last, we observe that each of $\mathrm{Re}(\kappa)$, $\mathrm{Im}(\kappa)$ manifests in both θ, η signals. The reason is that, as predicted by our theoretical model, $\sigma_c \propto \hat{\sigma}_{ee}\hat{\sigma}_{mm}$, where $\hat{\sigma}_{ee}$, $\hat{\sigma}_{mm}$ are complex in general. Hence, a purely real or a purely imaginary κ results in a complex-valued σ_c, therefore leading to nonzero θ and η. These observations are in contrast to traditional polarimetric measurements, where $\mathrm{Re}(\kappa)$ manifests only in θ ($\eta = 0$) and $\mathrm{Im}(\kappa)$ only in η ($\theta = 0$) [9]. In other words, the far-field θ and η signals express the response of the combined metasurface–chiral inclusion and, therefore, should be interpreted with caution in experiments. In essence, for all examined disk thicknesses we observe that $\theta(i\kappa) = -\eta(\kappa)$, $\eta(i\kappa) = +\theta(\kappa)$, which is a direct consequence of the fact that θ, η are the real and imaginary parts, respectively, of the same quantity $\Phi = t_c/t$, as predicted by our sheet model—because $t_c \propto \kappa\hat{\sigma}_{ee}\hat{\sigma}_{mm}$, the substitution $\kappa \rightarrow i\kappa$ leads to $t_c \rightarrow it_c$ and therefore $\Phi \rightarrow i\Phi$; as $\mathrm{Re}(i\Phi) = -\mathrm{Im}(\Phi)$, $\mathrm{Im}(\Phi) = \mathrm{Re}(\Phi)$ for any complex number Φ, this leads to $\theta(i\kappa) = -\eta(\kappa)$, $\eta(i\kappa) = +\theta(\kappa)$. This distinct behavior enables us to distinguish $\mathrm{Re}(\kappa)$, $\mathrm{Im}(\kappa)$, and their signs, as we will demonstrate next.

9.3.4 Complete measurement of unknown chirality

Inspection of equation (9.29) reveals that, because κ is dispersive, σ_c exhibits the dispersive features of both the chiral inclusions and the metasurface. Therefore, it is crucial to be able to isolate the metasurface's contribution to the total chiroptical signal, in order to deconvolve and unambiguously determine the unknown chirality. While several chiral sensing schemes based on CD measurements are able to detect and differentiate enantiomers [15, 22, 24, 27–29], a scheme that is able to distinguish $\mathrm{Re}(\kappa)$, $\mathrm{Im}(\kappa)$ is missing.

Due to the linearity of σ_c in κ (see equation (9.29)), θ, η are proportional to κ, justifying why measurements of θ, η are sensitive to both $\mathrm{sgn}(\kappa)$ and $|\kappa|$, and to both $\mathrm{Re}(\kappa)$ and $\mathrm{Im}(\kappa)$. According to equations (9.49) and (9.50), θ and η are the real and imaginary parts, respectively, of the same quantity, t_c/t, and, therefore, we may combine them and solve for t_c, to obtain

$$t_c = -(\theta + i\eta)t. \tag{9.51}$$

This result is of central importance; it enables the unambiguous determination of an unknown chirality, solely from the chiroptical signals θ, η. To demonstrate this possibility, let us consider the chiroptical signals θ_A, η_A and θ_B, η_B from simulations or experiments with inclusions A and B, respectively, which correspond to the individual transmission amplitudes t_c^A and t_c^B, according to equation (9.43). Taking into account that $\hat{\sigma}_{ee}$, $\hat{\sigma}_{mm}$ do not depend on κ (they are properties of the metasurface) and that $\hat{\sigma}_c \propto \kappa$, we can divide t_c^A and t_c^B to eliminate the common t term and express the unknown κ_A in terms of the reference κ_B, as

$$\kappa_A = \kappa_B \frac{\theta_A + i\eta_A}{\theta_B + i\eta_B}. \tag{9.52}$$

To demonstrate how we can use this formula to accurately determine the chirality of the unknown chiral inclusion A, i.e. the true value of κ_A, using a chiral inclusion of known chirality parameter, κ_B, we use dispersive chirality parameters of the form [7, 40]

$$\kappa_{A/B}(\omega) = \frac{\omega_{\kappa,A/B}\,\omega}{\omega_{0\kappa,\,A/B}^2 - \omega^2 + i\gamma_{\kappa,A/B}\,\omega}, \tag{9.53}$$

with $\omega_{0\kappa,A} = 2\pi \times 420$ THz, $\gamma_{\kappa,A} = 2\pi \times 10$ THz and $\omega_{\kappa,A} = 2\pi \times 2 \times 10^{-3}$ THz for inclusion A, and $\omega_{0\kappa,B} = 2\pi \times 380$ THz, $\gamma_{\kappa,B} = 2\pi \times 15$ THz and $\omega_{\kappa,B} = 2\pi \times 1.5 \times 10^{-3}$ THz for inclusion B. To emphasize the strength of the proposed scheme, these parameters are chosen to undergo changes in both sign and magnitude within the examined frequency range, and to provide chiralities close to those of realistic chiral molecules, such as biomolecules [15, 24, 29] or aqueous solutions of monosaccharides [41, 42]. The calculated chiroptical signals θ, η for the metasurface of figure 9.6 with disk thickness $d = 130$ nm are shown in the top row panels of figure 9.8, and the chirality parameters of the two inclusions are shown in the bottom row panels of figure 9.8 (cyan lines). Combining all signals with the reference κ_B in equation (9.52), we successfully retrieve the unknown κ_A across the

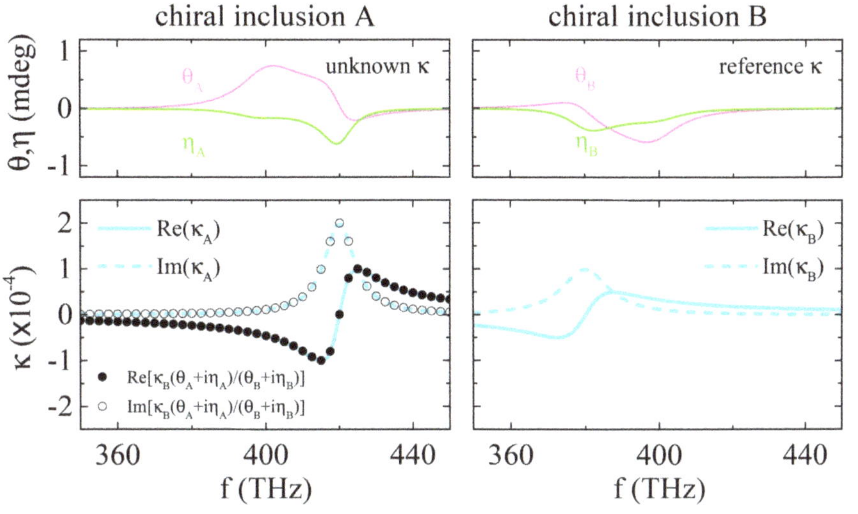

Figure 9.8. Retrieval of the unknown chirality parameter κ_A of inclusion A, using a reference chirality parameter κ_B of inclusion B, and the chiroptical signals θ, η. Top row: numerically calculated chiroptical signals θ, η of the transmitted wave for each inclusion, for the achiral isotropic metasurface of figure 9.6 with $d = 130$ nm. Bottom row: chirality parameters κ_A, κ_B of the two inclusions (cyan lines) and retrieved κ_A (black circles) of inclusion A. Reprinted (figure) with permission from [30], Copyright (2020) by the American Physical Society.

entire spectral range (complete measurement), as depicted in the bottom-left panel of figure 9.8 (black circles).

9.4 Chiral sensing with anisotropic metasurfaces

Anisotropic metasurfaces offer additional degrees of freedom with respect to their isotropic counterparts, enabling detection schemes beyond probing with linearly polarized beams. In this section we use an achiral anisotropic metasurface to detect a chiral inclusion and retrieve its unknown chirality parameter κ. We derive analytically, and verify numerically, formulas that provide insight into the sensing mechanism and explain how anisotropic metasurfaces can be used to (i) achieve enhanced chiroptical signals by more than two orders of magnitude, (ii) perform complete measurements of the total chirality (magnitude and sign of both its real and imaginary parts) and (iii) enable chirality measurements in an absolute manner, i.e. without the need for sample removal, using a crucial signal reversal (excitation with reversed polarization). To understand the various sensing mechanisms, we first need to establish a connection between the scattering amplitudes and the surface conductivities that express the chirality-induced coupling of the metasurface's electric and magnetic modes.

9.4.1 Scattering amplitudes

In the case of anisotropic metasurfaces, the conductivity tensor of equation (9.41) can be generalized to account for the polarization anisotropy as

$$\overleftrightarrow{\sigma} = \begin{pmatrix} \sigma_{ee}^{xx} & 0 & \sigma_c^{xx} & 0 \\ 0 & \sigma_{ee}^{yy} & 0 & \sigma_c^{yy} \\ -\sigma_c^{xx} & 0 & \sigma_{mm}^{xx} & 0 \\ 0 & -\sigma_c^{yy} & 0 & \sigma_{mm}^{yy} \end{pmatrix}. \tag{9.54}$$

Using equation (9.54) to solve the system of equations (9.31)–(9.34) for x- and y-linearly polarized incident waves, we find that the transmission and reflection amplitudes are expressed in terms of the surface conductivities as

$$t_{xx} = \frac{1 - \hat{\sigma}_{ee}^{xx}\hat{\sigma}_{mm}^{yy}}{(1 + \hat{\sigma}_{ee}^{xx})(1 + \hat{\sigma}_{mm}^{yy})}, \tag{9.55}$$

$$t_{yy} = \frac{1 - \hat{\sigma}_{ee}^{yy}\hat{\sigma}_{mm}^{xx}}{(1 + \hat{\sigma}_{ee}^{yy})(1 + \hat{\sigma}_{mm}^{xx})}, \tag{9.56}$$

$$t_c \equiv t_{xy} = -t_{yx} = \frac{\sigma_c^{xx}}{2(1 + \hat{\sigma}_{ee}^{xx})(1 + \hat{\sigma}_{mm}^{xx})} + \frac{\sigma_c^{yy}}{2(1 + \hat{\sigma}_{ee}^{yy})(1 + \hat{\sigma}_{mm}^{yy})}, \tag{9.57}$$

$$r_{xx} = \frac{\hat{\sigma}_{mm}^{yy} - \hat{\sigma}_{ee}^{xx}}{(1 + \hat{\sigma}_{ee}^{xx})(1 + \hat{\sigma}_{mm}^{yy})}, \tag{9.58}$$

$$r_{yy} = \frac{\hat{\sigma}_{mm}^{xx} - \hat{\sigma}_{ee}^{yy}}{(1 + \hat{\sigma}_{ee}^{yy})(1 + \hat{\sigma}_{mm}^{xx})}, \tag{9.59}$$

$$r_c \equiv r_{xy} = r_{yx} = \frac{\sigma_c^{xx}}{2(1 + \hat{\sigma}_{ee}^{xx})(1 + \hat{\sigma}_{mm}^{xx})} - \frac{\sigma_c^{yy}}{2(1 + \hat{\sigma}_{ee}^{yy})(1 + \hat{\sigma}_{mm}^{yy})}, \tag{9.60}$$

where, for simplicity, $\epsilon_1 = \epsilon_2 \equiv \epsilon$, $\mu_1 = \mu_2 \equiv \mu$, and we have introduced the dimensionless conductivities $\hat{\sigma}_{ee}^{\alpha\alpha} = \zeta\sigma_{ee}^{\alpha\alpha}/2$ and $\hat{\sigma}_{mm}^{\alpha\alpha} = \sigma_{mm}^{\alpha\alpha}/2\zeta$, with $\alpha = \{x, y\}$ and $\zeta = \sqrt{\mu_0\mu/\epsilon_0\epsilon}$ ($\sigma_c^{\alpha\alpha}$ is dimensionless by definition).

9.4.2 Enhancement of chiral–matter interaction

To understand how the magneto-electric coupling is enhanced in anisotropic metasurfaces, we solve equations (9.55)–(9.60) in terms of the conductivities

$$\hat{\sigma}_{ee}^{xx} = \frac{1 - r_{xx} - t_{xx}}{1 + r_{xx} + t_{xx}}, \tag{9.61}$$

$$\hat{\sigma}_{ee}^{yy} = \frac{1 - r_{yy} - t_{yy}}{1 + r_{yy} + t_{yy}}, \tag{9.62}$$

$$\hat{\sigma}_{mm}^{xx} = \frac{1 + r_{yy} - t_{yy}}{1 - r_{yy} + t_{yy}}, \tag{9.63}$$

$$\hat{\sigma}_{mm}^{yy} = \frac{1 + r_{xx} - t_{xx}}{1 - r_{xx} + t_{xx}}, \tag{9.64}$$

$$\sigma_c^{xx} = \frac{4(t_c + r_c)}{(1 + r_{xx} + t_{xx})(1 - r_{yy} + t_{yy})}, \tag{9.65}$$

$$\sigma_c^{yy} = \frac{4(t_c - r_c)}{(1 - r_{xx} + t_{xx})(1 + r_{yy} + t_{yy})}. \tag{9.66}$$

As an example, we consider a realistic anisotropic metasurface that supports resonant conductivities along the y-axis only, similar to the one examined in [43] for chiral sensing in the near infrared. The metasurface is composed of a 100 nm thin dielectric slab on which a 50 nm chiral layer with refractive index $n = 1.5 - 0.001i$ and chirality parameter $\kappa = 10^{-5}$ is placed (see figure 9.9(a)). The dielectric slab has refractive index $n = 3.4$ and supports TE (components H_x, E_y, H_z) and TM (components E_x, H_y, E_z) waveguide modes, which we use to implement the electric/magnetic moment pair. The slab is periodically interrupted by metallic wires with periodicity $\alpha = 840$ nm and the whole system is placed on a glass substrate of refractive index $n = 1.5$. For simplicity, the space on the opposite side (adjacent to the chiral layer) is index-matched with the substrate. The purpose of the metallic wires is to (i) spatially quantize the continuous TE/TM slab waveguide modes and provide discrete sets of resonant states and (ii) perturb the symmetry of the fields across each unit cell, essentially leading to residual moments in the dominant field components of each mode, i.e. the electric (magnetic) dipole moment in the E_y (H_y) component of the TE (TM) mode. The metal is a Drude silver (Ag) of permittivity

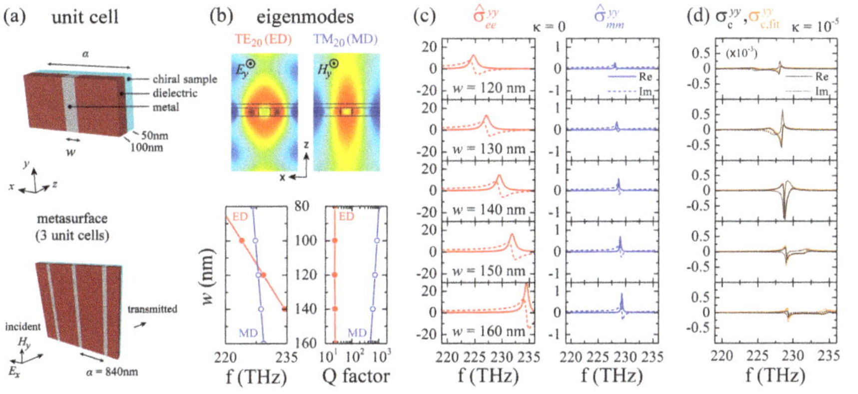

Figure 9.9. Enhanced chiral sensing using an achiral anisotropic metasurface. (a) Schematic illustration of a single unit cell and the metasurface (three unit cells shown here). (b) Eigenmodes, one of electric (ED—electric dipole) and one of magnetic (MD—magnetic dipole) type, that are utilized for sensing. Top panel: field distribution of the electric type (TE_{20}) and magnetic type (TM_{20}) modes. Bottom panel: spectral detuning (left panel) and Q factors (right panel) of TE_{20} and TM_{20} as a function of the metal width, w. (c) Retrieved conductivities $\hat{\sigma}_{ee}^{yy}$, $\hat{\sigma}_{mm}^{yy}$ as a function of the metal width, w, for $\kappa = 0$. (d) Retrieved conductivity σ_c^{yy} and its analytical fit $\sigma_{c,\text{fit}}^{yy}$ using equation (9.29), as a function of the metal width, w, for $\kappa = 10^{-5}$.

based on Johnson and Christy data [44] and its width, w, is used for tuning the spectral separation between the two modes, which we label as TE_{20} (electric dipole or ED) and TM_{20} (magnetic dipole or MD), in accordance with [25, 43]. The spatial distribution of their dominant field components that we use to implement the electric and magnetic dipole moments, E_y and H_y, respectively, is shown in figure 9.9(b), where their frequency tuning and individual Q factors are also shown.

To retrieve the effective surface conductivities of the metasurface, first, with $\kappa = 0$, we send x- and y-linearly polarized waves separately, which excite the metasurface along its anisotropic axes. From the numerically calculated reflection and transmission amplitudes, t_{xx}, t_{yy}, r_{xx}, r_{yy}, we retrieve the surface conductivities $\hat{\sigma}_{ee}^{xx}$, $\hat{\sigma}_{ee}^{yy}$, $\hat{\sigma}_{mm}^{xx}$, $\hat{\sigma}_{mm}^{yy}$ using equations (9.61)–(9.64). In figure 9.9 (c) we present $\hat{\sigma}_{ee}^{yy}$, $\hat{\sigma}_{mm}^{yy}$ for metal widths, w, ranging from 120 to 160 nm. Then, we set $\kappa = 10^{-5}$ and repeat the two experiments with x- and y-linearly polarized waves. Due to chirality, the cross-polarized terms t_c, r_c now become nonzero, giving rise to the magneto-electric conductivities σ_c^{xx}, σ_c^{yy}, which we retrieve with equations (9.65)–(9.66). Due to the absence of modes with dipole moments along the x-direction, our calculations verify that $\sigma_c^{xx} = 0$, i.e. the chiroptical effects are mediated entirely by σ_c^{yy}. The contrast between σ_c^{xx}, σ_c^{yy} also verifies the strong inherent anisotropy of the metasurface; clearly, as we tune the metal width w, we detune the resonant frequencies of $\hat{\sigma}_{ee}^{yy}$, $\hat{\sigma}_{mm}^{yy}$, in turn leading to a distinct change in σ_c^{yy}, as shown in figure 9.9 (d), while σ_c^{xx} remains practically zero. We observe that σ_c^{yy} is maximized when the detuning between $\hat{\sigma}_{ee}^{yy}$ and $\hat{\sigma}_{mm}^{yy}$ is minimized, in accord with our previous observations in figure 9.4, which occurs for $w = 140$ nm.

In figure 9.9(d), we also plot $\sigma_{c,\,\mathrm{fit}}^{yy} = \kappa C_{0,\mathrm{fit}}(\hat{\sigma}_{ee}^{yy} - i\beta_e\omega)(\hat{\sigma}_{mm}^{yy} - i\beta_m\omega)$ (orange lines), the analytical fit of σ_c^{yy}, using equation (9.29) with a similar modification to that introduced earlier for isotropic metasurfaces to account for background contributions from nearby modes. In this expression, $\hat{\sigma}_{ee}^{yy}$, $\hat{\sigma}_{mm}^{yy}$ are the numerically retrieved conductivities shown in figure 9.9(c), $\kappa = 10^{-5}$ is the chirality parameter used in the simulations, $C_{0,\mathrm{fit}} = -23$, and $\beta_e = (2\pi)^{-1} \times 5.8$ fs and $\beta_m = (2\pi)^{-1} \times 0.3$ fs. The agreement between the numerically retrieved σ_c^{yy} and its fit is again excellent.

9.4.3 Enhanced detection of unknown chirality with TE/TM linearly polarized beams

Following the approach that we established with isotropic metasurfaces, we can use our theoretical framework to express the chiroptical signals θ and η in terms of the transmission amplitudes. Inserting equations (9.55)–(9.57) in the Stokes parameters, we find that the chiroptical signals are expressed in terms of t_{xx}, t_{yy}, and t_c as

$$\theta_{\mathrm{TM/TE}} = \mathrm{Re}\left(-\frac{t_c}{t_{xx/yy}}\right), \tag{9.67}$$

$$\eta_{\mathrm{TM/TE}} = \mathrm{Im}\left(-\frac{t_c}{t_{xx/yy}}\right), \tag{9.68}$$

where we have used $t_c \ll t_{xx}, t_{yy}$ to approximate $\tan^{-1}(\Phi) \sim \Phi$. We refer to the illumination with $\mathbf{E}\|\hat{x}$ ($\mathbf{E}\|\hat{y}$) as TM (TE) illumination, to emphasize the fact that this particular polarization couples with the TM_{20} (TE_{20}) mode, as also indicated in the subscripts of θ, η. Equations (9.67) and (9.68) dictate that the chiroptical signals θ, η are in principle different for x- and y-linearly polarized waves, contrary to isotropic metasurfaces, where the chiroptical signals are the same for both polarizations and inversely proportional to the common co-transmission amplitude. Importantly, because anisotropy allows for the independent control of the conductivities along the x and y directions, we can tune the co- and cross-transmission amplitudes independently and, likewise, the chiroptical signals. For example, t_{xx} (equation (9.55)) changes only with $\hat{\sigma}_{ee}^{xx}$, $\hat{\sigma}_{mm}^{yy}$, whereas all $\hat{\sigma}_{ee}^{xx}$, $\hat{\sigma}_{mm}^{yy}$, $\hat{\sigma}_{ee}^{yy}$, $\hat{\sigma}_{mm}^{xx}$ are involved in the cross-polarized term t_c (equation (9.57)); the conductivities $\hat{\sigma}_{ee}^{yy}$, $\hat{\sigma}_{mm}^{xx}$ provide an additional degree of freedom to control the chiroptical signals independently from the transmittance t_{xx}, which does not change. Hence, using the anisotropy, the chiroptical signals can be maximized for a certain achievable transmittance. Similar conclusions hold for systems with resonant conductivities along a single direction, e.g. the y-axis, as in our example; upon TM illumination, t_{xx}, is tuned via $\hat{\sigma}_{mm}^{yy}$, while t_{yx} is tuned via both $\hat{\sigma}_{mm}^{yy}$ and $\hat{\sigma}_{ee}^{yy}$.

To demonstrate the above findings, in figure 9.10 we examine the metasurface of figure 9.9 for both TM and TE illumination, respectively, as a function of the metal width, w. The results demonstrate the additional degree of freedom offered by anisotropy in terms of the transmitted power, $|t_{xx}|^2$ or $|t_{yy}|^2$, and the polarization conversion strength, $|t_c| \equiv |t_{yx}| = |t_{xy}|$ (the slight discrepancy between $|t_{yx}|$, $|t_{xy}|$ is due to other nearby modes, contributing a constant background, which depends on the incident polarization). For example, one can choose between a strong output beam with moderately enhanced chiroptical signals θ_{TM}, η_{TM} (figure 9.10 (a)) or stronger chiroptical signals θ_{TE}, η_{TE} on a relatively weak output beam (figure 9.10

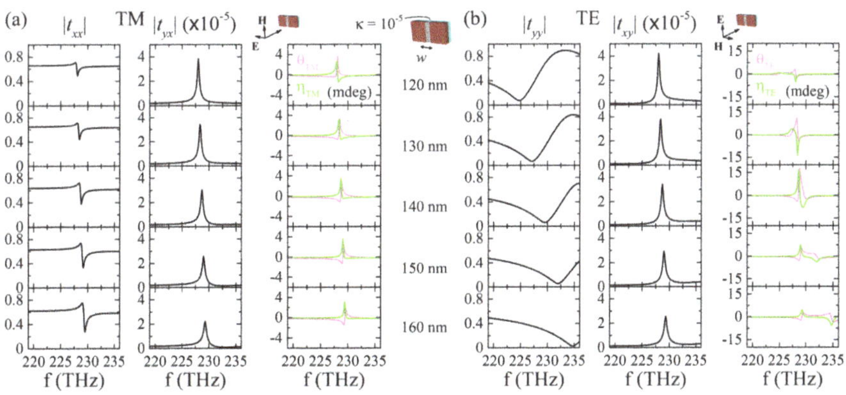

Figure 9.10. Enhanced detection of chiral inclusion, using the achiral anisotropic metasurface of figure 9.9. Magnitude of co- and cross-transmission amplitudes together with chiroptical signals θ (magenta lines) and η (green lines) for $\kappa = 10^{-5}$, under (a) TM illumination (x-polarized incident E-field) and (b) TE illumination (y-polarized incident E-field). The numerical results are in agreement with the analytical expressions given by equations (9.67) and (9.68).

(b)). We emphasize that this cannot be achieved with isotropic systems. For both choices of illumination, the chiroptical signals are 30–100 times stronger than those from the chiral inclusions alone, as for a 50 nm thin chiral film with $\kappa = 10^{-5}$ we find $\theta = 0.14$ mdeg and $\eta = 0$ at 230 THz, in the absence of the metasurface. The numerically calculated chiroptical signals are also in agreement with their analytical form given by equations (9.67) and (9.68), as we verify by calculating the ratios t_c/t_{xx} and t_c/t_{yy} separately, and comparing their real and imaginary parts with $\theta_{\mathrm{TM/TE}}$, $\eta_{\mathrm{TM/TE}}$.

9.4.4 Complete measurement of unknown chirality with TE/TM linearly polarized beams

We can now extend the approach we used with isotropic metasurfaces to unambiguously retrieve an unknown chirality A, with the aid of a reference chirality B. Following the same reasoning as with isotropic metasurfaces, we are led to $t_c = -(\theta_{\mathrm{TM/TE}} + i\eta_{\mathrm{TM/TE}})t_{xx/yy}$. Taking into account that $t_c \propto \kappa$, we can eliminate the $t_{xx/yy}$ term and express the unknown κ_A in terms of the reference κ_B, as

$$\kappa_A = \kappa_B \frac{\theta^A_{\mathrm{TM/TE}} + i\eta^A_{\mathrm{TM/TE}}}{\theta^B_{\mathrm{TM/TE}} + i\eta^B_{\mathrm{TM/TE}}}. \tag{9.69}$$

In figure 9.11 we demonstrate how we can accurately determine the chirality of the unknown chiral inclusion A, i.e. the true value of κ_A, using a reference chirality κ_B and comparing their respective chiroptical signals, obtained using transmission measurements under TE/TM illumination. For the chirality of the two inclusions we use equation (9.53), with $\omega_{\kappa,A} = 2\pi \times 1 \times 10^{-3}$ THz, $\omega_{0\kappa,A} = 2\pi \times 230$ THz, $\gamma_{\kappa,A} = 2\pi \times 1$ THz for inclusion A, and $\omega_{\kappa,B} = 2\pi \times 1 \times 10^{-3}$ THz,

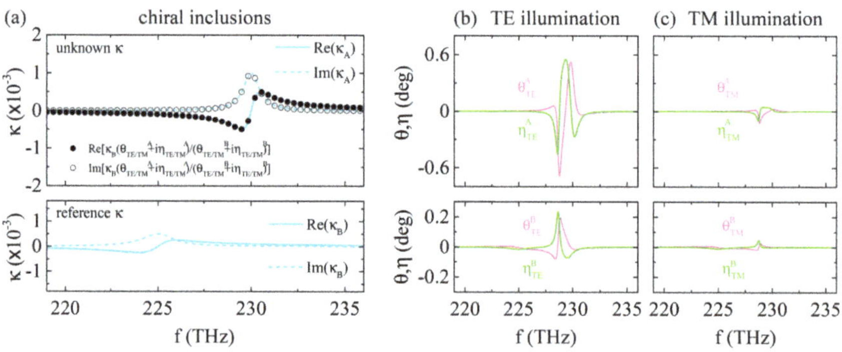

Figure 9.11. Retrieval of the chirality parameter κ_A of an unknown inclusion A, using a reference chirality parameter κ_B of inclusion B under TE and TM illumination. (a) Chirality parameters κ_A, κ_B of the two inclusions (cyan lines) and retrieved κ_A (black circles) of inclusion A. The numerically calculated chiroptical signals θ, η for each inclusion are shown in (b) for TE illumination and in (c) for TM illumination. The chiral inclusion is embedded in the metasurface of figure 9.9 with $w = 140$ nm. Reprinted (figure) with permission from [43], Copyright (2021) by the American Physical Society.

$\omega_{0\kappa,B} = 2\pi \times 225$ THz, $\gamma_{\kappa,B} = 2\pi \times 2$ THz for inclusion B. In figure 9.11(a) we present the chirality parameters of the two inclusions (cyan lines), and the accurately retrieved chirality for inclusion A using the chiroptical signals under either TE or TM illumination (black circles). Using the metasurface of figure 9.9 with $w = 140$ nm, the chiroptical signals for the two, known and unknown, chiral inclusions are presented in figure 9.11(b) for TE illumination and in figure 9.11(c) for TM illumination. Hence, with the aid of equation (9.69), we successfully deconvolve the spectral response of the chiral inclusion from the dispersion of the metasurface, and obtain the pure κ of the unknown chirality (complete measurement), even for cases of high chirality values (here of the order of $\sim 10^{-3}$ on resonance).

9.4.5 Complete measurement of unknown chirality with elliptically polarized beams

Next, we can take further advantage of the inherent anisotropy of the metasurface to gain access to even stronger chiroptical signals. This is possible with elliptically polarized incident waves, an additional functionality that is not feasible in isotropic systems [25]. Let us first parametrize the elliptically polarized incident wave as

$$\mathbf{E}_{\text{inc}} = \begin{pmatrix} E_{0x}\hat{x} \\ E_{0y}\hat{y} \end{pmatrix} = \begin{pmatrix} E_0 \cos \varphi_{\text{rot}} \hat{x} \\ E_0 \sin \varphi_{\text{rot}} e^{i\varphi_{\text{lag}}} \hat{y} \end{pmatrix}, \tag{9.70}$$

where the angle φ_{rot} is the angle between the incident wave's E-field and the x-axis, and φ_{lag} tunes the phase-lag between the x-, y-wave components; E_0 is a complex constant. By tuning φ_{rot} and φ_{lag} we can achieve any desired polarization, from linear to circular polarization. Of course, because the metasurface is anisotropic (birefringent), incident waves that are not parallel or vertical to the metal wires (equivalent to the fast/slow axis of typical anisotropic systems) will induce nonzero chiroptical signals θ, η, even for $\kappa = 0$. Therefore, for $\kappa \neq 0$, the θ, η signals may contain achiral background contributions from the metasurface itself, and have to be eliminated in order to achieve accurate measurement of the unknown chirality. Subtraction of the (achiral) background contribution can be realized using two different approaches: (a) by performing measurements with and without the chiral layer and (b) by implementing a signal reversal that allows us to isolate the signal from the chiral inclusions. We will examine both possibilities next.

9.4.5.1 Measurements in transmission with background subtraction

In figure 9.12 we employ the metasurface of figure 9.9 with $w = 140$ nm and for the chiral inclusion we use a constant chirality parameter $\kappa = 10^{-5}$. We fix the polarization ellipse orientation using $\varphi_{\text{rot}} = 80°$ and we tune the polarization ellipticity by scanning φ_{lag}. For each set of φ_{rot}, φ_{lag} we measure θ and η with and without the chiral inclusion, i.e. by setting $\kappa = 0$. As we scan φ_{lag} (figure 9.12 (a)), we find that the difference between the two measurements (with and without the chiral inclusion) increases and becomes maximum for $\varphi_{\text{lag}} = 50°$. For this illumination we obtain enhanced chiroptical signals by a factor of more than ~ 650 compared to the

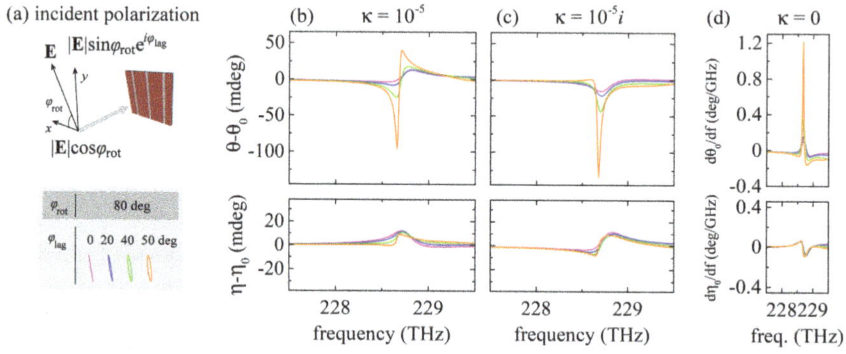

Figure 9.12. Enhanced chiral sensing using elliptically polarized incident wave. (a) Polarization ellipse of incident wave for fixed $\varphi_{\text{rot}} = 80°$ and tunable φ_{lag}. Subtraction of far-field measurements of rotation and ellipticity in the presence and absence of the chiral layer, i.e. $\theta - \theta_0$ and $\eta - \eta_0$, for (b) $\kappa = 10^{-5}$ and (c) $\kappa = 10^{-5}i$. (d) Calculated derivatives $d\theta_0/df$ and $d\eta_0/df$ for $\kappa = 0$, demonstrating that the chiroptical signals are maximized when these derivatives are also maximized. For $\varphi_{\text{rot}} = 80°$, $\varphi_{\text{lag}} = 50°$, the rotation signal is enhanced by a factor of more than ~650 compared to the respective signal of a pure chiral film with $\kappa = 10^{-5}$. Reprinted (figure) with permission from [43], Copyright (2021) by the American Physical Society.

respective signal from the pure chiral film (~0.14 mdeg), as we show in figure 9.12 (b). In figure 9.12 (c) we repeat the calculations using (purely imaginary) $\kappa = 10^{-5}i$.

To understand the mechanism of the enhancement, we expand $\theta \equiv \theta(f, \kappa)$ (a function of the frequency f and the chirality parameter κ), in terms of κ. If we keep the first two terms, we find $\theta - \theta_0 \approx \partial\theta/\partial\kappa\,|_{\kappa=0} \cdot \kappa$ or $\theta - \theta_0 \approx (\partial\theta/\partial f)(\partial f/\partial\kappa)|_{\kappa=0} \cdot \kappa$, where $\theta_0 = \theta(f, 0)$ is the calculated chiroptical signal θ in the absence of chirality ($\kappa = 0$) [25]. Because θ, η scale linearly with κ, we can write $\partial\theta/\partial f \propto d\theta_0/df$, which leads to $\theta - \theta_0 \sim (d\theta_0/df)(df/d\kappa) \cdot \kappa$ (similarly we obtain $\eta - \eta_0 \sim (d\eta_0/df)(df/d\kappa) \cdot \kappa$). This result implies that for a certain κ the far-field differential signals ($\theta - \theta_0$, $\eta - \eta_0$) are expected to be enhanced at frequency ranges where the derivatives of θ_0, η_0 become large, i.e. the metasurface's anisotropy changes abruptly. To independently verify this, we remove the chiral layer and measure the signals θ_0 and η_0. In figure 9.12(d) we present the calculated derivatives $d\theta_0/df$ and $d\eta_0/df$, which clearly follow $\theta - \theta_0$ and $\eta - \eta_0$, respectively. Thus, the chiroptical enhancement is mediated by the strongly dispersive metasurface's birefringence. This is in contrast to most contemporary approaches, which are based on the excitation of superchiral near fields and the background birefringence is undesired [22–24, 28].

9.4.5.2 Absolute measurements in transmission with signal reversals

Removing the chiral inclusion to subtract the background signals may result in unintentional changes in the measurement setup, possibly affecting measurement accuracy. Therefore, it would be desirable to have an absolute measurement scheme, i.e. a sensing scheme where chirality measurements are performed without the need to remove the chiral sample. Absolute measurements can be performed by means of a crucial signal reversal, with which one can directly isolate the (enhanced) chiroptical signal, a unique feature enabled by the anisotropy of the metasurface:

excitation with reversed polarization yields separate polarization effects of opposite sign, enabling the isolation of signals originating from the chiral inclusion only. This can be realized by illuminating the metasurface with $\pm\varphi_{\text{rot}}$, yielding two sets of chiroptical signals that we label as $\theta_{\pm}$ and $\eta_{\pm}$. Their mean value, $2\Delta\theta_{\text{rev}} \equiv \theta_{+} + \theta_{-}$ and $2\Delta\eta_{\text{rev}} \equiv \eta_{+} + \eta_{-}$, cancels the contributions from the metasurface, as well as any other potential achiral backgrounds, while doubling the pure chiroptical signal, which is even under this polarization reversal. The average signals $\Delta\theta_{\text{rev}}$ and $\Delta\eta_{\text{rev}}$ are expressed analytically as

$$\Delta\theta_{\text{rev}} = \text{Re}\left(-t_c \frac{c_{\text{rot}}^2 t_{xx} + s_{\text{rot}}^2 t_{yy} e^{2i\phi_{\text{lag}}}}{c_{\text{rot}}^2 t_{xx}^2 + s_{\text{rot}}^2 t_{yy}^2 e^{2i\phi_{\text{lag}}}}\right), \tag{9.71}$$

$$\begin{aligned}\Delta\eta_{\text{rev}} = \text{Im}\Bigg(&-t_c(c_{\text{rot}}^2 t_{xx}^* + s_{\text{rot}}^2 t_{yy}^*)\frac{|c_{\text{rot}}^2 t_{xx}^2 + s_{\text{rot}}^2 t_{yy}^2 e^{2i\phi_{\text{lag}}}|}{(c_{\text{rot}}^2 |t_{xx}|^2 + s_{\text{rot}}^2 |t_{yy}|^2)^2} \\ &+ \frac{\text{Re}\left\{t_c(t_{xx} - t_{yy})(c_{\text{rot}}^2 e^{-i\phi_{\text{lag}}} t_{xx}^2 + s_{\text{rot}}^2 e^{i\phi_{\text{lag}}} t_{yy}^2)^*\right\} 2c_{\text{rot}}^2 s_{\text{rot}}^2 e^{-i\phi_{\text{lag}}} t_{xx} t_{yy}^*}{(c_{\text{rot}}^2 |t_{xx}|^2 + s_{\text{rot}}^2 |t_{yy}|^2)^2 |c_{\text{rot}}^2 t_{xx}^2 + s_{\text{rot}}^2 e^{2i\phi_{\text{lag}}} t_{yy}^2|}\Bigg),\end{aligned} \tag{9.72}$$

where $c_{\text{rot}} = \cos(\varphi_{\text{rot}})$, $s_{\text{rot}} = \sin(\varphi_{\text{rot}})$ and t_{xx}, t_{yy}, t_c are the transmission amplitudes as given by equations (9.55)–(9.57). Because t_{xx}, t_{yy}, t_c do not depend on φ_{rot}, φ_{lag}, the form of equation (9.71) implies that there is always a combination of φ_{rot}, φ_{lag} that can minimize and possibly eliminate the denominator entirely, enhancing $\Delta\theta_{\text{rev}}$, as demonstrated in figure 9.12. On the other hand, because the denominator in equation (9.72) involves sums of positive quantities, a similar conclusion cannot be drawn directly for $\Delta\eta_{\text{rev}}$.

In figure 9.13 we present the calculated θ and η signals from simulations with two elliptically polarized incident waves, characterized by $\varphi_{\text{rot}} = \pm 80°$ and $\varphi_{\text{lag}} = 50°$.

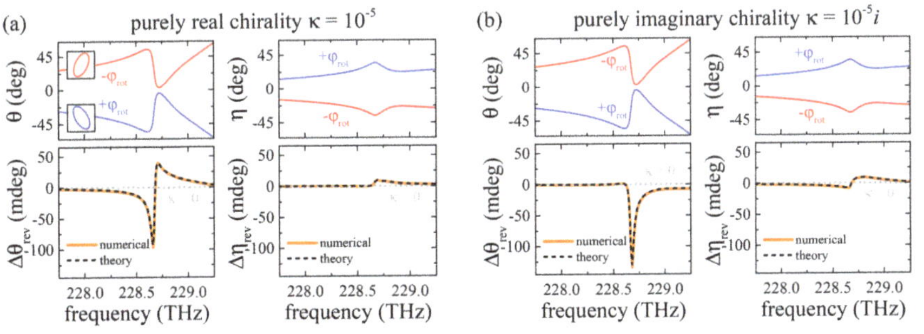

Figure 9.13. Absolute measurement of κ with polarization reversal of elliptically polarized incident wave for (a) $\kappa = 10^{-5}$, and (b) $\kappa = 10^{-5}i$. Measurements of θ, η for $\varphi_{\text{lag}} = 50°$ with $\varphi_{\text{rot}} = \pm 80°$ (blue and red lines, respectively, top row panels—see inset in (a)) are averaged to yield pure chiral signals $\Delta\theta_{\text{rev}}$, $\Delta\eta_{\text{rev}}$ (orange lines, bottom row panels). The semi-analytical plots of equations (9.71) and (9.72) for the respective cases are shown as dashed black lines. Gray dotted lines: $\kappa = 0$. Reprinted (figure) with permission from [43], Copyright (2021) by the American Physical Society.

The chirality parameter used in the simulations is purely real ($\kappa = 10^{-5}$) in figure 9.13(a) and purely imaginary ($\kappa = 10^{-5}i$) in figure 9.13(b). The θ and η signals for $\pm\varphi_{\text{rot}}$ (solid blue/red lines) are averaged to yield the $\Delta\theta_{\text{rev}}$, $\Delta\eta_{\text{rev}}$ signals, which originate solely from the chiral inclusions (solid orange lines). The numerically retrieved $\Delta\theta_{\text{rev}}$, $\Delta\eta_{\text{rev}}$ signals agree with the analytical predictions of equations (9.71) and (9.72) (dashed black lines), for which we have used t_{xx}, t_{yy} with $\kappa = 0$ and t_c with $\kappa = 10^{-5}$, from the simulations. These results are also in agreement with the results of figure 9.12.

9.4.5.3 Complete measurement of unknown chirality

We can now devise a measurement scheme to unambiguously retrieve an unknown chirality A, with the aid of a reference chirality B, using elliptically polarized beams with signal reversals. Note that in the case of TM/TE illumination, the simple form of equation (9.69) was derived on the basis that the rotation (θ) and ellipticity (η) are the real and imaginary parts of the same complex quantity, which is proportional to κ (see equations (9.67) and (9.68)). Although, under elliptical illumination, the relevant signals $\Delta\theta_{\text{rev}}$ and $\Delta\eta_{\text{rev}}$ refer to different complex quantities (see equations (9.71) and (9.72)), it is still possible to obtain the unknown chirality, as

$$\kappa_A = \kappa_B \frac{\Delta\theta^A_{\text{rev}} + g(\varphi_{\text{rot}}, \varphi_{\text{lag}}, t_{xx}, t_{yy})\Delta\eta^A_{\text{rev}}}{\Delta\theta^B_{\text{rev}} + g(\varphi_{\text{rot}}, \varphi_{\text{lag}}, t_{xx}, t_{yy})\Delta\eta^B_{\text{rev}}}, \tag{9.73}$$

where $g(\varphi_{\text{rot}}, \varphi_{\text{lag}}, t_{xx}, t_{yy})$ is a function of the angles φ_{rot}, φ_{lag} characterizing the elliptical polarization, and the co-transmission amplitudes t_{xx}, t_{yy} obtained previously with TE/TM illumination, and is given explicitly by

$$g(\varphi_{\text{rot}}, \varphi_{\text{lag}}, t_{xx}, t_{yy}) = i\frac{\dfrac{(c^2_{\text{rot}}|t_{xx}|^2 + s^2_{\text{rot}}|t_{yy}|^2)^2}{|c^2_{\text{rot}}t^2_{xx} + s^2_{\text{rot}}t^2_{yy}e^{2i\phi_{\text{lag}}}|}\left(\dfrac{c^2_{\text{rot}}t_{xx} + s^2_{\text{rot}}t_{yy}e^{2i\phi_{\text{lag}}}}{c^2_{\text{rot}}t^2_{xx} + s^2_{\text{rot}}t^2_{yy}e^{2i\phi_{\text{lag}}}}\right)^*}{(c^2_{\text{rot}}t_{xx} + s^2_{\text{rot}}t_{yy}) + 2ic^2_{\text{rot}}s^2_{\text{rot}}\dfrac{\text{Im}(e^{-i\phi_{\text{lag}}}t_{xx}t^*_{yy})(t^*_{xx} - t^*_{yy})(c^2_{\text{rot}}t^2_{xx}e^{-i\phi_{\text{lag}}} + s^2_{\text{rot}}t^2_{yy}e^{i\phi_{\text{lag}}})}{|c^2_{\text{rot}}t^2_{xx} + s^2_{\text{rot}}t^2_{yy}e^{2i\phi_{\text{lag}}}|^2}}. \tag{9.74}$$

The result of equation (9.73) generalizes equation (9.69) for any arbitrary incident polarization. Indeed, for $\varphi_{\text{rot}} = 0$ (TM illumination) or $\varphi_{\text{rot}} = \pi/2$ (TE illumination), it is straightforward to show that $g(\varphi_{\text{rot}}, \varphi_{\text{lag}}, t_{xx}, t_{yy}) = i$, reducing equation (9.73) to equation (9.69). As an example, in figure 9.14 we use equation (9.73) to retrieve the unknown chirality κ_A in terms of the reference chirality κ_B using elliptically polarized illumination with $\varphi_{\text{rot}} = \pm 80°$, $\varphi_{\text{lag}} = 50°$. For κ_A, κ_B we use the parameters previously used in figure 9.11 to demonstrate the retrieval procedure with TE/TM illumination.

With the introduction of an additional reference chirality κ_C, we may eliminate the parameter $g(\varphi_{\text{rot}}, \varphi_{\text{lag}}, t_{xx}, t_{yy})$. Repeating equation (9.73) for molecules A, C, and combining the results for both sets of molecule pairs (A, B and A, C) we obtain

$$\kappa_A = \kappa_B \frac{\Delta\theta^A_{\text{rev}}\Delta\eta^C_{\text{rev}} - \Delta\theta^C_{\text{rev}}\Delta\eta^A_{\text{rev}}}{\Delta\theta^B_{\text{rev}}\Delta\eta^C_{\text{rev}} - \Delta\theta^C_{\text{rev}}\Delta\eta^B_{\text{rev}}} + \kappa_C \frac{\Delta\theta^B_{\text{rev}}\Delta\eta^A_{\text{rev}} - \Delta\theta^A_{\text{rev}}\Delta\eta^B_{\text{rev}}}{\Delta\theta^B_{\text{rev}}\Delta\eta^C_{\text{rev}} - \Delta\theta^C_{\text{rev}}\Delta\eta^B_{\text{rev}}}, \tag{9.75}$$

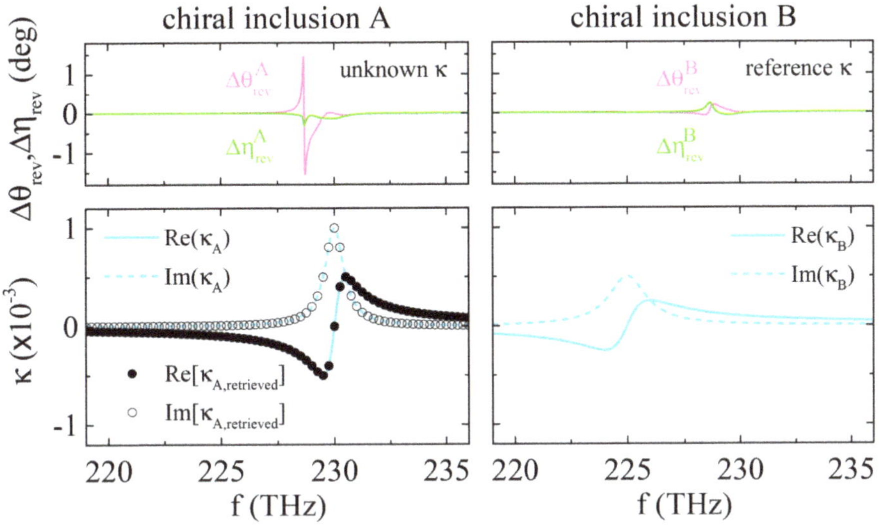

Figure 9.14. Retrieval of the unknown chirality parameter κ_A of an unknown inclusion A, using a reference chirality parameter κ_B of inclusion B under illumination with an elliptically polarized beam with $\varphi_{\text{rot}} = \pm 80°$, $\varphi_{\text{lag}} = 50°$. Top row: numerically calculated chiroptical signals $\Delta\theta_{\text{rev}}$, $\Delta\eta_{\text{rev}}$ for each inclusion. The chiral inclusions are embedded in the metasurface of figure 9.9 with $w = 140$ nm. Bottom row: chirality parameters κ_A, κ_B of the two chiral inclusions (cyan lines) and retrieved κ_A (black circles) of inclusion A, using the reference values of κ_B and the chiroptical signals $\Delta\theta_{\text{rev}}$, $\Delta\eta_{\text{rev}}$. Reprinted (figure) with permission from [43], Copyright (2021) by the American Physical Society.

where the superscript A, B or C refers to the chiroptical signals measured with the metasurface using molecule A, B or C, respectively.

Hence, with the aid of two references chiralities B, C, we can retrieve the unknown chirality A entirely in terms of the $\Delta\theta_{\text{rev}}$, $\Delta\eta_{\text{rev}}$ signals of the three molecules, without involving the parameter g.

9.4.6 Enhanced detection from thin films to monolayers

As the incident beam passes through the metasurface, it interacts directly with the chiral film and indirectly through its coupling with the electric and the magnetic mode. Therefore, the chiroptical signal originates from the interaction of the chiral film with both propagating (far-field) and evanescent (near-field) waves, respectively. This can be easily verified by probing the metasurface with TM illumination, while gradually separating the chiral layer from the metasurface and measuring the far-field signal, which undergoes a distance-dependent exponential decay, as shown in figure 9.15 (a). Note how, for large separation, the chiroptical signal converges to that of the chiral film alone (the minus sign results from analysing the waves from the source's view). This is a clear indication that the enhancement is practically mediated by the near fields of the electric and the magnetic mode, rather than the far-field contribution from the incident wave. In this example we have used the metasurface of figure 9.12 ($w = 140$ nm) and a 50 nm chiral film with $\kappa = +10^{-5}$, at 228.7 THz. The calculated θ for large δz is in agreement with the theoretically

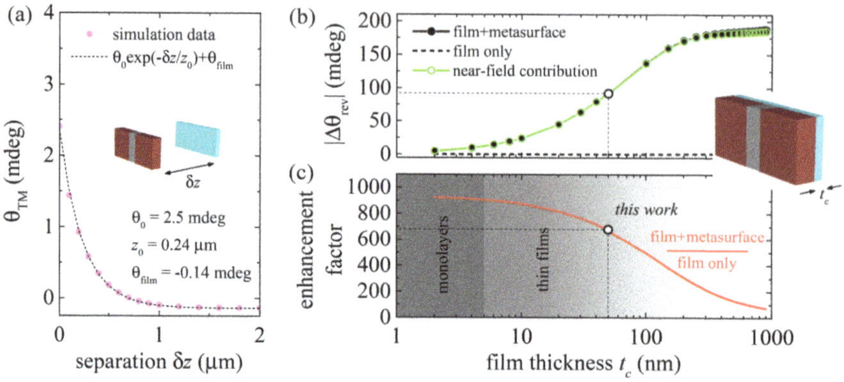

Figure 9.15. Mechanism of near-field enhancement for the metasurface of figure 9.12 ($w = 140$ nm) and chiral film with $\kappa = +10^{-5}$, at 228.7 THz. (a) Measured rotation θ for TM illumination, as a function of δz, the separation between the metasurface and the chiral film of thickness 50 nm. (b) Near- vs far-field contribution to the total chiroptical signal $|\Delta\theta_{\text{rev}}|$ for illumination with elliptical polarization, as a function of the chiral film thickness, t_c. (c) Ratio of $|\Delta\theta_{\text{rev}}|$ over the equivalent signal from the chiral film in the absence of the metasurface. The open dot denotes the film thickness studied throughout this work, i.e. 50 nm.

expected value $\theta = k_0 L\,\text{Re}(\kappa)$ for a pure chiral film [9], where k_0 is the free space wavenumber and L is the length of the chiral medium.

We can also fix the position of the chiral film and gradually increase its thickness, t_c. In figure 9.15 (b) we illuminate the metasurface with elliptically polarized light and repeat the measurements of figure 9.13 for variable film thickness, t_c. As expected, the magnitude of $\Delta\theta_{\text{rev}}$ becomes stronger with increasing film thickness, because the interaction volume between the fields and the chiral film increases. If we subtract the contribution of the chiral film without the metasurface to isolate the near-field contribution to the total signal, we observe that the near-field contribution dominates for film thicknesses up to $\sim\lambda$.

This is also observable in the enhancement factor shown in figure 9.15 (c). Even for chiral films with thickness of ~5 nm, the predicted signal is of ~10 mdeg when the equivalent OR polarimetric signal from a transmission measurement at 228.7 THz is ~14 μ deg. It is crucial to emphasize that the predicted chiroptical signals are well within the sensitivity range of commercial spectro-polarimeters and, in particular, of custom polarimetric instruments that can operate close to the fundamental polarimetric limits (i.e. photon shot noise limits). Therefore, anisotropic metasurfaces are ideal platforms for enhanced detection of chiral films, particularly of subwavelength thickness.

9.5 Using gain to enhance chiral sensing

Gain media are known to provide amplification. Therefore, it is intuitively expected that enhanced chiral sensing could be possible simply by introducing gain in the vicinity of a chiral medium. However, this would require gain media that can couple directly to the chiral medium. In this section we demonstrate that gain can be coupled to the chiral medium via a metasurface. We show that the achievable enhancement is larger than that achieved by the chiral medium alone or when combined with the same metasurface without gain. We explore the different coupling paths and we demonstrate that the chiroptical signal enhancement depends on how strongly gain couples with the metasurface, identifying two distinct regimes of operation, namely (a) background amplification and (b) loss compensation. The regime of background amplification occurs for weak coupling between the metasurface and gain, and is characterized by amplification of the incoming waves, but without chiroptical signal enhancement. The regime of loss compensation occurs for strong coupling between the metasurface and gain, and is characterized by undamping of the metasurface resonances, which accompanies the amplification of the incoming waves and the enhancement of the chiroptical signals.

9.5.1 Coupling paths in gain-assisted metasurfaces with chiral inclusions

To explore the different coupling paths between the chiral medium and gain, we use (here and throughout the entire section) the achiral anisotropic metasurface previously presented in figure 9.9, with $w = 140$ nm, under TM illumination. To provide gain, we use the four-level gain medium used in [45, 46]. To maximize the coupling between the gain medium, the metasurface and the chiral medium, it is necessary to tune the resonant modes of the metasurface, as well as the gain emission spectrum, at the frequency range where the chiral medium is to be probed. Depending on the target frequency, different gain media can be selected. For example, quantum wells and quantum dots are ideal candidates for sensing in the infrared, and dyes for sensing in the visible. The anisotropic metasurface used here is formed by a dielectric slab, which can be directly replaced by a quantum well, to serve as the dielectric slab with gain [47, 48].

Let us start by removing the metallic wires from the metasurface, embedding the gain medium in the remaining slab, and allowing the chiral film to move along the z-axis. We can now study the distance-dependent interaction of the chiral film with a pure gain slab. First, without pumping the gain material, we launch an x-linearly polarized wave, and we analyse the polarization of the transmitted wave in terms of its rotation θ and ellipticity η, as a function of the separation δz. In figure 9.16(a) we plot θ, which remains constant and equal to $\sim$0.14 mdeg, regardless of the separation δz (the minus sign results from analysing the waves from the source's view). The numerically calculated θ also agrees with its theoretically predicted value, provided by $\theta = \mathrm{Re}(\kappa)k_0 L$ [9], using $\kappa = 10^{-5}$ and $L = 50$ nm. We subsequently pump the gain at a constant pump rate of $R_p = 3 \times 10^7\ \mathrm{s}^{-1}$ and repeat the procedure. We observe no change in the polarization state of the transmitted wave, indicating that gain does not couple to the chiral medium, as shown schematically in the inset.

Next, we remove the gain material from the dielectric host and bring back the metallic wires in the dielectric slab to restore the metasurface. We introduce a gain slab of the same thickness (100 nm) far from the metasurface's near fields, as shown

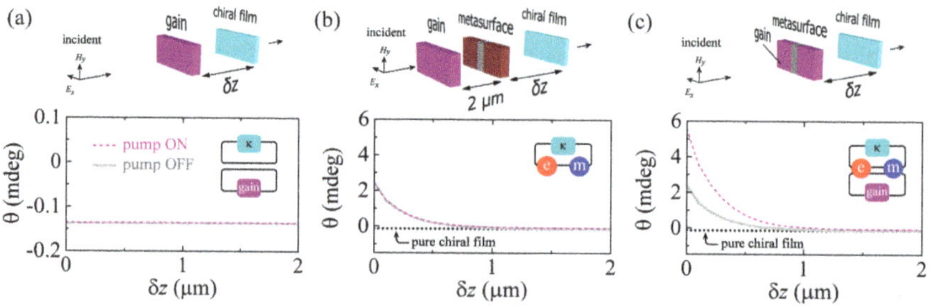

Figure 9.16. Coupling regimes between a chiral film and gain. (a) Chiral film and gain material uncoupled: the rotation θ of a linearly polarized incident wave does not change upon pumping. (b) Chiral film and gain material uncoupled, chiral film coupled to metasurface: enhanced θ with respect to the pure chiral film, but with no further enhancement upon pumping, due to weak coupling of the metasurface with the gain material. (c) Chiral film coupled to gain material via the metasurface: enhancement of θ upon pumping. For $\delta z \gg \lambda$, the rotation θ converges to that of the pure chiral film, because the chiral film is decoupled from the metasurface. Insets: schematic illustration of near-field interactions between the chiral medium, the metasurface and the gain medium. The coupling path between the chiral medium and gain material is mediated by the electric e and magnetic m dipole moments, provided by the metasurface. In all examples, the gain is pumped at constant pump rate ($R_p = 3 \times 10^7$ s^{-1}) and the frequency of the incident wave is $f = 228.6$ THz, where the electric and the magnetic modes of the metasurface are efficiently excited. Reprinted with permission from [46] © Optica Publishing Group.

in figure 9.16(b). With the pump off, the chiroptical signal is enhanced by a factor of $\sim$18 for $\delta z = 0$, compared to the same signal obtained with the pure chiral film (figure 9.16(a)) and, with increasing distance, the enhanced θ reduces, until it converges to that of the pure chiral film. These results are in accord with the findings of the previous section in figure 9.15(a). This is a clear indication of near-field coupling between the chiral film and the metasurface, as shown schematically in the inset. When we switch the pump on, no change occurs in the measured θ of the transmitted wave for the same δz range. This is a direct consequence of the fact that the overlap of gain with the near fields of the metasurface is weak and, therefore, gain is decoupled from the metasurface; what we observe is only the effect of near-field coupling between the chiral film and the metasurface.

Last, to increase the spatial overlap of gain with the modes of the metasurface, we embed the gain medium in the dielectric host of the metasurface. With the pump off we observe the same enhancement as previously in figure 9.16(b). However, when we switch the pump on, we observe that θ increases further, reaching an enhancement factor of $\sim$41 for $\delta z = 0$. Clearly, in this case the chiroptical signal enhancement is mediated by the near-field coupling between the gain medium and the metasurface, i.e. the chirality couples with the gain medium via the metasurface, as shown schematically in the inset. Again, with increasing distance, θ gradually converges to that of the pure chiral film of figure 9.16(a).

These three characteristic regimes convey two important messages: (a) the near-field coupling of the chiral film with the metasurface is necessary to enhance the chiroptical signals (θ reduces with increasing δz in all examples) and (b) the chiral medium couples to gain via the metasurface, i.e. when the metasurface is individually coupled to the chiral medium and the gain medium. However, since the gain medium amplifies the incoming waves, why does it only enhance the chiroptical signals in certain cases? This is what we investigate next.

9.5.2 Regimes of background amplification vs loss compensation

To understand the mechanism of chiroptical signal enhancement in the presence of gain, we restore the chiral film at the metasurface backplane to maximize the coupling of the metasurface with the chiral medium and examine two characteristic configurations in more detail: (a) the gain slab is placed sufficiently far from the metasurface (figure 9.17(a)) and (b) the gain is embedded in the metasurface (figure 9.17(b)). Indeed, the calculations across a wide spectral range verify that in both cases the magnitudes of t_{yx}, t_{xx} are enhanced, i.e. the transmitted waves are amplified. In particular, chiral effects manifest through the cross-polarized transmittance, t_{yx}, which expresses the strength of polarization conversion. However, although t_{yx} is amplified, θ, η are only enhanced in the example of figure 9.17(b), where the gain is embedded in the metasurface.

Retrieval of the effective conductivity $\sigma_{\text{eff}} \sim J/E$ of an equivalent, infinitely thin, current sheet reveals that chiroptical signal enhancement occurs when gain acts to undamp the resonances of the metasurface (σ_{eff} is obtained by dividing $J = i\omega P$, the polarization current oscillating in the dielectric slab, by the driving E-field, both probed at the same location). Note how, at the onset of pumping, σ_{eff} remains practically unchanged in figure 9.17(a), while it becomes spectrally reshaped in figure 9.17(b). Based on this observation, we may identify two distinct regimes of operation, the background amplification and the loss compensation regime, respectively [45]. In the background amplification regime the gain is coupled weakly to the metasurface, and consequently it does not impose changes on the characteristics of the resonators comprising the metasurface; the resonators first homogenize and then

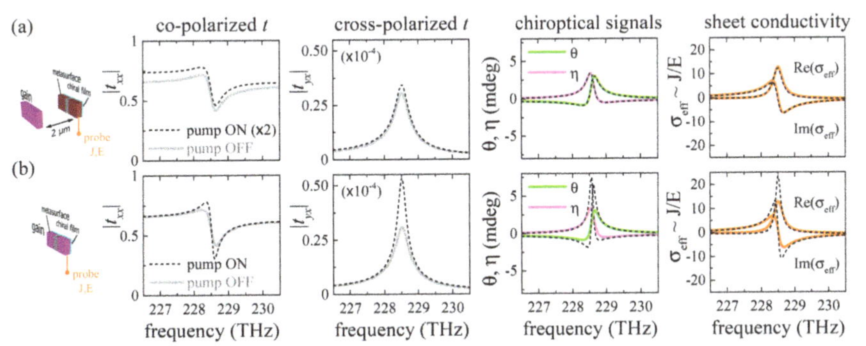

Figure 9.17. Mechanism of chiroptical signal enhancement in the presence of gain. (a) Regime of background amplification: gain coupled weakly to the metasurface leads to amplification of the transmitted wave (t_{xx}, t_{yx}), but with no enhancement of θ, η. (b) Regime of loss compensation: gain coupled strongly to the metasurface leads to enhancement of both θ, η, as a result of resonance undamping, as verified with the retrieved sheet conductivity, σ_{eff}. The gain is pumped at a constant pumping rate $R_p = 3 \times 10^7\ \text{s}^{-1}$ and in (a) the pump is $2 \times R_p$ to emphasize any subtle effects. Reprinted with permission from [46] © Optica Publishing Group.

couple to gain. In the loss compensation regime the gain is coupled strongly to the metasurface, and it acts to undamp (narrow) the resonances and reshape the spectral response of the metasurface; in this case the resonators first couple to gain and subsequently homogenize.

According to equations (9.67) and (9.68), the chiroptical signal enhancement is associated with the ratio of t_{yx} and t_{xx}, rather than their individual values; a proportional change in t_{yx} and t_{xx} ensures that their ratio is constant, leading to unchanged θ and η, as also discussed in the previous sections. Indeed, as we also verify by numerically calculating the ratio t_{yx}/t_{xx}, in the background amplification regime, t_{yx} and t_{xx} change proportionally, and consequently θ and η do not change, although gain is present. In the loss compensation regime, however, t_{yx} changes disproportionately with respect to t_{xx}, with the latter being associated with changes in the impedance mismatch between the metasurface and its surroundings, due to the resonance undamping. As a result, the enhancement of θ, η is mediated by the resonance undamping of the metasurface, i.e. gain-enhanced chiral sensing occurs in the loss compensation regime [46]. In this regime, a weak chirality can induce large, detectable changes to the polarization state of the transmitted wave.

9.5.3 Chiroptical signal enhancement from loss compensation to lasing

With more gain, it is intuitively expected that the observed chiroptical signals can be further enhanced. However, with increasing pump, gain will gradually compensate the radiative and dissipative losses, and when these losses become exactly balanced the metasurface will start to lase. Although once lasing is reached, chirality detection becomes meaningless, right before the lasing threshold, the metasurface lies deep in the loss compensation regime, opening up the possibility for large chiroptical signal enhancement.

To study the chiroptical signal enhancement as a function of gain, we analyse the polarization state of the transmitted wave under constant pump, within the range $R_p = 10^5\ \text{s}^{-1} - 10^9\ \text{s}^{-1}$. Because the enhancement is mediated by the electric (TE_{20}) and the magnetic (TM_{20}) resonant modes of the metasurface, it is also instructive to monitor how the associated dipole moments evolve with increasing gain. To this end, we use equations (9.62) and (9.64) to retrieve the effective surface conductivities $\hat{\sigma}_{ee}^{yy}$ and $\hat{\sigma}_{mm}^{yy}$, which are presented in figure 9.18(a). To monitor the transition to lasing, we also probe $J = i\omega P$, the polarization current oscillating in the dielectric slab. When the system passes on to lasing, gain balances the radiative and dissipative losses and, consequently, the resonant current J becomes spectrally a delta function. Hence, the transition to lasing manifests as spectral narrowing of J.

Lasing occurs once the pump rate exceeds $R_p = 7.9 \times 10^7 s^{-1}$, as can be observed in the full-width half-maximum (FWHM) of J in the top panel of figure 9.18(b). Beyond this pumping rate the system becomes a self-sustained oscillator and outgoing waves are produced regardless of the incoming waves; as a result, the transmittance becomes meaningless and is, therefore, marked with a shaded area. Before the lasing threshold, as the pump increases, each mode reaches the full loss compensation regime at different pump rates, as observed in the dramatic narrowing

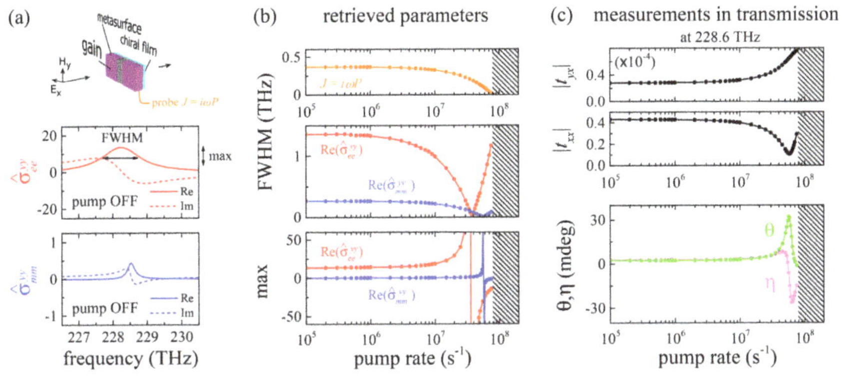

Figure 9.18. Gain-enhanced chiral sensing, from loss compensation to lasing. (a) Effective surface conductivities of the metasurface for $\kappa = 0$, retrieved using equations (9.61)–(9.66) without pumping ($R_p = 0$). (b) FWHM evolution of probed polarization current J and retrieved surface conductivities, as a function of the pumping rate R_p. (c) Magnitude of co-polarized (t_{xx}), cross-polarized (t_{yx}) transmission amplitudes and chiroptical signals θ, η for TM illumination, as a function of the pump rate R_p. The shaded region marks the transition to lasing. Reprinted with permission from [46] © Optica Publishing Group.

of $\text{Re}(\hat{\sigma}_{ee}^{yy})$ and $\text{Re}(\hat{\sigma}_{mm}^{yy})$, respectively (at $R_p = 3.8 \times 10^7\ \text{s}^{-1}$ for TE_{20} and at $R_p = 5.6 \times 10^7\ \text{s}^{-1}$ for TM_{20}). The observed narrowing occurs to an effective material property and, therefore, is not associated with the transition to lasing, rather indicates that the mode from lossy becomes amplifying (overcompensation), as shown in figure 9.18(b), where $\max[\text{Re}(\hat{\sigma}_{ee}^{yy})]$ and $\max[\text{Re}(\hat{\sigma}_{mm}^{yy})]$ undergo sign reversal. After the point of full loss compensation, each resonance begins to broaden, until the gain exactly balances all radiative and dissipative losses and the metasurface starts to lase; this occurs first for TM_{20}.

With increasing pumping, right before the onset of lasing, the transmission amplitudes t_{yx}, t_{xx} undergo qualitatively different changes; while t_{yx} increases monotonically, t_{xx} undergoes non-monotonic changes, due to the impedance mismatch that changes as the gain increases, causing the resonance to undamp. This is demonstrated in figure 9.18(c), where the chiral inclusion is probed at 228.6 THz. As a result, with increasing gain the chiroptical signals θ, η start to increase, but undergo non-monotonic changes close to full loss compensation. The observed enhancement in both signals reaches a factor of ~200, with respect to the corresponding signals from a pure chiral film ($\theta \sim 0.14$ mdeg, see figure 9.16(a)). Taking into account that the demonstrated enhancement is achieved with linear polarization, the combination of gain with elliptically polarized beams could boost the detected signals to even higher levels, enabling absolute measurements of unknown chiralities with unprecedented enhancement.

9.6 Discussion

It is well known from traditional polarimetric measurements of chirality that, for pure chiral media, $\text{Re}(\kappa)$ is determined entirely in terms of θ, and $\text{Im}(\kappa)$ in terms of η. This property can be derived directly from the analytical expressions of equations

(9.49) and (9.50), which for waves propagating through a bulk chiral medium take the simple form $\theta = \mathrm{Re}(\kappa)k_0L$ and $\eta = \mathrm{Im}(\kappa)k_0L$ [9], respectively; L is the length of the chiral medium and k_0 is the free-space wavenumber. Therefore, for pure chiral media, an unknown κ can be always be retrieved unambiguously by means of the signals θ, η.

In nanophotonic-based, chiral sensing approaches, however, when a chiral medium is introduced, e.g. in a metasurface, the incident wave probes both the chiral medium and the metasurface, instead of the chiral medium alone. Hence, the transmission results from the dispersion of both κ and the metasurface, i.e. from an effective chirality parameter that corresponds to the combined metasurface–chiral medium. As a result, the one-to-one correspondence between θ and $\mathrm{Re}(\kappa)$ and η and $\mathrm{Im}(\kappa)$ breaks.

A direct consequence is that a purely real κ may lead to nonzero η and a purely imaginary κ to nonzero θ, as demonstrated in the various examples in this chapter. For complex κ, the measurement of an unknown chirality can be even more demanding, because the chiroptical signals result from both the real and imaginary parts of κ. This is why, in nanophotonic-based chiral sensing schemes, it is necessary to deconvolve the unknown κ from the dispersion of the metasurface, in order to be able to detect and differentiate enantiomers correctly. Note that many popular nanophotonic platforms for chiral sensing are based on measurements of differential transmission between left and right circularly polarized waves. Such measurements must be interpreted with care, as they are intuitively expected to be sensitive to $\mathrm{Im}(\kappa)$ only, but they are sensitive to $\mathrm{Re}(\kappa)$, as well. By contrast, the techniques proposed in this work enable the unambiguous, complete and absolute measurement of an unknown chirality, with large enhancement factors.

9.7 Concluding remarks

In this chapter we theoretically examined chiral sensing using achiral metasurfaces. We derived analytically, and verified numerically, expressions that provide insight into the enhancement mechanism of the magneto-electric coupling and explained why the chiroptical signals of rotation θ and ellipticity η can both arise from both the real and the imaginary parts of the chirality parameter κ. We demonstrated how to deconvolve the unknown chirality from the background dispersion of the metasurface, and we proposed practical measurement schemes for enhanced detection, unambiguous enantiomer differentiation and complete measurement of the total chirality (magnitude and sign of both its real and imaginary parts). We also explained how anisotropic metasurfaces offer additional degrees of freedom with respect to their isotropic counterparts, offering chiroptical signals with unprecedented enhancement, and the possibility for absolute chirality measurements, i.e. without the need for sample removal, using a crucial signal reversal (excitation with reversed polarization). Last, we explored the possible coupling paths between the chiral inclusion, the metasurface and gain, and we demonstrated how gain can be used to further enhance chiral sensing.

Acknowledgment

The author would like to thank L Bougas for discussions on experimental aspects of chiral sensing.

References

[1] Fortson E N and Lewis L L 1984 Atomic parity nonconservation experiments *Phys. Rep.* **113** 289–344

[2] Fasman G D 2010 *Circular Dichroism and the Conformational Analysis of Biomolecules* (New York: Springer)

[3] Kelly S M, Jess T J and Price N C 2005 How to study proteins by circular dichroism *Biochim. Biophys. Acta* **1751** 119–39

[4] Nordén B, Rodger A and Dafforn T 2010 *Linear Dichroism and Circular Dichroism* (London: RSC)

[5] Hutt A J and Tan S C 1996 Drug chirality and its clinical significance *Drugs* **52** 1–12

[6] Nguyen L A, He H and Pham-Huy C 2006 Chiral drugs: an overview *Int. J. Biomed. Sci.* **2** 85–100

[7] Condon E U 1937 Theories of optical rotatory power *Rev. Mod. Phys.* **9** 432–57

[8] Lindell I V, Sihvola A H, Tretyakov S A and Viitanen A J 1994 *Electromagnetic Waves in Chiral and Bi-Isotropic Media* (Norwood, MA: Artech House)

[9] Barron L D 2004 *Molecular Light Scattering and Optical Activity* 2nd edn (Cambridge: Cambridge University Press)

[10] Busch K W and Busch M A 2006 *Chiral Analysis* (Amsterdam: Elsevier)

[11] Vaccaro P H 2011 *Optical Rotation and Intrinsic Optical Activity* (New York: Wiley) ch 11 pp 275–323

[12] Pellegrini G, Finazzi M, Celebrano M, Duò L and Biagioni P 2017 Chiral surface waves for enhanced circular dichroism *Phys. Rev.* B **95** 241402

[13] Droulias S and Bougas L 2019 Surface plasmon platform for angle-resolved chiral sensing *ACS Photon* **6** 1485–92

[14] Govorov A O, Fan Z, Hernandez P, Slocik J M and Naik R R 2010 Theory of circular dichroism of nanomaterials comprising chiral molecules and nanocrystals: plasmon enhancement, dipole interactions, and dielectric effects *Nano Lett.* **10** 1374–82

[15] Abdulrahman N A, Fan Z, Tonooka T, Kelly S M, Gadegaard N, Hendry E, Govorov A O and Kadodwala M 2012 Induced chirality through electromagnetic coupling between chiral molecular layers and plasmonic nanostructures *Nano Lett.* **12** 977–83

[16] Davis T J and Hendry E 2013 Superchiral electromagnetic fields created by surface plasmons in nonchiral metallic nanostructures *Phys. Rev.* B **87** 085405

[17] Maoz B M, Chaikin Y, Tesler A B, Elli O B, Fan Z, Govorov A O and Markovich G 2013 Amplification of chiroptical activity of chiral biomolecules by surface plasmons *Nano Lett.* **13** 1203–9

[18] Schäferling M, Dregely D, Hentschel M and Giessen H 2012 Tailoring enhanced optical chirality: design principles for chiral plasmonic nanostructures *Phys. Rev.* X **2** 031010

[19] Tullius R, Karimullah A S, Rodier M, Fitzpatrick B, Gadegaard N, Barron L D, Rotello V M, Cooke G, Lapthorn A and Kadodwala M 2015 Superchiral' spectroscopy: detection of protein higher order hierarchical structure with chiral plasmonic nanostructures *J. Am. Chem. Soc.* **137** 8380–3
[20] Tullius R *et al* 2017 Superchiral plasmonic phase sensitivity for fingerprinting of protein interface structure *ACS Nano* **11** 12049–56
[21] Zhang W, Wu T, Wang R and Zhang X 2017 Amplification of the molecular chiroptical effect by low-loss dielectric nanoantennas *Nanoscale* **9** 5701–7
[22] Mohammadi E, Tsakmakidis K L, Askarpour A N, Dehkhoda P, Tavakoli A and Altug H 2018 Nanophotonic platforms for enhanced chiral sensing *ACS Photon.* **5** 2669–75
[23] Mohammadi E, Tavakoli A, Dehkhoda P, Jahani Y, Tsakmakidis K L, Tittl A and Altug H 2019 Accessible superchiral near-fields driven by tailored electric and magnetic resonances in all-dielectric nanostructures *ACS Photon.* **6** 1939–46
[24] García-Guirado J, Svedendahl M, Puigdollers J and Quidant R 2020 Enhanced chiral sensing with dielectric nanoresonators *Nano Lett.* **20** 585–91 PMID: 31 851 826
[25] Droulias S and Bougas L 2020 Absolute chiral sensing in dielectric metasurfaces using signal reversals *Nano Lett.* **20** 5960–6
[26] Tang Y and Cohen A E 2010 Optical chirality and its interaction with matter *Phys. Rev. Lett.* **104** 163901
[27] Zhao Y, Askarpour A N, Sun L, Shi J, Li X and Alù A 2017 Chirality detection of enantiomers using twisted optical metamaterials *Nat. Commun.* **8** 14180
[28] Solomon M L, Hu J, Lawrence M, García-Etxarri A and Dionne J A 2019 Enantiospecific optical enhancement of chiral sensing and separation with dielectric metasurfaces *ACS Photon.* **6** 43–9
[29] Kelly C, Khorashad L K, Gadegaard N, Barron L D, Govorov A O, Karimullah A S and Kadodwala M 2018 Controlling metamaterial transparency with superchiral fields *ACS Photon* **5** 535–43
[30] Droulias S 2020 Chiral sensing with achiral isotropic metasurfaces *Phys. Rev.* B **102** 075119
[31] Hendry E, Carpy T, Johnston J, Popland M, Mikhaylovskiy R V, Lapthorn A J, Kelly S M, Barron L D, Gadegaard N and Kadodwala M 2010 Ultrasensitive detection and characterization of biomolecules using superchiral fields *Nat. Nanotechnol.* **5** 783
[32] Tang Y and Cohen A E 2011 Enhanced enantioselectivity in excitation of chiral molecules by superchiral light *Science* **332** 333–6
[33] Karimullah A S, Jack C, Tullius R, Rotello V M, Cooke G, Gadegaard N, Barron L D and Kadodwala M 2015 Disposable plasmonics: plastic templated plasmonic metamaterials with tunable chirality *Adv. Mater.* **27** 5610–6
[34] Luo Y, Chi C, Jiang M, Li R, Zu S, Li Y and Fang Z 2017 Plasmonic chiral nanostructures: chiroptical effects and applications *Adv. Opt. Mater.* **5** 1700040
[35] Tassin P, Zhang L, Zhao R, Jain A, Koschny T and Soukoulis C M 2012 Electromagnetically induced transparency and absorption in metamaterials: the radiating two-oscillator model and its experimental confirmation *Phys. Rev. Lett.* **109** 187401
[36] Haus H A and Huang W 1991 Coupled-mode theory *Proc. IEEE* **79** 1505–18
[37] Pelet P and Engheta N 1990 Coupledmode theory for chirowaveguides *J. Appl. Phys.* **67** 2742–5
[38] Fietz C and Shvets G 2010 Homogenization theory for simple metamaterials modeled as one-dimensional arrays of thin polarizable sheets *Phys. Rev.* B **82** 205128

[39] Tassin P, Koschny T and Soukoulis C M 2012 Effective material parameter retrieval for thin sheets: theory and application to graphene, thin silver films, and single-layer metamaterials *Physica* B **407** 4062–5 Proceedings of the conference—Wave Propagation: From Electrons to Photonic Crystals and Metamaterials

[40] Zhao R, Zhou J, Koschny T H, Economou E N and Soukoulis C M 2009 Repulsive casimir force in chiral metamaterials *Phys. Rev. Lett.* **103** 103602

[41] Sofikitis D, Bougas L, Katsoprinakis G E, Spiliotis A K, Loppinet B and Rakitzis T P 2014 Evanescent-wave and ambient chiral sensing by signal-reversing cavity ringdown polarimetry *Nature* **514** 76

[42] Bougas L, Sofikitis D, Katsoprinakis G E, Spiliotis A K, Tzallas P, Loppinet B and Peter Rakitzis T 2015 Chiral cavity ring down polarimetry: chirality and magnetometry measurements using signal reversals *J. Chem. Phys.* **143** 104202

[43] Droulias S and Bougas L 2021 Chiral sensing with achiral anisotropic metasurfaces *Phys. Rev.* B **104** 075412

[44] Johnson P B and Christy R W 1972 Optical constants of the noble metals *Phys. Rev.* B **6** 4370–9

[45] Droulias S, Koschny T, Kafesaki M and Soukoulis C M 2019 On loss compensation, amplification and lasing in metallic metamaterials *Nanomater. Nanotechnol.* **2019** 9

[46] Droulias S 2021 Enhanced chiral sensing using achiral metasurfaces with gain *J. Opt. Soc. Am.* B **38** C210–6

[47] Droulias S, Jain A, Koschny T and Soukoulis C M 2017 Fundamentals of metasurface lasers based on resonant dark states *Phys. Rev.* B **96** 155143

[48] Droulias S, Mohamed S, Rekola H, Hakala T K, Soukoulis C M and Koschny T 2022 Experimental demonstration of dark-state metasurface laser with controllable radiative coupling *Adv. Opt. Mater.* **10** 2102679